ALLUVIAL FANS

Frontispiece

Milner Creek fan, White Mountains,
California. Photo: C. B. Beaty, 16 June, 1987.

ALLUVIAL FANS

A Field Approach

Edited by

Andrzej H. Rachocki

Institute of Geography,
University of Gdańsk

and

Michael Church

Department of Geography,
University of British Columbia

JOHN WILEY & SONS

Chichester · New York · Brisbane · Toronto · Singapore

Copyright © 1990 by John Wiley & Sons Ltd,
Baffins Lane, Chichester,
West Sussex PO19 1UD, England

All rights reserved.

No part of this book may be reproduced by any means,
or transmitted, or translated into a machine language
without the written permission of the publisher.

Other Wiley Editorial Offices

John Wiley & Sons, Inc., 605 Third Avenue,
New York, NY 10158-0012, USA

Jacaranda Wiley Ltd, G.P.O. Box 859, Brisbane,
Queensland 4001, Australia

John Wiley & Sons (Canada) Ltd, 22 Worcester Road,
Rexdale, Ontario M9W 1L1, Canada

John Wiley & Sons (SEA) Pte Ltd, 37 Jalan Pemimpin 05-04,
Block B, Union Industrial Building, Singapore 2057

Library of Congress Cataloging-in-Publication Data:
Alluvial fans : a field approach / edited by Andrzej H. Rachocki and
  Michael Church.
       p.   cm.
    Includes bibliographical references.
    ISBN 0-471-91694-3
    1. Alluvial fans.   I. Rachocki, Andrzej.   II. Church, Michael
Anthony, 1942–
GB591.A45   1990
551.4′56—dc20   89–22435
                                                          CIP

British Library Cataloguing in Publication Data:
Alluvial fans: a field approach.
  1. Alluvial fans. Formation.
  I. Rachocki, Andrzej   II. Church, Michael
  551.3′55

ISBN 0-471-91694-3

Typeset by Acorn Bookwork, Salisbury, Wiltshire
Printed in Great Britain by Courier International Ltd., Tiptree, Essex

Contents

Contributing Authors — vii

Preface — ix

INTRODUCTION

1 The Alluvial Fan Problem — 3
   S. A. Lecce

SECTION I ALLUVIAL FANS IN RECENT ENVIRONMENTAL SETTINGS

2 Quaternary Alluvial Fans in the Karakoram Mountains — 27
   E. Derbyshire and L. A. Owen

3 Ice Marginal Ramps and Alluvial Fans in Semiarid Mountains: Convergence and Difference — 55
   M. Kuhle

4 Anatomy of a White Mountains Debris-Flow—The Making of an Alluvial Fan — 69
   Ch. B. Beaty

5 Alluvial Fans in Japan and South Korea — 91
   Y. Ono

6 Humid Fans of the Appalachian Mountains — 109
   R. Craig Kochel

7 The Chandigarh Dun Alluvial Fans: An Analysis of the Process–Form Relationship — 131
   A. B. Mukerji

8 Morphology of the Kosi Megafan — 151
   K. Gohain and B. Parkash

9 The Portage la Prairie 'Floodplain Fan' — 179
   W. F. Rannie

10 Fan Deltas—Alluvial Fans in Coastal Settings 195
W. A. Wescott and F. G. Ethridge

11 The Yallahs Fan Delta: A Coastal Fan in a Humid Tropical Climate 213
W. A. Wescott

SECTION II INTERPRETATION OF THE ENVIRONMENT OF ALLUVIAL FANS

A. Alluvial Fans in Studies of Palaeogeomorphology

12 Evolution of the Alluvial Fans of the Alföld 229
Z. Borsy

13 Factors Influencing Quaternary Alluvial Fan Development in Southeast Spain 247
A. M. Harvey

14 Long-term Palaeochannel Evolution During Episodic Growth of an Exhumed
Plio-Pleistocene Alluvial Fan, Oman 271
J. Maizels

15 The Leba River Alluvial Fan and its Palaeogeomorphological Significance 305
A. H. Rachocki

B. Alluvial Fans: A Scene of Human Activity

16 Development of Alluvial Fans in the Foothills of the Darjeeling Himalayas and their
Geomorphological and Pedological Characteristics 321
S. R. Basu and S. Sarkar

17 Hazard Management on Fans, with Examples from British Columbia 335
R. Kellerhals and M. Church

18 Artificial Recharge of Aquifers in Alluvial Fans in Mexico 355
N. Otero San Vicente

19 Geomorphic Appraisals for Development on Two Steep, Active Alluvial Fans,
Mt Cook, New Zealand 369
I. E. Whitehouse and M. J. McSaveney

Index 385

Contributing Authors

S. R. Basu	*Department of Geography and Applied Geography, The University of North Bengal, Siliguri 734 430, Darjeeling, West Bengal, India.*
Ch. B. Beaty	*Department of Geography, The University of Lethbridge, 4401 University Drive, Lethbridge, Alberta, Canada T1K 3M4.*
Z. Borsy	*Institute of Geography, Lajos Kossuth University, H-4010 Debrecen Pf.9, Hungary.*
M. Church	*Department of Geography, The University of British Columbia, 217–1984 West Mall, Vancouver, B.C., Canada V6T 1W5.*
R. Craig Kochel	*Department of Geology, Southern Illinois University at Carbondale, Carbondale, Illinois 62901, U.S.A.*
E. Derbyshire	*Department of Geography, University of Leicester, Leicester LE1 7RH, U.K.*
F. G. Ethridge	*Department of Earth Resources, Colorado State University, Ft. Collins, Colorado 80523 U.S.A.*
K. Gohain	*Geological Department, Oil India Ltd., Duliajan, Assam 786 602, India.*
A. M. Harvey	*Department of Geography, University of Liverpool, Roxby Building, Liverpool L69 3BX, U.K.*
R. Kellerhals	*P.O. Box 250, Heriot Bay, British Columbia, Canada V0P 1H0.*
M. Kuhle	*Geographisches Institut der Universität Göttingen, Goldschmiedstrasse 5, D-3400 Göttingen, BRD (F.R.G.).*
S. A. Lecce	*Department of Geography, University of Wisconsin, Madison, Wisconsin 53706, U.S.A.*

J. Maizels	*Department of Geography, University of Aberdeen, Elphinstone Road, Aberdeen AB9 2UF, U.K.*
M. J. McSaveney	*New Zealand Geological Survey, University of Canterbury, Private Bag, Christchurch, New Zealand.*
A. B. Mukerji	*Department of Geography, Panjab University, Chandigarh 160 014, India.*
Y. Ono	*Graduate School of Environmental Science, Hokkaido University, Sapporo 060, Japan.*
N. Otero San Vicente	*Rio Jordan 888, Colonia Estrella, Torreon, Coahuila, Mexico 27010.*
L. A. Owen	*Department of Geography, University of Leicester, Leicester LE1 7RH, U.K.*
B. Parkash	*Department of Earth Sciences, University of Roorkee, Roorkee U.P.–247 667, India.*
A. H. Rachocki	*Institute of Geography, University of Gdańsk, Marchlewskiego St.16a, 80-952 Gdańsk 6, Poland.*
W. F. Rannie	*Department of Geography, University of Winnipeg, Winnipeg, Canada R3B 2E9.*
S. Sarkar	*Department of Geography and Applied Geography, North Bengal University, Siliguri 734 430, Darjeeling, India.*
W. A. Wescott	*Amoco Production Company, 501 West Lake Park Blvd. P.O. Box 3092, Houston, Texas 77253, U.S.A.*
I. E. Whitehouse	*Division of Land and Soil Sciences, Department of Scientific and Industrial Research, P.O. Box 29199, Christchurch, New Zealand.*

Preface

No one knows whether Frederick Drew's descriptions of alluvial fans in the Himalayan foothills, published in 1873, were the first. One may suppose so. In the ensuing period, these landforms only occasionally attracted the attention of earth scientists, as is witnessed by the number of publications. Dispersed in various more or less specialized periodicals, they hardly exceeded one hundred in the next century. This situation changed rapidly about twenty-five years ago. The increasing interest of geomorphologists in the contemporary processes shaping the surface of the earth and the progressive reorientation of sedimentologists toward better understanding the mechanisms of sediment deposition, as well as recognition that recent sedimentary environments provide a template for studying ancient ones, triggered an avalanche of sometimes very detailed studies in the early 1960s. One may note the rebirth of studies on alluvial fan morphology and sedimentology from that time.

Aside from their intrinsic interest, alluvial fans represent a relatively convenient sedimentary feature for study: they are of relatively limited size, are relatively accessible, and are rather transparently linked to their sediment source regions by evident processes of material transfer. Modern studies of alluvial fans differ greatly in approach from the classical studies, such as those of Eckis, Blissenbach, Solch, Czajka, or Horowitz. Whereas the earlier studies were based upon Davisian principles of landscape evolution, those written by Bull, Hooke, Lustig, and many others attempted to illuminate interrelationships between the geometric size and shape of catchments – the sources of material for fan creation – and the fans themselves, on the assumption that some more or less stable 'equilibrium' form is reached. These relationships quite often have been expressed in the form of statistical or mathematical formulae linking basins and fans in an inseparable functional unit. The main emphasis shifted, of course, according to the personal inclinations of individual researchers. Surely better recognition of depositional and erosional processes on alluvial fans, and relationships between the magnitude and frequency of effective forces and shape, size and depositional structures of alluvial fans were important achievements of that time.

A distinctive group of investigations was done in indoor (Hooke, Kadar) or outdoor (Rachocki, Weaver) laboratories. These experimental studies were expected to shed new light upon details of the mechanisms of fan deposition, and upon the origin of specific features of surficial morphology of fans. Although sometimes very interesting from the theoretical point of view, these studies could not incorporate, in scales of space and time, all the various factors existing in the natural environment. After two decades or so, the first enthusiasm for these laboratory studies has to some degree run down. If at the present stage of atomization in geomorphology one can still speak about any tendencies at all, and especially any which could be termed 'general', some 'return to nature' might be recorded in recent years. The aim of this book is to gather together results from the return to alluvial fans in the field.

The alluvial fans that have been studied are scattered over nearly the entire globe. They vary in size from the megafan of Kosi River in the Himalayan foreland, to a quite small fan on the Leba River in a pradolina of northern Poland. Equally variable are their histories and ages, which range from decades to hundreds of thousands of years.

As different as the alluvial fans themselves are the approaches essayed to the problems presented by their morphology, depositional history and contemporary development. One can note two main attitudes toward the study of alluvial fans. In the first, the entire attention of the researchers is focussed strictly on the alluvial fans. In the second, the fans are treated as a useful artefact enabling one to discover and reconstruct palaeoenvironmental changes in the region where they occur.

Two features of this book are particularly pleasing for us. Our group of collaborators ranges far beyond the usual North Atlantic academic mafia. In appraising the papers in this collection, one must remember that as various as are the alluvial fans and their environmental context, so also are the conditions in which contributors work. Just as the landforms studied each has its own morphodynamical milieu, each of the authors has a particular intellectual, political and economic environment exerting an influence on real possibilities. This often is reflected in the objectives of studies, in the approach, and in the mode of presentation of the results. This perhaps has generated the second salutary outcome: a significant proportion of the papers report applied investigations. They vary from water resources assessments, through land use studies, to hazard appraisals. Given the frequent association between alluvial fans and high mountains, where useable land is restricted, these are important concerns. Alluvial fans often are preferred sites for settlement. Hazards on active fans may be severe and they are not like usual hazards associated with stream courses. Defining and characterizing resources and hazards is an important task for the maturing subject of geomorphology. It will also generate new insights for basic study. So the mixture of papers in this book augurs well for future work.

We wish to thank a number of individuals and organizations for special help. The reviewers of the papers provided substantial commentaries for the authors, and the papers have benefitted greatly. Paul Jance, cartographer at the University of British Columbia, has performed remarkable repairs on many of the drawings: it is a measure of his skill that most authors will not detect any intrusion upon their individual styles. Mrs. Helen Bailey, earth science editor at John Wiley, helped editors to launch the book and has been unfailingly patient and helpful during its growth. The Oxford–Poland Academic Visitor Programme allowed us to meet in Oxford to coordinate our editorial tasks. We acknowledge the spirit of cooperation shown by one group of the authors approached rather late in the day, when we decided that some specific topics should be sought in order to bring balance to the overall collection: their delivery of interesting papers on a very tight schedule has both enhanced the book and allowed us to maintain reasonable times for all the authors. Finally, we salute all of the authors: a Polish–Canadian editorial team must be one of the more unlikely hazards they ever will encounter.

Andrzej H. Rachocki and Michael Church
February 20, 1989

INTRODUCTION

CHAPTER 1

The Alluvial Fan Problem

Scott A. Lecce
Arizona State University, Tempe

Abstract

The 'alluvial fan problem' is defined as the difficulty of formulating a general model of alluvial fan development. The paradigm concept provides a useful framework for evaluating trends in alluvial fan research. The predominant paradigm since the early 1960s has been the fan dynamics paradigm, in which the major research question of the paradigm was to determine the process involved in long-term fan development. Equilibrium and evolutionary approaches in the fan dynamics paradigm were proposed to explain fan development. Difficulty in successfully applying time-independent equilibrium concepts to the alluvial fan problem required the integration of equilibrium and evolutionary approaches. Reconciliation of these two competing concepts was achieved by considering the time span over which equilibrium is viewed. The resulting integrated approach to the fan dynamics paradigm involves the investigation of depositional processes in different environments and the historical analysis of long-term fan evolution.

Field data collection and laboratory fan models have been essential components of fan research. Fluvial theory in geomorphology has benefited from emphasis on empirical data collection; however, field site selection has been spatially biased, favouring the arid to semiarid regions of the American Southwest.

Introduction

It is not unusual in geomorphology for difficult questions related to the origin and formation of geomorphic features to be regarded as 'problems'. The 'pediment problem' (Oberlander, 1974) and the 'arroyo problem' (Cooke and Reeves, 1976; Graf, 1983) are examples of the treatment of questions in this manner. The 'alluvial fan problem' discussed in this paper is defined as the difficulty of establishing a general model of alluvial fan development. The formation of alluvial fans has attracted the attention of geomorphologists since the exploratory surveys of the American West in the late 1800s, yet the alluvial fan problem has generated significant interest only since the early 1960s. Assessment of the progress towards resolution of the alluvial fan problem is accomplished by discussing the development of paradigms in alluvial fan research. The purpose of this paper is to review the trends in alluvial fan research by focusing on research paradigms and to examine the influence that research on alluvial fans has had in the development of fluvial theory.

The significance of alluvial fan research within

Alluvial Fans: A Field Approach edited by A. H. Rachocki and M. Church
Copyright © 1990 John Wiley & Sons Ltd.

the general field of geomorphology is threefold. While there may have been a shift of interest in the past several decades to process studies at the expense of the study of landforms, research on alluvial fans serves to stress a fundamental goal in geomorphology—to understand the significance of process–form relationships. Second, the investigation of the processes responsible for the construction of alluvial fans helps to improve understanding of sediment transport in other fluvial systems, particularly in arid and semiarid environments (Bull, 1968). Third, the geomorphic response of fan morphology to changes in fluvial processes, tectonic activity, climate, and drainage basin variables are concerns common to the science of geomorphology as a whole.

Alluvial Fans

An alluvial fan consists of stream deposits, the surface of which forms a segment of a cone that radiates downslope from the point where the stream emerges from the mountain area (Bull, 1963, 1968) (Figure 1.1). Some authors have used the term 'alluvial cone' to describe small fans steeper than 20 degrees that are formed by both fluvial deposition and mass wasting (Bull, 1977), while others have used it as a synonym for an 'alluvial fan' (Knopf, 1918). This paper will restrict the use of the term alluvial fan to features deposited by fluvial processes and debris-flow processes (thereby excluding other forms of mass wasting). An alluvial fan may consist of debris-flow deposits, water-laid sediments, or both. The highest point on the fan, where the stream leaves the confines of the mountain, is the fan apex. The overall radial profile of an ideal alluvial fan surface (longitudinal profile) is concave, while the cross-fan profile (parallel to the mountain front) is convex. In plan view the deposit is typically fan shaped, with contours that are convex outward from the mountain front (Bull, 1977).

Fans often have stream channels that are incised into the fan surface. The particular case known as 'fanhead trenching' generally is deepest

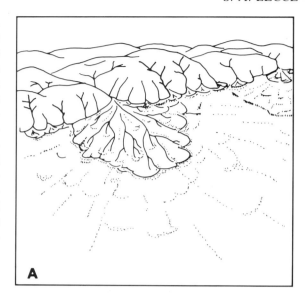

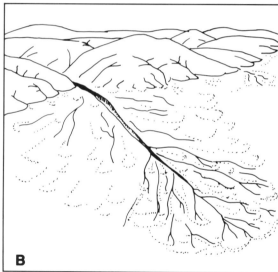

Figure 1.1. Two phases in the development of alluvial fans under the influence of tectonic uplift (A) Area of deposition adjacent to the mountain front; (B) Area of deposition shifted downfan due to stream channel entrenchment. (Reproduced by permission from Bull, 1968)

at the fan apex and progressively decreases downfan until the stream channel and the fan surface intersect. This point on the fan surface is called the intersection point (Wasson, 1974).

Alluvial fans may coalesce along a mountain

front to form a depositional piedmont called an alluvial apron or bajada (Eckis, 1928). Coalescent fans are frequently mistaken for pediments. However, alluvial fans are distinguishable as depositional features, while pediments are erosional landforms. Pedimented landscapes commonly contain bedrock knobs that protrude for the ramplike erosional surface, whereas alluvial fans with such knobs projecting through the surface are seldom observed. Doehring (1970) provided a method for discriminating between fans and pediments using topographic map analysis. In general, a fan may be distinguished from a pediment when the thickness of the deposit is greater than 1/100 of the length of the landform (Bull, 1977).

While the term 'alluvial fan' was first coined by Drew (1873) in reference to features in the upper Indus basin, Surell (1841, from Bull, 1977; Anstey, 1965) appears to have been the first to discuss fans. Alluvial fans are commonly found in arid and semiarid regions with tectonically active mountains where there is an abundant supply of sediment, but their occurrence is by no means limited to such environments. Alluvial fans also occur in humid–temperate, subtropical, arctic, and alpine environments (Table 1.1).

Table 1.1 Alluvial fan studies in non-arid environments*

Study area	References
Australia	Wasson, 1977a,b
Canada	
Alberta	Winder, 1965
	McPherson and Hirst, 1972
	Roed and Wasylyk, 1973
	Kostaschuk et al., 1986
British Columbia	Ryder, 1971a,b,
	Church and Ryder, 1972
Northwest Territories	Leggett et al., 1966
Costa Rica	Kesel, 1985
Greece	Schroeder, 1971
Honduras	Schramm, 1981
Iceland	Boothroyd and Nummedal, 1978
India	Drew, 1873
	Gole and Chitale, 1966
	Mukerji, 1976
Japan	Murata, 1931a,b; 1966
	Yazawa et al., 1971
	Matsuda, 1974
	Iso et al., 1980
	Saito, 1980
New Zealand	Carryer, 1966
Poland	Rachocki, 1981
Sweden	Hoppe and Ekman, 1964
United Kingdom	Wells and Harvey, 1987
United States	
Alaska	Anderson and Hussey, 1962
	Boothroyd, 1972
	Ritter and Ten Brink, 1986
Colorado	Blair, 1987
Virginia	Kochel and Johnson, 1984

*From Nilsen and Moore (1984), with additions.

Paradigms

KUHN'S PARADIGM CONCEPT

Before considering the trends in alluvial fan studies, it is valuable to review Kuhn's (1970) perspective on the sociology of science, because it provides a useful conceptual framework for examining trends in scientific research. According to Kuhn, the development of science is characterized by long periods of problem-solving when knowledge is accumulated in a steady manner, interrupted periodically by brief episodes of intellectual crisis. The practitioners of a science share a common approach and philosophy to research and use an agreed-upon methodology (Gregory, 1985). Kuhn defined a paradigm as 'universally recognized scientific achievements that for a time provide model problems and solutions to a community of practitioners' (Kuhn, 1970: viii). The paradigm defines whether or not specific research is relevant to the current view of the science. 'Normal science' is research based upon past scientific achievements reached within the domain of the current paradigm that form the foundation for subsequent investigations (Kuhn, 1970). The objective of researchers is to solve problems that are identified within the theoretical framework provided by the paradigm in a manner that accumulates generalizations and extends the range of the theory (Johnston, 1983). By working within the structure of the paradigm, practitioners of normal science are provided with (1) an accepted body of knowledge, (2) a general indication of the problems that remain to be solved, and (3) a methodology designed to solve those problems (Johnston, 1983).

As normal science proceeds, occasional anomalies surface that are not in keeping with the assumptions of the paradigm. Minor changes in the paradigm may be necessary to assimilate these discrepancies, but if anomalies persist and accumulate, normal science will be interrupted by an intellectual crisis. At this time some scientists will search for a new paradigm that is able to account for the anomalies. The successful outcome of this 'extraordinary' research presents an alternative paradigm to the scientific community, which, if accepted, marks a scientific revolution whereby the prevailing paradigm is replaced by a new one.

Although Kuhn's work has stimulated criticism (Shapere, 1964; Masterman, 1970; Watkins, 1970) and interest in alternative approaches (Popper, 1970; Lakatos, 1970), his philosophy appears to have particular relevance to the historical development of geomorphology (Graf, 1983).

PARADIGMS IN GEOMORPHOLOGY

Following the early exploratory surveys of the American West, the science of geomorphology experienced a period of normal science dominated by the Davisian concept of landscape evolution which continued until the early 1950s (Tinkler, 1985). While Davis viewed landforms as a function of 'structure', 'process', and 'stage', he placed primary emphasis on stage. Consequently, the investigation of process was virtually ignored (Hart, 1986). Dissatisfaction with Davis's evolution paradigm led to an intellectual crisis in geomorphology. Alternatives (Penck, 1924; King, 1953, 1962; Büdel, 1977) and criticisms (Strahler, 1950) accumulated until a scientific revolution occurred and the evolution paradigm was replaced. Although the search for an alternative to the evolution paradigm had begun early with the work of Leighly (1932), Horton (1932), Hjulström (1935), and Bagnold (1935, 1937, 1941), the shift to a new process paradigm was delayed until Horton published his influential paper in 1945.

A new period of normal science emphasized equilibrium concepts and a quantitative approach to the investigation of surface morphology and process using principles developed in physics (Tinkler, 1985). Research under this process paradigm continued until the early 1970s when geomorphology appeared to enter another crisis period which persists to the present. Problems that are solvable by the equilibrium approach appear to be fewer and some researchers have recognized the advantage of integrating useful aspects of the Davisian and equilibrium approaches (Schumm, 1977). A new conceptual framework emphasizing spatial- and temporal-dependent solutions has been advocated and appears promising for new theory development (Kennedy, 1977; Graf, 1983).

Although paradigms usually pertain to scientific disciplines, it is also important to consider paradigms at the smaller scale of the specialized research field. The development of paradigms in alluvial fan studies generally followed events in the broader science; however, significant differences can be observed which makes the analysis of paradigms at the smaller scale a reasonable proposition. The following section reviews the pre-paradigm period that emphasized description and classification. Next, a more elaborate review is provided of the paradigm of fan dynamics, which is concerned with process and fan development. The third section discusses an approach that combines useful parts of equilibrium and evolutionary hypotheses which illustrate the relevance of the alluvial fan problem to research conducted in the paradigm of fan dynamics. The objective of these sections is to analyse the prevailing paradigms in alluvial fan studies and their influence on the progress of fan research. It is recognized that organizing intellectual history according to paradigms is a subjective exercise at best; nevertheless, it is a useful way to make sense of a seemingly random collection of information and to gain some 'knowledge of the nature of intellectual debate' (Hart, 1986:199). Because paradigms provide the means for answering specific research questions, the goal here will be to identify how paradigms failed or succeeded in resolving the fundamental research questions associated with the paradigm. Finally, fan research is evaluated in the context of fluvial theory development.

Pre-paradigm Research

While the broader field of geomorphology was growing quickly during a Davisian period of normal science, interest in alluvial fans as subjects for research was developing quite slowly. The number of researchers and studies lacked the 'critical mass' necessary to develop a well-defined paradigm for alluvial fans. This period was characterized by lack of a common methodology and literature, resulting in relatively disjointed research efforts. Interest was generally confined to the geologic description of fan deposits, the establishment of fan classification criteria and terminology, and speculation about the depositional processes that form alluvial fans. Few researchers collected quantitative data on surface processes or morphology.

The dominance of Davisian concepts in geomorphology had a relatively limited impact on alluvial fan studies, yet it was through Davis's ideas that the alluvial fan problem was first addressed by the consideration of evolutionary stages (Eckis, 1928). Eckis proceeded under the notion that fans indicated conditions of youth in the geographic cycle, and are therefore only temporary features. He further suggested that trenching signalled maturity and the eventual destruction of the fan, and noted that the significant factors to be considered in the trenching of fans are tectonic uplift, climatic change, reduced sediment load, base-level change, and 'dissection in the normal course of the cycle' (Eckis, 1928:237). Eckis proposed that the last factor, trenching due to the progressive downwearing of the stream profile above the fanhead, was responsible for fan entrenchment in the San Gabriel Mountains of southern California. Evolutionary change was conceptualized in a closed system which neglected periodic inputs of energy due to tectonic uplift.

Following the work of Eckis, Blackwelder (1928) recognized the importance of debris-flows in fan construction and Chawner (1935) described the sediments deposited following a major flood on an alluvial fan in Los Angeles County. In 1952, Blissenbach found that a close relationship exists between the rates of change of particle size and surface slope. He later summarized the current knowledge about fan sedimentology and morphology (Blissenbach, 1954).

Although this period exhibited some of the characteristics of the evolutionary paradigm in the broader field of geomorphology, the work of two researchers (Eckis, 1928; Blissenbach, 1954) was hardly sufficient to be described as a paradigm for alluvial fan research. The concern with fan evolution was but a small component of research that instead emphasized descriptive studies with fans as secondary topics of interest (Tolman, 1909; Trowbridge, 1911; Lawson, 1913; Knopf, 1918; Blackwelder, 1928; Chawner, 1935). These researchers made few efforts to relate what they learned about alluvial fans to broader concepts of landscape evolution.

The Paradigm of Fan Dynamics and Hypotheses of Fan Development

The early 1960s marked a transition in alluvial fan research. Following the pre-paradigm period of description and classification in which there was no generally accepted view or body of knowledge upon which to build, one paradigm became dominant over the other competing views. The paradigm of fan dynamics was similar in many ways to the process paradigm in the parent field of geomorphology. Introduction of the equilibrium concept and systems theory encouraged quantitative studies that placed primary emphasis on fan processes and morphology. The major research question for the paradigm of fan dynamics was one of determining the processes involved in fan development, i.e. the alluvial fan problem. Interpretations of how fans developed varied considerably. To support their claims, many researchers applied allometric models to describe the rates of change between fan and drainage basin characteristics (Bull, 1964b; Melton, 1965; Hooke, 1968a; Hooke and Rohrer, 1977; Denny, 1967; Beaumont, 1972) (Figure 1.2). Fan–basin relationships were used to suggest factors that control fan development. In addition, research investigating the causes of fan entrenchment produced useful evidence to support fan development hypotheses.

The two primary hypotheses proposed to explain fan development were the evolutionary hypothesis and the equilibrium hypothesis. Climatic and tectonic effects are considered as factors influencing fan development in an equilibrium hypothesis. The evolutionary hypothesis received much less attention than the equilibrium hypothesis due to widespread dissatisfaction with historical approaches bearing any resemblance to

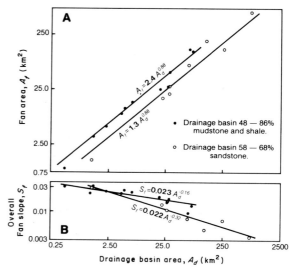

Figure 1.2. Relation between drainage basin area and (A) fan area and (B) fan slope. (A) Fans with basins containing mostly erodible mudstone and shale (solid dots) are proportionately larger than their counterparts built of material from basins composed predominantly of more resistant sandstone; (B) Fan slope decreases as basin area increases. (From Bull, 1964b)

Davis's cycle of erosion. However, it is important to recognize the evolutionary hypothesis early in the discussion of the fan dynamics paradigm because later it will play a significant role in the restructuring of the paradigm when difficulties arise in successfully applying equilibrium concepts to the alluvial fan problem.

THE EVOLUTIONARY HYPOTHESIS

The evolutionary hypothesis was first expressed in the work of Eckis (1928) during the pre-paradigm period in alluvial fan research. The influence of process and its variation with time was virtually ignored while emphasis was placed on qualitative assessments of 'stage' in Davis's (1905) arid cycle. Conditions of youth were indicated by small, actively growing fans fed by streams flowing from V-shaped canyons in uplifted mountain ranges (Lustig, 1965). As fans evolved they were thought to grow until they filled the depositional basin with sediment,

THE ALLUVIAL FAN PROBLEM

burying remnants of the mountain range in alluvium.

Beaty (1970:71) rejected the possibility that the fans on the west side of the White Mountains in California are in a steady-state condition or in a state of active dissection and volume reduction. Instead, he concluded that conditions of 'active growth' and a 'Davisian stage of youth' are applicable to the west-side fans, while the older fans on the east side of the range are physiographically mature. However, Beaty also suggested that these fans would eventually reach some sort of equilibrium condition if given enough time. The evolutionary hypothesis in a strictly Davisian (cyclic, stage-oriented) sense was not likely to provide the answers sought by geomorphologists regarding fan development. A different conceptual approach was needed.

THE EQUILIBRIUM HYPOTHESIS

The concept of equilibrium soon became the guiding principle in the effort to resolve the alluvial fan problem. Initially, researchers chose to view fans in steady state equilibrium—as time-independent landforms (Figure 1.3) (see Schumm, 1977, pp. 4–13 for a review of equilibrium terminology). Denny (1965, 1967) contended that a state of dynamic equilibrium exists between alluvial fans and their source areas, an idea first proposed for alluvial fans by Hack (1960). The rate of deposition on the fan was considered equal to the rate of erosion from the fan to the floodplain. To maintain such a steady state Denny envisioned a fan divided into areas of erosion and deposition, where areas of deposition are converted to areas of erosion by repeated stream piracies (Figure 1.4).

In a different view of equilibrium, Hooke (1968a) noted that in Denny's system erosional material must be removed entirely from the basin, otherwise the playa would encroach onto the fan. In such a scenario fan area would decrease, causing an increase in the rate of deposition per unit area on the fan. The rate of erosion from the fan would decrease because erosional processes would have less area in which to oper-

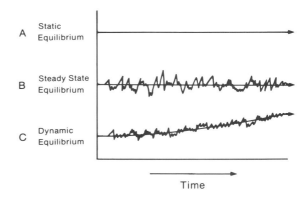

Figure 1.3. Types of equilibrium. (A) A system in static equilibrium has no change in the system variables over time; (B) Systems in steady-state equilibrium show variations about a mean condition that remains constant over time; (C) Dynamic equilibrium means that system variables fluctuate around a mean condition that changes through time. (Reproduced by permission from Schumm, 1977, after Chorley and Kennedy, 1971)

ate (Hooke, 1968a). Hooke suggested that a steady-state condition exists (in a closed system) between the amounts of deposition on the fan and on the playa. These equilibrium adjustments result in equal rates of fan and playa aggradation. Thus, according to Hooke, the main flaw in Denny's analysis is that it will work only in open systems where sediment from the playa can be removed, whereas most basins of interior drainage are more easily viewed as closed systems.

Consideration of alluvial fans as time-independent landforms may seem to suggest that the equilibrium approach was not useful for resolving the alluvial fan problem. However, Schumm and Lichty (1965) showed that a system in steady-state equilibrium (during a graded time span) does not necessarily preclude evolutionary change (over a cyclic time span) (Figure 1.5). They reconciled the apparent contradiction of considering steady-state equilibrium in a model of progressive change by specifying the temporal scale involved. The graded time span therefore refers to a short period of cyclic time when dynamic equilibrium exists. Because evolutionary change takes place over long time spans (cyclic time) but is observed over shorter time spans (graded time or steady time), the system appears

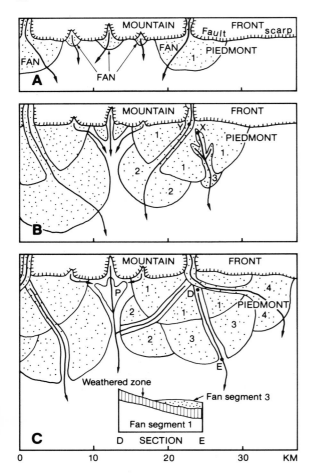

Figure 1.4. Diagram showing development of alluvial fans and shifts in the location of erosion and deposition. (A) Small fans at base of recently elevated mountain front; (B) Wash has dissected original fan segment (1) and is constructing a new fan segment (2); (C) Stream piracy has caused the abandonment of segments (2) and (3) and construction of another new fan segment (4). Stratigraphic relations between segments (1) and (3) in gully D–E. (Reproduced by permission of the American Journal of Science from Denny, 1967)

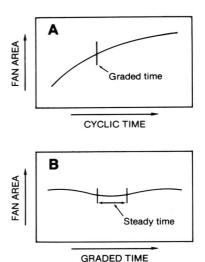

Figure 1.5. The concepts of cyclic, graded, and steady time as reflected in changes of fan area through time. (A) Progressive increase of fan area during cyclic time. Graded time is a small fraction of cyclic time, so fan area remains relatively constant during the graded time interval; (B) Fluctuations of fan area around a mean value during graded time. Fan area is constant when observed during the brief period of steady time. (Modified by permission of the American Journal of Science from Schumm and Lichty, 1965)

to be in steady-state equilibrium when, in fact, it is not. The variables in the system fluctuate about a changing mean condition (dynamic equilibrium). So, the equilibrium approach remains viable so long as the role of time is taken into account. In essence, this represents the integration of the evolutionary and the equilibrium concepts. The implications of this new approach in the fan dynamics paradigm will be considered below, but first it is necessary to comment in some detail on climatic and tectonic complications.

CLIMATIC AND TECTONIC COMPLICATIONS OF THE EQUILIBRIUM HYPOTHESIS

Schumm and Lichty (1965) argued that it may be impossible to exclude time from the analysis of landforms. Although Hack (1960) suggested that a steady-state balance exists between form and process that is independent of time, he qualified his argument by stating that if erosional energy changes through time, then form must also change. The two most important factors causing this change in erosional energy through time are climatic change and tectonics. Thus, the consideration of climatic and tectonic complications of the equilibrium hypothesis served to reintroduce the dimension of time to alluvial fan research, thereby refocusing attention on the alluvial fan problem.

The Climatic Factor

Climatic change influences the development of alluvial fans by inducing variability in the magnitude and frequency of fluvial processes that alter alluvial fan features. Lustig (1965) concluded that fans tend toward equilibrium in adjustment to different climatic conditions. Lustig's evidence included: (1) a shift in the locus of deposition downfan; (2) deep fanhead trenching at the fan apex; (3) presence of desert varnish on abandoned fan surfaces above the intersection point; and (4) greater estimated tractive force in active channels than on the fan surface.

Lustig believed that fan aggradation occurs regardless of climatic regime. During wet periods aggradation takes place within the drainage basin and below the mountain front (Figure 1.6A). During dry conditions trenching occurs in the upper portion of the fan as debris-flows become the dominant erosional process. This causes fanhead entrenchment and the shift of deposition downfan, but the fan still builds out from the mountain front (Figure 1.6B). Because fan entrenchment was so widespread in the American Southwest, Lustig cited the need for a regional explanation. He favoured a regional change in climate rather than a local explanation such as tectonic uplift. The type of climatic change required to accomplish entrenchment might be an increase in either storm frequency or intensity, in total precipitation, or a decrease in total precipitation that coincides with an increase in storm intensity (Cooke and Warren, 1973). Denny (1967:104) criticized Lustig's 'overreliance on climatic change as an explanation for features whose mode of origin may be in doubt,' noting that many of the features indicating climatic change could be explained by stream piracy.

Another problem with Lustig's model is that, in an arid phase, it requires debris-flows to be the dominant erosional agents due to their greater tractive force. Hooke (1967) and Wasson (1977b) doubted that debris-flows have the erosive ability to accomplish this. Based on observations of fans

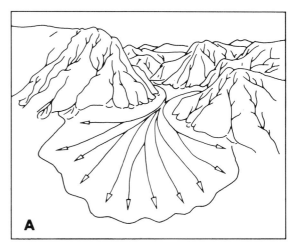

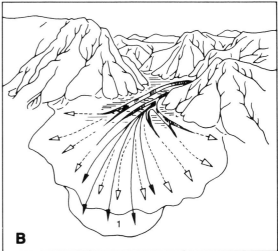

Figure 1.6. The formation of alluvial fans under the control of climate in an equilibrium hypothesis. (A) Aggradation during more humid or pluvial periods; (B) Trenching during a subsequent drier climate period. Intersection point moves downfan and mudflows and other infrequent density flows transport sediment to the lower reaches of the fan, building it outward into the valley (A). (From Lustig, 1965)

in the White Mountains of California, Beaty (1963, 1970) suggested that fans are built, not eroded by debris-flows. Others saw the present regime as a destructive phase in which no new material is being deposited (Hunt and Mabey, 1966:A97). Fans may also be relict features from different climatic conditions during the Pleistocene, therefore modern processes may not be indicative of events that influenced fan construction (Nilsen and Moore, 1984). Fan aggradation may be caused by climatically controlled increases in sediment production followed by fan trenching initiated when the temporarily abundant sediment supply is depleted. Bull (1964a, 1964b) cited variation in rainfall intensity as the principal cause of temporary fanhead trenching in Fresno County, California. As an alternative to climatic or tectonic explanations for stream entrenchment, Schumm (1973) suggested that the complex response of drainage systems may not be related to external influences. Instead a fan may grow until it exceeds some threshold slope that causes trenching to occur. Trenching was observed on an experimental fan where oversteepening of the fanhead was the result of exceeding an intrinsic geomorphic threshold (Weaver and Schumm, 1974; Schumm, 1980).

The Tectonic Factor

Although alluvial fans may form in areas where tectonic uplift is not an important factor (Drew, 1873; Carryer, 1966; Ryder, 1971a, 1971b; Church and Ryder, 1972; McPherson and Hirst, 1972; Roed and Wasylyk, 1973), they are especially prominent where uplift of mountainous regions provides a continual supply of fresh debris from steep drainage basins (Beaty, 1970). Regional analysis of landforms in the American West has revealed that the tectonically active Basin and Range Province has an abundance of alluvial fans, while in the tectonically stable areas of south–central Arizona the pediment is the dominant landform (Bull, 1977). In areas where tectonic activity is decreasing, alluvial fans may be replaced by pediments as the typical landform. The tectonic influence on the development of alluvial fans is considered in terms of fan entrenchment, fan segmentation, and the sedimentology, shape, and thickness of both modern and ancient alluvial fan deposits.

The rate of tectonic uplift in mountain areas relative to the rate of stream channel downcutting largely determines the locus of deposition and the thickness of the alluvial fan deposit (Bull, 1972, 1977). Bull proposed two phases of development for alluvial fans by considering differential uplift of the mountain block with respect to the valley block (Figure 1.1). The first type of fan develops where the rate of uplift is greater than the rate of stream channel downcutting, resulting in deposition adjacent to the mountain front (Figure 1.1A). Continued tectonic uplift results in the accumulation of thick alluvial fan deposits.

The second type occurs where the rate of channel downcutting at the mountain front exceeds the rate of uplift of the mountain mass; the fanhead becomes entrenched and the locus of deposition moves downslope from the fan apex (Figure 1.1B). This situation may be associated with a decrease in the rate of uplift. If the rate of tectonic uplift does not increase so that it exceeds the rate of stream channel downcutting, the fanhead area will be removed from active deposition and soil development will take place on that portion of the fan surface.

Although the overall radial profile of an alluvial fan is concave (Blissenbach, 1954), it is not always a smooth exponential curve. Instead, the connection of a series of distinct straight or, less commonly, concave segments that have progressively lower gradients downslope form what Bull (1961, 1964b) termed 'segmented fans' (Figure 1.7). Two types of segmented fans have been identified (Figure 1.8) and are related to changes in stream channel slope upstream from the fan apex due to tectonic uplift, climatic change, or base-level change (Bull, 1961).

The first type occurs in areas characterized by rapid, intermittent uplift. Mountain stream gradients will steepen and, as a result, a new and steeper fan segment will be constructed upslope of the previous segment. The opposite situation occurs where either a reduced rate of tectonic uplift or a climatic change induces accelerated stream incision and rapid downfan movement of the locus of deposition, resulting in a new fan segment with a lesser gradient. Therefore, in the first situation fan segments are steeper and younger in the upfan direction (Figure 1.8C), whereas in the second situation, fan segments are younger and gentler in the downfan direction (Figure 1.8D). Recognition of fan segmentation of the

THE ALLUVIAL FAN PROBLEM

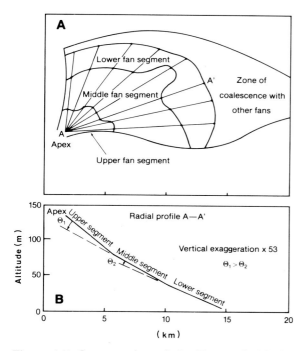

Figure 1.7. Segmentation of the Tumey Gulch fan, western Fresno County, California. (A) The segment boundaries are more strongly concentric than the contours, showing that fan segments are depositional forms rather than purely tectonic features. Each of the eight radial profiles has three straight line segments (B). (From Bull, 1964b)

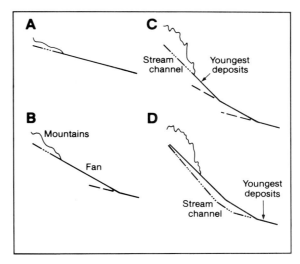

Figure 1.8. Segmented alluvial fan development in western Fresno County, California. (A) Fan and channel have developed a common gradient; (B) Fan gradient adjusts to steeper stream gradient due to uplift in mountains; (C) New phase of uplift causes younger segment, above older segment; (D) Increase in amount and intensity of precipitation causes stream channel entrenchment, moving the locus of deposition downfan. (From Bull, 1961)

first type is useful for reconstruction of the tectonic history of a mountain range (Bull, 1961).

From research conducted in Death Valley, California, Beaty (1961) provided an interpretation which conflicts with that proposed by Bull for the response of alluvial fans to differential uplift across a mountain front. Beaty suggested that uplift of the Grapevine Mountains had caused rejuvenation of the stream above the fan apex. This led to the incision of the fanhead and the construction of a new fan segment with a more gentle slope below the old fan surface. Further evidence of the influence of tectonic effects on fan entrenchment was provided by Denny (1967), who described how relative sinking of the valley floor due to normal faulting may cause fanhead trenching. Hooke (1968a, 1972) and Denny (1965, 1967) noted that deep trenching of the fans on the west side of Death Valley has been produced by the eastward tilting of the valley.

The tectonic history of a mountain range may be deciphered over longer time periods if one considers the geometry of the alluvial fan deposit. The accumulation of thick alluvial deposits in the geologic past has been noted frequently in the geologic literature (Nilsen, 1969; Miall, 1970; Steel, 1974). Bull (1972) identified three basic types of fan deposits using cross-sections along radial profiles. The first is wedge shaped with the thickest part near the mountain front and the thinnest part away from it (Figure 1.9A). The tectonic interpretation is one of major uplift in the mountains prior to fan deposition. The second type consists of lens-shaped sedimentary bodies that are thin both near the mountain front and far away from it, reflecting uplift that has continued during fan deposition (Figure 1.9B). The third type is represented by wedge-shaped deposits that are thin next to the mountain front and thick away from it. These indicate the cessation of

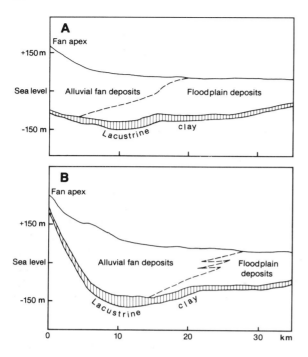

Figure 1.9. Longitudinal cross-sections of fan deposits. (A) Fan deposits that are thickest adjacent to the mountain front; (B) Fan deposits that are lenticular in shape. (Reproduced by permission from Bull, 1972 after Magleby and Klein, 1965, Plates 4 and 5)

tectonic activity and the erosion and redistribution of fan material away from the mountain front, which may eventually result in pediment formation.

Tectonic uplift affects the thickness of fan deposits and cycles of progradation, thus influencing sedimentary sequences. Two types of vertical fan sequence have generally been recognized. The first is a coarsening and thickening upward sequence that has been interpreted as indicative of fan progradation (outbuilding) at tectonically active basin margins (Heward, 1978). A fining and thinning upward sequence has been considered to be the result of fan retrogradation (retreat) and/or a reduction of source area relief associated with less tectonic activity at basin margins (Galloway and Hobday, 1983). Rust (1978) believed that the ideal sedimentary response to tectonic activity would be a coarsening upward sequence followed by a fining upward sequence as the fan system attempted to return to an equilibrium state.

CRISIS IN THE PARADIGM OF FAN DYNAMICS

Schumm and Lichty's effort to demonstrate that the equilibrium concept can be incorporated in an evolutionary approach took time to diffuse through geomorphology and be recognized explicitly by alluvial fan researchers (Wells, 1977). The investigation of climatic and tectonic complications to the equilibrium hypothesis, however, had a more direct influence on the approach taken in alluvial fan research.

The success of the equilibrium approach to the paradigm of fan dynamics was limited because geomorphologists had difficulty providing answers to the fundamental questions posed by the paradigm. After a decade of intensive study in the 1960s, researchers were unable to resolve the alluvial fan problem: to explain the evolutionary development of alluvial fans. Lustig (1967:96) concluded that 'the basic problem of fan formation is not yet resolved'. Furthermore, he suggested that future progress on this problem would require more data on rates of sedimentation and erosion on fans, and rates of bedrock weathering. The preceding discussion of fan development hypotheses illustrates the anomalies associated with each hypothesis and the overall diversity of opinions that prevailed. Although much progress was made towards understanding depositional processes, fan–basin relationships, and fan sedimentology, dissatisfaction with progress in resolving the alluvial fan problem signalled a shift in scientific efforts away from normal science and into a crisis stage.

According to Kuhn (1970), a crisis stage in the intellectual development of a scientific discipline is followed by a scientific revolution, unless the paradigm is able to adjust to accommodate the causes of the crisis. Further progress in the fan dynamics paradigm was largely dependent upon a more complete reconciliation of the evolutionary and equilibrium approaches. It was necessary to recognize that fans may tend toward some equilibrium state, but that it is difficult to evaluate what this state is or when it is reached without taking a historical perspective. This reconciliation was first suggested by Beaty (1970), but explicitly stated by Wasson (1977a). They have advocated construction of a general theory capable of accommodating the variety of conflicting conclusions reached thus far regarding the contemporary condition of alluvial fans, thereby integrating the time-independent and the time-bound aspects of alluvial fans into one comprehensive theory of alluvial fan development. Wasson (1977a) proposed a strategy for development of such a theory that would integrate investigation of modern processes on fans in different environments, with the historical analysis of alluvial fan evolution on active or relict fans. While some have doubted whether a general theory is achieveable (Rachocki, 1981), it does appear that recent fan research has adhered closely to this mandate. The fan dynamics paradigm has continued to remain viable by combining the successful parts of the evolutionary and equilibrium concepts into an approach that deals explicitly with the historical development of fans and with sedimentary processes in various environments.

An Integrated Approach to the Paradigm of Fan Dynamics

Several recent trends in the fan literature show promise to advance the field closer towards the goal of constructing a general model to explain fan development. The predominant trends observed throughout this period have been: (1) The detailed investigation of depositional processes and the relative importance of various processes in different environments; (2) Consideration of the effects of catastrophic events on alluvial fan processes and morphology; and (3) The use of sedimentology and numerical dating techniques to analyse the historical development of alluvial fans.

DEPOSITIONAL PROCESSES

Although the basic depositional processes operating on alluvial fans were recognized early during investigation of the fan dynamics paradigm (Bull, 1963, 1972; Beaty, 1963; Hooke, 1967), substantial progress has been made recently in examining the relative roles played by these processes in various environments. Because alluvial fans are commonly inactive (in some environments) during historical time periods, sedimentological analysis is usually the method used to interpret depositional processes.

Blair (1987) pointed out that previous studies have failed to provide detailed documentation of vertical stratification sequences and the sedimentary processes that produced them. He found that the Roaring River alluvial fan (created by a catastrophic flood in Rocky Mountain National Park, Colorado) was formed by the processes of sheetflooding and noncohesive sediment–gravity flow (NCSGF) deposition. An NCSGF is similar to but differentiated from a debris-flow (cohesive sediment–gravity flow) by the near absence of silt and clay in the matrix. The dominant sedimentary process responsible for building the fan was sheetflooding, an unconfined water flow moving down a slope (Hogg, 1982). Blair's study demonstrated the importance of source area sediment characteristics in the production of an NCSGF and water flows instead of debris-flows. Fanhead incision was attributed to the progressive downslope movement of the intersection point as fan progradation occurred. In addition, because he was able to obtain aerial photographs of the fan during deposition and because postflood modifications of the fan were carefully monitored, Blair suggested that the sedimentary processes inferred from examination of surficial fan deposits are not necessarily the same processes that build alluvial fans. Postflood modifications (i.e. development of braided distributary channels on top of the sheetflood deposits) on the Roaring River fan have eliminated almost all evidence that sheetflooding was the dominant depositional process responsible for fan aggradation.

The relative importance of different depositional processes on fans has become a major topic of interest (Harvey, 1984b; Kostaschuk et al., 1986; Wells and Harvey, 1987). A more traditional methodological approach using fan–basin morphometric relationships was applied to a group of alluvial fans in a recently glaciated, subhumid, mountainous region in the Bow River Valley of Alberta, Canada (Kostaschuk *et al.*, 1986). The fans were classified as fluvially dominated or debris-flow dominated. Allometric analysis of fan–basin morphometric relationships for these two distinct types of fans showed that large low gradient fans dominated by fluvial processes were associated with large, less rugged drainage basins. On the other hand, small, steep fans dominated by debris-flow processes were found to be associated with small rugged basins.

In a field study of small alluvial fans in northwestern England, Wells and Harvey (1987) showed that the spatial variation of the type of depositional process was controlled by variables other than tectonism or climatic change. They attributed the dominance of a particular depositional process to drainage basin size, channel gradient, the percentage of the drainage basin eroded, the type of sediment available from the drainage basin, and the location of the storm cell. Debris-flow processes were associated with small steep basins peripheral to the centre of the storm cell with a large percentage of the basin area yielding sediment. Intrinsic geomorphic thresholds were recognized as influencing depositional process types and the resulting spatial distribution of facies. Harvey (1984b) obtained similar results from a study of Spanish Quaternary deposits, but he also showed the importance of source area geology. Debris-flow fans were associated with small steep basins on sedimentary and low-grade metamorphic rocks, while fluvial fans were associated with high-grade metamorphic rocks.

Wasson (1977a,b) pointed out the importance of studying the differences in process in various environments. The primary factor responsible for the differences in fan morphology and sedimentology that exist between fans that form in different climates is the nature of the dominant depositional process on the fan (Kochel and Johnson, 1984). Detailed analysis of fan sediments in the Lower Derwent Valley, Tasmania, indicated that the processes operating on the Derwent fans were similar in most respects to those on the arid fans of the American Southwest (Wasson, 1977b). However, comparison with Hooke's (1967) maps showed that the Death Valley fans demonstrated little surface evidence of sheet-like flows, while the Derwent fans showed few indications of leveed debris-flows. Wasson (1977b) concluded that these kinds of comparisons could be improved only if processes of fan accumulation were catalogued for fans with different geologic, climatic, and topographic environments.

As a result, previously neglected alluvial fans found in humid-temperate and wet–tropical regions have received recent attention. While many humid-temperate fans have been considered as relict landforms, Kochel and Johnson (1984) demonstrated that a group of fans in central Virginia are still active today and were formed primarily by mass-wasting processes (debris-flows and debris-avalanches) generated by catastrophic precipitation events. Alluvial fans in Costa Rica provide further insight into the processes associated with a wet-tropical environment where an episodic sediment supply and highly variable stream discharge controls fan morphology and sedimentology (Kesel, 1985). These fans are dominated by relatively brief periods of active streamflow deposition in braided channels separated by long periods when deposition is minor and the growth of tropical vegetation gradually alters the braided, active channel to a more sinuous, inactive channel pattern.

CATASTROPHIC EVENTS

Previous study of the magnitude and frequency of fluvial processes (Wolman and Miller, 1960) has suggested that relatively frequent events of moderate intensity accomplish the greatest amount of geomorphic work—a defence of the principle of uniformitarianism. In contrast, based on evidence from the White Mountains of California and Nevada, Beaty (1974) reported that 'normal' stream processes contribute little in the way of sediment to the fans—no more than 10–15% by volume. He argued that catastrophic events (producing debris-flows) of low frequency and high magnitude are the usual geomorphic process in the western Great Basin. Although uniformitarianism may prevail over long timescales, catastrophism is more applicable over short time spans at the small scale of the individual drainage system (Beaty, 1974:50).

Determination of a dominant discharge has the potential to answer some of the questions posed by the catastrophism–uniformitarianism debate, but 'identification of the dominant event in controlling landforms has been an elusive problem' (Graf et al., 1980:282). Fan research has presented ideas on both sides of the argument: control by extreme events, or control by moderate events. The concept of dominant discharge as adapted to alluvial fan studies is defined as 'that discharge which, if it alone occurred, would produce a fan having the same slope as a fan built with a distribution of discharges' (Hooke and Rohrer, 1979:151). In the absence of a way to determine the magnitude and frequency of the dominant discharge on alluvial fans, Hooke and Rohrer used data from laboratory fans. They concluded that these fans were adjusted to forces of moderate magnitude and not to catastrophic events. The dominant discharge was also found to increase with increasing debris size, reaffirming Wolman and Miller's (1960) observation that catastrophic events become increasingly important as the threshold stress required to move material increases.

Although the geomorphic role of infrequent, large magnitude events is controversial, much recent research in the integrated approach to the fan dynamics paradigm has focused on catastrophic events. Radiocarbon dating of fan stratigraphy in central Virginia provided a preliminary estimate of a recurrence interval of 3000 to 6000 years for events with a magnitude similar to Hurricane Camille (which initiated the most recent activity on these fans in 1969) (Kochel and Johnson, 1984). In another humid–temperate region (England), Wells and Harvey (1987) concluded, on the effects of a storm with a return interval greater than 100 years, that high magnitude events of low frequency may be equally important in accomplishing geomorphic work as relatively frequent events of moderate magnitude.

In an attempt to describe the spatial variation of erosion and deposition during flood events, Harvey (1984c) looked at stream channel adjustment produced by a flood with a return interval of 25–100 years on an alluvial fan in a semiarid region in southern Spain. Even with continuous flow from this extreme event, erosion predominated in the proximal areas of the fan (fanhead trench) and deposition took place downfan. It was concluded that the spatial distribution of flow frequency influenced channel differences between the proximal and distal parts of the fan. The fanhead is modified by frequent events while downfan locations are affected only by extreme events. The creation of the Roaring River alluvial fan, formed during a catastrophic flood caused by the failure of a dam in Colorado, further demonstrates the importance of extreme events. However, Blair (1987) also pointed out that postflood, noncatastrophic events significantly modified the fan surface, obscuring evidence of the depositional process primarily responsible for fan formation (sheetflooding).

HISTORICAL ANALYSIS

Research dealing with the historical development of alluvial fans clearly demonstrates continued interest in the alluvial fan problem. The concurrent development and maturity of several dating techniques and sedimentological analyses has been the most important means for attacking this question. Studies of fan evolution fit well with research dealing with dominant depositional processes because progressive change in fan morphology also involves progressive change in the dominant process (Harvey, 1984a).

Wasson (1977a) provided a historical account of the evolution of a group of relict fans in Tasmania by using stratigraphic methods to determine the age of the fans. He suggested controls on their accumulation and dissection by examining changing sedimentary conditions in the drainage basin. Fan construction was associated with debris-flows caused by increased sediment production from periglacial and nivational processes of the last glacial period. Fan dissection occurred as drainage basins were revegetated following deglaciation, causing a decrease in the ratio of sediment yield to discharge.

In a series of papers from research conducted on Spanish alluvial fans, Harvey (1978, 1984a,b) first described their sedimentary and geomorphic characteristics and then identified two main phases of development. These were, first, a fan aggradation phase during a period of net excess sediment supply, and second, a fan dissection phase of net sediment deficiency. He then proposed that fan development reflected long-term progressive change (dynamic equilibrium over cyclic time) complicated by short-term response to climatic fluctuations and spatially variable trenching thresholds (Harvey, 1984b). Harvey (1984a) concluded that the Spanish fans were not equilibrium forms as proposed by Hooke (1968a).

The development of alluvial fans in or near glaciated regions is often strongly controlled by the cyclic alternation of glacial and nonglacial conditions, presenting geomorphologists with a potentially useful method for reconstructing Quaternary history. 'Fan formation continues until the system reaches a quasi-equilibrium condition or the supply of fan sediment is arrested' (Ritter and Ten Brink, 1986:621). In this perspective fans are seen as time-dependent landforms. Similarly, the development of wet-tropical fans is closely related to an episodic sediment supply, in this case, caused by volcanic activity or earthquakes in the source basin (Kesel, 1985).

Rachocki (1981) suggested a model of fan evolution in which fans develop through time until the source area can no longer provide sediment to the fan. The model (his figure 132) is similar to Figure 1.5A showing fan area (a surrogate measure of fan volume) increasing with time at a decreasing rate until fan growth levels off and the depositional landform (Bull, 1977) becomes an erosional surface which decreases in volume.

The establishment of depositional chronologies for fan deposits would be an important step towards construction of an acceptable model of fan development, but dated fan surfaces and knowledge of their associated environmental conditions are rarely available (Cooke and Warren, 1973; Lustig, 1967), particularly in arid regions where organic material for radiocarbon dating is lacking. The stable carbon isotopic analysis of organic matter in rock varnish provides information on the relative aridity of the environment in which varnish is formed, and when used with cation–ratio dating, it can provide a chronology of fan development related to climatic fluctuations (Dorn et al., 1987). Reconstruction of the history of two alluvial fans in Death Valley has led Dorn et al. (1987) to identify three aggradation–entrenchment cycles. They concluded that these fans were deposited during humid periods with greater basin vegetation cover and soil development. Climatic change to more arid conditions was thought to have promoted fanhead trenching which was made permanent by the eastward tilting of the Panamint Range. The use of rock varnish has the potential to resolve some of the conflicting interpretations of fan development arrived at by early researchers in the Death Valley region.

The integrated approach to the fan dynamics paradigm has been relatively successful in providing a focus for alluvial fan research. Researchers interested in studying alluvial fans are not yet ready to reject the basic precepts of the fan dynamics paradigm. Instead, the reconciliation of equilibrium and evolutionary approaches has led to the detailed investigation of process with significant attention paid to long-term development.

Alluvial Fans and Fluvial Theory

In reviewing the intellectual history of alluvial fan research it is useful to consider how research on alluvial fans has contributed to theory development in the broader field of geomorphology. Following dissatisfaction with concepts of long-term landform evolution associated with Davis's cycle of erosion, geomorphologists shifted to the smaller temporal and spatial scales of process investigation. In order to describe the direction of research in contemporary fluvial geomorphology, Knighton (1984) emphasized the distinction between empirical and theoretical approaches. The empirical approach is largely inductive and involves establishing relationships between form and process variables obtained from the analysis of field data. Its goals are prediction using a 'black-box' approach rather than explanation of causal process links (Chorley, 1978). Statistical relationships and use of power functions are typical of this mode of inquiry. This approach treats external artifacts that result from the operation of geomorphic systems rather than the systems themselves (Chorley, 1978). The deductive nature of the theoretical approach involves the construction of deterministic and stochastic models. It places more emphasis on explanation than the identification of functional relationships. A 'white-box' approach is used in the detailed investigation of transfers of mass and energy. Deductive reasoning leads to statements that constitute theories (Church and Mark, 1980). Due to the lack of established fluvial theory in geomorphology, ideas have been borrowed from related fields such as hydraulic engineering and then modified for use in geomorphic problem-solving. These theories have their foundation in the laws of chemistry and physics.

Most research on alluvial fans has followed the empirical approach (Beaty, 1963; Bull, 1964b; Denny, 1965; Hooke, 1968a). Bull (1964b) proposed

$$A_f = cA_d^b \qquad (1.1)$$

to relate characteristics of alluvial fans to characteristics of their drainage basins, by the use of the power function where A_f is fan area, A_d is drainage basin area, and b and c are empirically derived coefficients. The exponent b has been found to be approximately equal to 0.9, while c varies geographically (Denny, 1967; Hooke, 1968a) (Figure 1.2). This relationship represents the correlation between fan and basin geometry. Church and Mark (1980) stated that these relationships are of scientific interest only if they can provide a better understanding of causal mechanisms. Even though such techniques are not deductively rigorous, they may serve as important preliminary steps in the development of geomorphic theories.

Thus, the contribution made by alluvial fan research to the development of fluvial theory has been to provide a preliminary step in theory formulation. This is not surprising in view of the general lack of theory in geomorphology (Knighton, 1984). Fan research has focused on the collection and analysis of empirical data and on the establishment of relationships between variables expressing geomorphic phenomena. The empirical approach is considered a necessary precursor to successful theory formulation since the testing of theoretical models with 'real world' data is required to verify their validity as accurate representations of natural phenomena. Herein lies the value of field research in theory development. Fan research is therefore an ideal example of the field approach in geomorphology.

Although field work has been an essential element in alluvial fan studies, generalizations arrived at empirically are influenced by the selection of one geographic region over another. Graf (1984) concluded that the size of the geomorphic research community has been a major factor influencing spatial bias in geomorphic theory development. By analogy, the smaller size of the alluvial fan research community would be expected to have promoted an even greater spatial bias in the selection of fan field sites. Indeed, this has been the case. Most field sites have been located in the arid and semiarid regions of the American Southwest, namely portions of California, Arizona, and Nevada that lie within the Basin and Range Province. If one identifies the number of publications that emanate from selected field locations, spatial bias in the selection of field sites becomes even more pronounced. For example, W. B. Bull (western Fresno County, California), R. L. Hooke (Death Valley, California), and C. B. Beaty (White Mountains, California and Nevada) have produced a disproportionately large number of publications from research conducted largely in just these three field locations (about 25% of the literature dealing with modern alluvial fans). More importantly, those publications considered as the seminal works in fan research come almost exclusively from the American Southwest.

In spite of recent attempts to study fans in different environments, striking gaps remain in site selection for alluvial fan field research. Review of Nilsen and Moore's (1984) *Bibliography of Alluvial Fan Deposits* (containing over 700 entries) revealed only two references each from the continents of Africa and South America and none from China or Southeast Asia. Language barriers undoubtedly explain part of the apparent spatial bias in field site selection. All too often research in languages other than English goes unnoticed. Efforts are needed to expose English-speaking fan researchers to work done in these other regions. Future investigators might consider selecting field sites in different environments in order to test hypotheses based on previous research in spatially biased localities.

Even though field research has played a predominant role in fan research, the construction of artificial alluvial fans in the laboratory has been attempted by several researchers (Hooke, 1965, 1967, 1968a, 1968b; Weaver and Schumm, 1974; Hooke and Rohrer, 1979; Rachocki, 1981; Schumm *et al.*, 1987) partly because the conditions required for their construction are easily simulated (Schumm, 1977). Processes of erosion, transportation, and deposition on natural alluvial fans can be observed only infrequently; therefore, experimental fans are useful for investigating temporal changes in process operation. Hooke (1968b) recognized the difficulty involved with scaling problems and chose to treat laboratory fans as small fans in their own right instead of as scale models of larger natural fans. His 'similarity of process' approach involved using fans constructed in the laboratory to answer questions of process and morphology posed by the investigation of fans in the field. Data obtained from laboratory fans were qualitative because scaling relationships for sediment transport by streams and debris-flows were not understood sufficiently to permit exact modelling. Hooke and Rohrer (1979) later used laboratory fans to show the effect of discharge and sediment size on fan slope, the effect of sediment size on the magnitude of dominant discharge, and the variation of slope with azimuth. Hooke's 'similarity of process' approach served to enhance the position of geomorphology as a field science because it made laboratory modelling dependent upon information obtained by field observation.

Conclusion

Review of the alluvial fan literature shows that research paradigms provide a useful means for evaluating trends in fan research and the progress made towards resolving the alluvial fan problem. Most research has been accomplished under the paradigm of fan dynamics. The fundamental research question posed by this paradigm addressed the determination of the long-term development of alluvial fans (i.e. the alluvial fan problem). The equilibrium hypothesis soon became the main approach that dealt with the question of fan development. The early assumption of the time-independence of alluvial fans was followed by the recognition that fan variables may fluctuate around a mean condition that changes with time. Investigation of the climatic and tectonic complications of the equilibrium hypothesis also demonstrated the need to incorporate the equilibrium concept in an evolutionary hypothesis. Dissatisfaction with the progress made towards resolution of the alluvial fan problem resulted in an intellectual crisis which was resolved by combining the equilbrium and evolutionary approaches. This integrated approach to the fan dynamics paradigm has attempted to combine the investigation of sedimentary processes on fans in different environments with the historical analysis of fan evolution for the purpose of formulating a general model of fan development.

The contribution made by fan research to the development of fluvial theory has been to establish empirical relationships between fan and basin variables that may be used to verify theoretical models. Field work has been an essential component of fan research; however, generalizations based on evidence gathered from past alluvial fan field sites are likely to be spatially biased. Future research efforts should consider the selection of field sites in different environmental settings. Laboratory fan models have also been used to answer questions of process and morphology derived from the study of fans in the field, and are heavily dependent upon empirical field data.

Acknowledgements

I wish to express gratitude to Jonathan D. Phillips (Arizona State University) for reading the final draft and providing many useful comments for completion of the manuscript, and William L. Graf (Arizona State University) for providing me with valuable advice and the opportunity to write. Review by Chester B. Beaty greatly improved the final product. A special note of gratitude is extended to Linda S. O'Hirok, Tomas A. Miller, and Judith K. Haschenburger for their continual support, helpful editorial comments, and thought-provoking discussion throughout the project. Thanks are also due Charlie Rader (Staff Cartographer, Arizona State University) for drafting the line figures.

References

Anderson, G. S. and Hussey, K. M. 1962. Alluvial fan development at Franklin Bluffs, Alaska. *Iowa Academy of Sciences Proceedings*, **69**, 310–322.

Anstey, R. L. 1965. Physical characteristics of alluvial fans. *U.S. Army Natick Laboratories, Technical Report*, **ES-20**, 109 pp.

Bagnold, R. A. 1935. The movement of desert sand. *Geographical Journal*, **85**, 342–369.

Bagnold, R. A. 1937. The transport of sand by wind. *Geographical Journal*, **89**, 409–438.

Bagnold, R. A. 1941. *The Physics of Blown Sand and Desert Dunes*. Methuen, London. 265 pp.

Beaty, C. B. 1961. Topographic effects of faulting: Death Valley, California. *Annals, Association of American Geographers*, **51**, 234–240.

Beaty, C. B. 1963. Origin of alluvial fans, White Mountains, California and Nevada. *Annals, Association of American Geographers*, **53**, 516–535.

Beaty, C. B. 1970. Age and estimated rate of accumulation of an alluvial fan, White Mountains, California, U.S.A. *American Journal of Science*, **268**, 50–70.

Beaty, C. B. 1974. Debris flows, alluvial fans, and a revitalized catastrophism. *Zeitschrift für Geomorphologie Supplementband*, **21**, 39–51.

Beaumont, P. 1972. Alluvial fans along the foothills of the Elburz Mountains, Iran. *Paleogeography, Paleoclimatology, Paleoecology*, **12**, 251–273.

Blackwelder, E. 1928. Mudflow as a geologic agent in semi-arid mountains. *Geological Society of America Bulletin*, **39**, 465–484.

Blair, T. C. 1987. Sedimentary processes, vertical stratification sequences, and geomorphology of the Roaring River alluvial fan, Rocky Mountain National Park. *Journal of Sedimentary Petrology*, **57**, 1–18.

Blissenbach, E. 1952. Relation of surface angle distribution to particle size distribution on alluvial fans. *Journal of Sedimentary Petrology*, **22**, 25–28.

Blissenbach, E. 1954. Geology of alluvial fans in semi-arid regions. *Geological Society of America Bulletin*, **65**, 175–190.

Boothroyd, J. C. 1972. Coarse-grained sedimentation on a braided outwash fan, northwest Gulf of Alaska. *University of South Carolina, Coastal Research Division, Department of Geology, Technical Report*, **6–CRD**, 127 pp.

Boothroyd, J. C. and Nummedal, D. 1978. Proglacial braided outwash—a model for humid alluvial fan deposits. In Miall, A. D. (Ed.), *Fluvial Sedimentology*. Canadian Society of Petroleum Geologists Memoir, **5**, 641–668.

Büdel, J. 1977. *Klima-Geomorphologie*. Gebrüder Borntraeger, Berlin. Translated Fischer, L. and Busche, D.

Büdel, J. 1982. *Climatic Geomorphology*. Princeton University Press, Princeton. 443 pp.

Bull, W. B. 1961. Tectonic significance of radial profiles of alluvial fans in western Fresno County, California. *U.S. Geological Survey Professional Paper*, **424–B**, 181–184.

Bull, W. B. 1963. Alluvial-fan deposits in western Fresno County, California. *Journal of Geology*, **71**, 243–251.

Bull, W. B. 1964a. History and causes of channel trenching in western Fresno County, California. *American Journal of Science*, **262**, 249–258.

Bull, W. B. 1964b. Geomorphology of segmented alluvial fans in western Fresno County, California. *U.S. Geological Survey Professional Paper*, **352–E**, 89–129.

Bull, W. B. 1968. Alluvial fans. *Journal of Geologic Education*, **16**, 101–106.

Bull, W. B. 1972. Recognition of alluvial fan deposits in the stratigraphic record. In Hamblin, W. K. and Rigby, J. K. (Eds), *Recognition of Ancient Sedimentary Environments*. Society of Economic Paleontologists and Mineralogists Special Publication, **16**, 63–83.

Bull, W. B. 1977. The alluvial fan environment. *Progress in Physical Geography*, **1**, 222–270.

Carryer, S. J. 1966. A note on the formation of alluvial fans, New Zealand. *Journal of Geology and Geophysics*, **9**, 91–94.

Chawner, W. D. 1935. Alluvial fan flooding. *Geographical Review*, **25**, 255–263.

Chorley, R. J. 1978. Bases for theory in geomorphology. In Embleton, C., Brunsden, D., and Jones, D. K. (Eds), *Geomorphology: Present Problems and*

Future Prospects. Oxford University Press, Oxford. 1–13.
Chorley, R. J. and Kennedy, B. A. 1971. *Physical Geography: A Systems Approach*. Prentice-Hall International, London, 375 pp.
Church, M. and Mark, D. 1980. On size and scale in geomorphology. *Progress in Physical Geography*, **4**, 342–390.
Church, M. and Ryder, J. M. 1972. Paraglacial sedimentation: a consideration of fluvial processes conditioned by glaciation. *Geological Society of America Bulletin*, **83**, 3059–3072.
Cooke, R. U. and Reeves, R. W. 1976. *Arroyos and Environmental Change in the American South-West*. Clarendon Press, Oxford. 213 pp.
Cooke, R. U. and Warren, A. 1973. *Geomorphology in Deserts*. United Kingdom, Batsford. 374 pp.
Davis, W. M. 1905. The geographical cycle in an arid climate. *Journal of Geology*, **13**, 381–407.
Denny, C. S. 1965. Alluvial fans in the Death Valley region, California and Nevada. *U.S. Geological Survey Professional Paper*, **466**, 1–62.
Denny, C. S. 1967. Fans and pediments. *American Journal of Science*, **265**, 81–105.
Doehring, D. O. 1970. Discrimination of pediments and alluvial fans from topographic maps. *Geological Society of America Bulletin*, **81**, 3109–3115.
Dorn, R. I., DeNiro, M. J., and Ajie, H. O. 1987. Isotopic evidence for climatic influence on alluvial-fan development in Death Valley, California. *Geology*, **15**, 108–110.
Drew, F. 1873. Alluvial and lacustrine deposits and glacial records of the Upper Indus basin. *Quarterly Journal of the Geological Society of London*, **29**, 441–471.
Eckis, R. 1928. Alluvial fans in the Cucamonga district, southern California. *Journal of Geology*, **36**, 111–141.
Galloway, W. E. and Hobday, D. K. 1983. *Terrigenous Clastic Depositional Systems*. Springer-Verlag, New York. 423 pp.
Gole, C. V. and Chitale, S. V. 1966. Inland delta building activity of the Kosi River. *American Society of Civil Engineers, Journal of the Hydraulics Division*, **92** (HY2), 111–126.
Graf, W. L. 1983. The arroyo problem—paleohydrology and paleohydraulics in the short term. In Gregory, K. J. (Ed.), *Background to Paleohydrology*, John Wiley and Sons, London. 279–302.
Graf, W. L. 1984. The geography of American field geomorphology. *Professional Geographer*, **36**, 78–82.
Graf, W. L., Trimble, S. W., Toy, T. J., and Costa, J. E. 1980. Geographic geomorphology in the eighties. *Professional Geographer*, **32**, 279–284.
Gregory, K. J. 1985. *The Nature of Physical Geography*. Arnold, London. 262 pp.
Hack, J. T. 1960. Interpretation of erosional topography in humid temperate regions. *American Journal of Science*, **258–A**, 80–97.
Hart, M. G. 1986. *Geomorphology: Pure and Applied*. George Allen and Unwin, London, 228 pp.
Harvey, A. M. 1978. Dissected alluvial fans in southeast Spain. *Catena*, **5**, 177–211.
Harvey, A. M. 1984a. Aggradation and dissection sequences on Spanish alluvial fans: Influence on morphological development. *Catena*, **11**, 289–304.
Harvey, A. M. 1984b. Debris flows and fluvial deposits in Spanish Quaternary alluvial fans: implications for fan morphology. In Koster, E. H. and Steel, R. J. (Eds), *Gravels and Conglomerates. Canadian Society of Petroleum Geologists Memoir*, **10**, 123–132.
Harvey, A. M. 1984c. Geomorphological response to an extreme flood: a case from southeast Spain. *Earth Surface Processes and Landforms*, **9**, 267–279.
Heward, A. P. 1978. Alluvial fan sequence and megasequence models: with examples from Westphalian D–Stephanian B coalfields, northern Spain. In Miall, A. D. (Ed.), *Fluvial Sedimentology. Canadian Society of Petroleum Geologists Memoir*, **5**, 669–702.
Hjulström, F. 1935. Studies of the morphological activity of rivers as illustrated by the River Fyris. *Bulletin of the Geological Institute of Uppsala*, **25**, 221–527.
Hogg, S. E. 1982. Sheetfloods, sheetwash, sheetflow, or . . .? *Earth Science Review*, **18**, 59–76.
Hooke, R. L. 1965. *Alluvial fans*. California Institute of Technology, Pasadena, Ph.D. Thesis. 192 pp.
Hooke, R. L. 1967. Processes on arid-region alluvial fans. *Journal of Geology*, **75**, 438–460.
Hooke, R. L. 1968a. Steady-state relationships of arid-region alluvial fans in closed basins. *American Journal of Science*, **266**, 609–629.
Hooke, R. L. 1968b. Model geology: prototype and laboratory streams: a discussion. *Geological Society of America Bulletin*, **79**, 391–394.
Hooke, R. L. 1972. Geomorphic evidence for Late-Wisconsin and Holocene tectonic deformation in Death Valley, California. *Geological Society of America Bulletin*, **83**, 2073–2098.
Hooke, R. L. and Rohrer, W. L. 1977. Relative erodibility of source-area rock types, as determined from second-order variations in alluvial-fan size. *Geological Society of America Bulletin*, **88**, 1177–1182.
Hooke, R. L. and Rohrer, W. L. 1979. Geometry of alluvial fans: effect of discharge and sediment size. *Earth Surface Processes*, **4**, 147–166.
Hoppe, G. and Ekman, S. R. 1964. A note on the alluvial fans of Ladtjovagge, Swedish Lapland. *Geografiska Annaler*, **46**, 338–342.
Horton, R. E. 1932. Drainage basin characteristics. *Transactions of the American Geophysical Union*, **13**, 350–361.
Horton, R. E. 1945. Erosional development of streams and their drainage basins: hydrophysical approach to

quantitative morphology. *Geological Society of America Bulletin*, **56**, 275–370.

Hunt, C. B. and Mabey, D. R. 1966. Stratigraphy and structure, Death Valley, California. *U.S. Geological Survey Professional Paper*, **494–A**, 162 pp.

Iso, N., Yamakawa, K., Yonezawa, H., and Matsubara, T. 1980. Accumulation rates of alluvial cones constructed by debris-flow deposits in the drainage basin of the Takahara River, Gifu prefecture, central Japan. *Geographical Review of Japan*, **53**, 699–720.

Johnston, R. J. 1983. *Geography and Geographers: Anglo-American Human Geography since 1945*. Arnold, London. 264 pp.

Kennedy, B. A. 1977. A question of scale? *Progress in Physical Geography*, **1**, 154–157.

Kesel, R. H. 1985. Alluvial fan systems in a wet-tropical environment, Costa Rica. *National Geographic Research*, **1**, 450–469.

King, L. C. 1953. Canons of landscape evolution. *Geological Society of America Bulletin*, **64**, 721–752.

King, L. C. 1962. *The Morphology of Earth*, Oliver and Boyd, Edinburgh. 726 pp.

Knighton, D. 1984. *Fluvial Forms and Processes*. Arnold, London. 218 pp.

Knopf, A. 1918. A geologic reconnaissance of the Inyo Range and the eastern slope of the Sierra Nevada, California. *U.S. Geological Survey Professional Paper*, **438**, 68 pp.

Kochel, R. C. and Johnson, R. A. 1984. Geomorphology, sedimentology, and depositional processes of humid-temperate alluvial fans in central Virginia, U.S.A. In Koster, E. and Steel, R. (Eds), *Gravels and Conglomerates, Canadian Society of Petroleum Geologists Memoir*, **10**, 109–122.

Kostaschuk, R. A., MacDonald, G. M., and Putnam, P. E. 1986. Depositional process and alluvial fan–drainage basin morphometric relationships near Banff, Alberta, Canada. *Earth Surface Processes and Landforms*, **11**, 471–484.

Kuhn, T. S. 1970. *The Structure of Scientific Revolutions*. University of Chicago Press, Chicago. 210 pp.

Lakatos, I. 1970. Falsification and the methodology of scientific research programs. In Lakatos, I. and Musgrave, H. (Eds), *Criticism and the Growth of Knowledge*. Cambridge University Press, London. 282 pp.

Lawson, A. C. 1913. The petrographic designation of alluvial fan formations. *University of California Publications in Geology*, **7**, 325–334.

Leggett, R. F., Brown, R. J., and Johnston, G. H. 1966. Alluvial fan formation near Aklavik, Northwest Territories, Canada. *Geological Society of America Bulletin*, **77**, 15–30.

Leighly, J. B. 1932. Towards a theory of the morphologic significance of turbulence in the flow of water in streams. *University of California Publications in Geography*, **6**, 1–22.

Lustig, L. K. 1965. Clastic sedimentation in Deep Springs Valley, California. *U.S. Geological Survey Professional Paper*, **352–F**, 131–192.

Lustig, L. K. 1967. *Inventory of research on geomorphology and surface hydrology of desert enviroments. Chapter IV*. Office of Arid Lands Research, University of Arizona, Tucson, Arizona. 189 pp.

Magleby, D. C. and Klein, I. E. 1965. Ground-water conditions and potential pumping resources above the Corcoran Clay—an addendum to the ground-water geology and resources definite plan appendix. *U.S. Bureau of Reclamation Open File Report*, 21 plates.

Masterman, M. 1970. The nature of a paradigm. In Lakatos, I. and Musgrave, A. (Eds), *Criticism and the Growth of Knowledge*. Cambridge University Press, Cambridge. 59–90.

Matsuda, I. 1974. Distribution of the recent deposits and buried landforms in the Kanto Lowland, central Japan. *Tokyo Metropolitan University, Geographical Report*, **9**, 1–36.

McPherson, H. J. and Hirst, F. 1972. Sediment changes on two alluvial fans in the Canadian Cordillera. *British Columbia Geographical Series*, **14**, 161–175.

Melton, M. A. 1965. The geomorphic and paleoclimatic significance of alluvial deposits in southern Arizona. *Journal of Geology*, **73**, 1–38.

Miall, A. D. 1970. Devonian alluvial fans, Prince of Wales Island, Arctic Canada. *Journal of Sedimentary Petrology*, **40**, 556–571.

Mukerji, A. B. 1976. Terminal fans of inland streams in Sutlej–Yamuna Plain, India. *Zeitschrift für Geomorphologie*, **20**, 190–204.

Murata, T. 1931a. Theoretical consideration on the shape of alluvial fans. *Geographical Review of Japan*, **7**, 569–586. (In Japanese)

Murata, T. 1931b. Relation between a fan and its surrounding mountains. *Geographical Review of Japan*, **7**, 649–663. (In Japanese)

Murata, T. 1966. A theoretical study of the forms of alluvial fans. *Geographical Reports of Tokyo Metropolitan University*, **1**, 33–43.

Nilsen, T. H. 1969. Old Red sedimentation in the Buelandet–Vaerlandet Devonian district, western Norway. *Sedimentary Geology*, **3**, 35–57.

Nilsen, T. H. and Moore, T. E. 1984. *Bibliography of Alluvial-Fan Deposits*. Geo Books, Norwich. 96 pp.

Oberlander, T. M. 1974. Landscape inheritance and the pediment problem in the Mojave Desert of southern California. *American Journal of Science*, **274**, 849–875.

Penck, W. 1924. *Die Morphologische Analyse*. Englehorn, Stuttgart. 283 pp.

Popper, K. R. 1970. Normal science and its dangers. In Lakatos, I. and Musgrave, A. (Eds), *Criticism and the Growth of Knowledge*. Cambridge University Press, London. 51–58.

Rachocki, A. 1981. *Alluvial Fans: An Attempt at an*

Empirical Approach. John Wiley and Sons, New York. 166 pp.

Ritter, D. F. and Ten Brink, N. W. 1986. Alluvial fan development in the glaciofluvial cycle, Nenana Valley, Alaska. *Journal of Geology*, **94**, 613–625.

Roed, M. A. and Wasylyk, D. G. 1973. Age of inactive alluvial fans—Bow River Valley, Alberta. *Canadian Journal of Earth Sciences*, **10**, 1834–1840.

Rust, B. R. 1978. Depositional models for braided alluvium. In Miall, A. D. (Ed.), *Fluvial Sedimentology*. Canadian Society of Petroleum Geologists Memoir, **5**, 605–625.

Ryder, J. M. 1971a. The stratigraphy and morphology of paraglacial fans in south–central British Columbia. *Canadian Journal of Earth Sciences*, **8**, 279–298.

Ryder, J. M. 1971b. Some aspects of the morphometry of paraglacial alluvial fans in south–central British Columbia. *Canadian Journal of Earth Sciences*, **8**, 1252–1264.

Saito, K. 1980. Classification of alluvial fans in Tohuku district based on cluster analysis. *Geographical Review of Japan*, **53**, 721–729.

Schramm, W. E. 1981. *Humid Tropical Alluvial Fans, Northwest Honduras*. M.S. thesis, Louisiana State University, Baton Rouge.

Schroeder, B. 1971. The age of alluvial fans east of Corinth, Greece. *Neues Jahrbuch für Geologie und Paleontologie*, **6**, 363–371.

Schumm, S. A. 1973. Geomorphic thresholds and complex response of drainage systems. In Morisawa, M. (Ed.), *Fluvial Geomorphology*. Proceedings of the 4th Annual Binghamton Geomorphology Symposium, Binghamton, 299–310.

Schumm, S. A. 1977. *The Fluvial System*. John Wiley and Sons, New York. 338 pp.

Schumm, S. A. 1980. Some applications of the concept of geomorphic thresholds. In Coates, D. R. and Vitek, J. D. (Eds), *Thresholds in Geomorphology*. Proceedings of the 9th Annual Binghamton Geomorphology Symposium, Oct. 1978, Binghamton. 473–485.

Schumm, S. A. and Lichty, R. W. 1965. Time, space and causality in geomorphology. *American Journal of Science*, **263**, 110–119.

Schumm, S. A., Mosley, M. P., and Weaver, W. E. 1987. *Experimental Fluvial Geomorphology*. Wiley–Interscience (John Wiley and Sons), New York. 413 pp.

Shapere, D. 1964. The structure of scientific revolutions. *Philosophical Review*, **73**, 383–394.

Steel, R. J. 1974. New Red Sandstone floodplain and piedmont sedimentation in the Hebridean Province, Scotland. *Journal of Sedimentary Petrology*, **44**, 336–357.

Strahler, A. N. 1950. Davis's concept of slope development viewed in the light of recent quantitative investigations. *Annals, Association of American Geographers*, **40**, 209–213.

Surell, A. 1841. *Etude sur les Torrents de Hautes Alpes*. 1st Edition, Paris; cited by Bull, W. B. 1977. loc. cit. ante.

Tinkler, K. J. 1985. *A Short History of Geomorphology*. Croom Helm, London and Sydney, 317 pp.

Tolman, C. F. 1909. Erosion and deposition in the southern Arizona bolson region. *Journal of Geology*, **17**, 136–163.

Trowbridge, A. C. 1911. The terrestrial deposits of the Owens Valley, California. *Journal of Geology*, **19**, 706–747.

Wasson, R. J. 1974. Intersection point deposition on alluvial fans: an Australian example. *Geografiska Annaler*, **56**, 83–92.

Wasson, R. J. 1977a. Catchment processes and the evolution of alluvial fans in the lower Derwent Valley, Tasmania. *Zeitschrift für Geomorphologie, Supplementband*, **21**, 147–168.

Wasson, R. J. 1977b. Late-glacial alluvial fan sedimentation in the lower Derwent Valley, Tasmania. *Sedimentology*, **24**, 781–799.

Watkins, R. J. 1970. Against normal science. In Lakatos, I. and Musgrave, A., (Eds), *Criticism and the Growth of Knowledge*. Cambridge University Press, London. 25–38.

Weaver, W. E. and Schumm, S. A. 1974. Fanhead trenching: an example of a geomorphic threshold. *Geological Society of America, Abstracts with Program*, **6**, 481.

Wells, S. G. 1977. Geomorphic controls of alluvial fan deposition in the Sonoran Desert, southwestern Arizona. In Doehring, D. O. (Ed.), *Geomorphology in Arid Regions*. State University of New York, Publications in Geomorphology, Binghamton. 27–50.

Wells, S. G. and Harvey, A. M. 1987. Sedimentologic and geomorphic variations in storm-generated alluvial fans, Howgill Fells, northwest England. *Geological Society of America Bulletin*, **98**, 182–198.

Winder, C. G. 1965. Alluvial cone construction by alpine mudflow in a humid-temperate region. *Canadian Journal of Earth Sciences*, **2**, 270–277.

Wolman, M. G. and Miller, J. P. 1960. Magnitude and frequency of forces in geomorphic processes. *Journal of Geology*, **68**, 54–74.

Yazawa, D., Toya, H., and Kaizuka, S. 1971. *Alluvial Fans*, Kokon Shoin, Tokyo. 318 pp. (In Japanese).

SECTION I

ALLUVIAL FANS IN RECENT ENVIRONMENTAL SETTINGS

CHAPTER 2

Quaternary Alluvial Fans in the Karakoram Mountains

Edward Derbyshire and Lewis A. Owen
University of Leicester, Leicester

Abstract

The Karakoram Mountains are one of the highest and most geomorphologically dynamic regions in the world, influenced by intense tectonism, high denudation, and a wide range of differing environments, altitudinally controlled but essentially juxtaposed. This provides a large range of landforms and sediment deposits: the fan is the most characteristic sedimentary geometry.

The techniques used to study these landforms are discussed and case studies from the Gilgit and Hunza valleys are used to provide models for fan development, emphasizing their polygenetic formation. The sediment fans in this area are essentially ancient, and few fans are still actively aggrading by fluvial or debris-flow processes. This is evident by marked fanhead entrenchment and fan-toe truncation. Resedimentation of till deposits by debris-flow processes has been the most important process in the formation of the sediment fans. A major phase of fan development occurred in the Gilgit area about 60 000 yr B.P., soon after the last major glaciation in the Karakoram Mountains. In the Hunza Valley, the fans are younger and developed after about 47 000 yr B.P., as glacial advances still dominated this area at that time.

Introduction

River valleys and intermontane basins within high mountain belts act as sinks for a large variety of sediments derived from the adjacent mountains including the catchment areas of fluvial, glacial, and aeolian systems. A variety of landforms results from the deposition of these sediments, including floodplains, terraces, and fans. These landforms may become modified by erosion and subsequent deposition, resulting in complex polygenetic landforms, many of which may have similar morphology.

Fans according to Denny (1967) are landforms having slopes of less than 20° with an apex at a topographical high from which the landform appears to radiate. Alluvial fans are a special type, but definitions are vague, involving to varying degrees genetic and morphological considerations (Rachocki, 1981). The 'Encyclopedia of Geomorphology' (Bull, 1968) defines an alluvial fan as: 'a body of stream deposits whose surface approximates a segment of a cone that radiates downslope from that point where the stream leaves a mountainous area.'

A working definition is provided by combining

Alluvial Fans: A Field Approach edited by A. H. Rachocki and M. Church
Copyright © 1990 John Wiley & Sons Ltd.

those of Patton *et al.* (1970) and Bull (1977):

'An alluvial fan is a low cone of gravels, sand and finer sediments that resembles an unfolded oriental fan in outline' (Patton *et al.*, 1970) 'that radiates downslope from the point where the stream leaves the source area' (Bull, 1977).

Drew (1873) first used the term 'alluvial fan' for forms in the Karakoram Mountains with reference to features in the basins and valleys of the Upper Indus. He described their morphology but only briefly mentioned their sedimentology. Influenced by the earlier works of Surell (1841) and Haast (1864, 1879) in the Alps and on the Canterbury Plains of New Zealand respectively, he attributed these landforms to fluvial processes, envisaging alluvial sedimentation by tributary valley streams entering larger, less confined valleys, and depositing their load as stream power decreased.

Since Drew's (1873) work, little attention has been given to these fans, or to similar features in adjacent parts of the Trans-Himalayan mountain range, until very recently. Detailed sedimentological study of these 'alluvial fans' within the Pakistan Karakoram, shows that they comprise a large variety of sediment types including debris-flow, mudflow, fluvial, till, lacustrine, aeolian, and glaciofluvial. The purpose of this chapter is to describe the complex variety of sediments and the modes of deposition characteristic of landforms which mimic closely the form of alluvial fans within high mountain valleys. The models presented may prove useful in studies of fan development in adjacent parts of the Himalaya and in other high mountain areas such as the Andes and the Southern Alps of New Zealand. However, since the majority of these fans are not strictly alluvial in origin, the epithet 'alluvial' is clearly inappropriate: the features are here referred to as sediment fans. Stream incision and fan-toe truncation are widespread features: thus isolated remnants of fan surfaces stand above the valleys, making up terrace series (*cf.* Rhind, 1968). The term 'fan terrace' is therefore proposed for such features.

The Karakoram Mountain Environment

The Karakoram Mountains (Figure 2.1) are situated at the western end of the Trans-Himalayan mountain belt in a pivotal position bordering the northwestern margin of the Himalayas and connecting the Hindu Kush and Pamir Mountains. Tectonically, the Karakoram Mountains represent the intercontinental collision of the Indian and Asian plates (Gansser, 1964; Dewey and Burke, 1973; Le Fort, 1975). Owen (1989) suggests that most of the uplift in the Karakoram mountains occurs as regional doming, with few zones of intense neotectonic deformation. It is one of the more rapidly rising mountain belts in the world, recent work suggesting that uplift across the whole of the mountain area averages 2 mm a^{-1} (Zeitler, 1985; Mehta, 1980; Gornitz and Seeber, 1981; Seeber and Gornitz, 1983; Lyon-Caen and Molnar, 1983; Ferguson, 1984). Locally, uplift may exceed 10 mm a^{-1}, especially in the Nanga Parbat–Haramosh area (Zeitler, 1985). This results in the Karakoram being one of the highest mountain belts on Earth, with peaks rising to between 7 000 and 8 000 m a.s.l. Denudation is intense, involving a formidable combination of processes including frost-shattering (Hewitt, 1968; Goudie *et al.*, 1984), chemical weathering by salts involving crystal growth (Goudie, 1984; Whalley *et al.*, 1984) and granular disintegration (Goudie *et al.*, 1984), glacial erosion (Goudie *et al.*, 1984; Li Jijun *et al.*, 1984), fluvial incision (Ferguson, 1984; Ferguson *et al.*, 1984), and mass movement processes (Brunsden and Jones, 1984; Brunsden *et al.*, 1984). In addition, the Karakoram mountains have undergone at least three extensive valley glaciations during Quaternary time (Derbyshire *et al.*, 1984). These resulted in intense glacial erosion and the production of extensive and thick till deposits.

The combination of such intense denudation and high uplift rates has produced the greatest relative relief on Earth with valley floors averaging 1 500 m in altitude and peaks lying between 7 000 and 8 000 m a.s.l. Valley sedimentary fills frequently exceed 700 m in thickness. The sediment loads transported by some of the world's greatest rivers (Indus, Gilgit, Hunza, Shyok) are

QUATERNARY ALLUVIAL FANS IN THE KARAKORAM MOUNTAINS

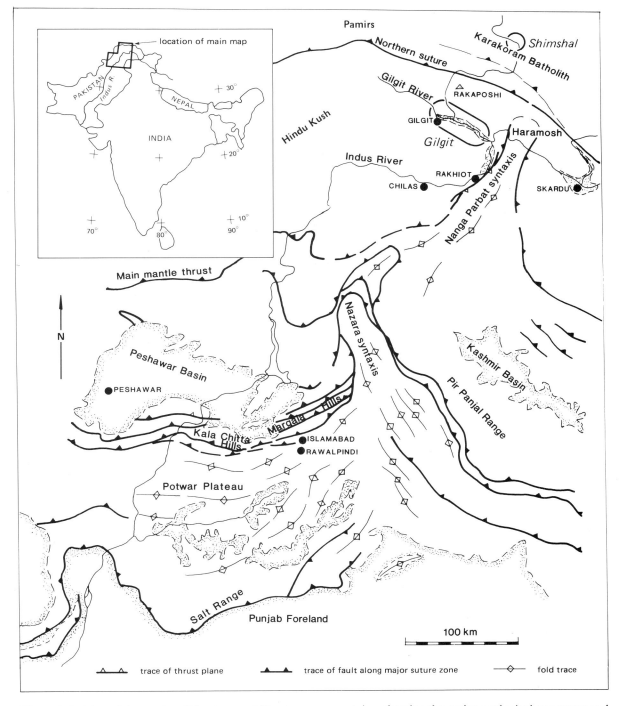

Figure 2.1. Map of the western Himalaya and Karakoram mountains, showing the major geological structures and Quaternary basins. Field areas are indicated (italics)

among the highest known, e.g. the Hunza River has a sediment yield of $4800 \text{ t km}^2 \text{ a}^{-1}$ (Ferguson, 1984). Moreover, the modern glaciers, covering at least 20% of the land area, are of the high activity type (Derbyshire, 1982; Derbyshire and Miller, 1981) and include some of the largest outside the polar regions.

The climate of the region is transitional between central Asian and monsoonal south Asian types, varying considerably with altitude, aspect, and local relief. The Karakoram valley floors are essentially deserts with a mean annual precipitation of less than 150 mm, most of this occurring over short periods in summer as heavy storms. Extreme diurnal maximum temperatures in summer exceed 38°, while winter temperatures fall below 0°C, even in the valleys (Goudie et al., 1984). Dust storms occur about once a week, enhanced in the summer by katabatic effects.

Vegetation is altitudinally controlled (Paffen et al., 1956). It is sparse along the valley, being of desert steppe type, and is replaced at higher levels, first by temperate coniferous trees and then by alpine meadow vegetation. Most sediment fans are found in the zone of desert–steppe vegetation. Man-made irrigation systems, dependent to a large degree on glacier meltwaters, are extensive on fan surfaces and form the basis of an agriculture in which wheat, corn, barley, potatoes, and deciduous orchards are important.

Catastrophic events are frequent in the Karakoram and Himalaya, notably large debris-flows, and flooding produced by the failure of natural dams associated with landslides and rapid glacier advances (Mason, 1929, 1935; Brunsden and Jones, 1984). Such events are capable of causing rapid and violent erosion and widespread resedimentation of unconsolidated deposits, dramatically reshaping the landscape.

The most common depositional landform is the sediment fan. These occur within beaded networks of small interconnecting intermontane basins that frequently coincide with the confluences of glaciated and deeply incised river valleys. Elucidation of the evolution of the sediment fans and their depositional environment within the Karakoram Mountains and similar high mountain belts requires an understanding of the nature of the interaction between tectonics, uplift, climate, Quaternary glaciations, and catastrophic land-forming events.

The Karakoram Sediment Fans

The Karakoram sediment fans vary from debris slopes (including rockfall screes 32–37°, scree fans 15–30° (Figure 2.2A), and mudflow fans 5–10°; Brunsden et al., 1984) to dominantly debris-flow ($< 15°$: Figure 2.2B), 'dominantly fluvial' ($< 10°$), and composite (5–30°) fans. The composite sediment fan is the norm in this region and may include debris-flow, mudflow, till, glaciofluvial, fluvial, lacustrine, and aeolian sediments (Figure 2.2C). Case studies will be described both to exemplify the variety and to form the basis of some models of sediment fan development. First, however, particular techniques will be described which have been applied in order to cope with the special problems of scale and facies variability.

Methods

MAPPING

Map coverage of the area is poor, the largest scale providing full coverage being the United States Military 1:250 000 U502 series. Maps at a scale of 1:50 000 are available in two sheets for the Nanga Parbat and Minapin areas only. The whole area is covered by aerial photographs, but Pakistan Government security policy prohibits access to them. However, none of these items provides sufficient data for detailed geomorphic mapping or for precise fixing of locations for sedimentological work. Areas of interest must be surveyed using reconnaissance techniques such as plane tabling (cf. Pugh, 1975) or terrestrial photogrammetric surveys. In the work discussed here, sketch maps (cf. Figure 14) have been constructed using a series of oblique photographs of sediment fan surfaces taken by hand from high peaks or high rock benches. Field sketches of geomorphic relationships have been used throughout, further details being added using photographs which may

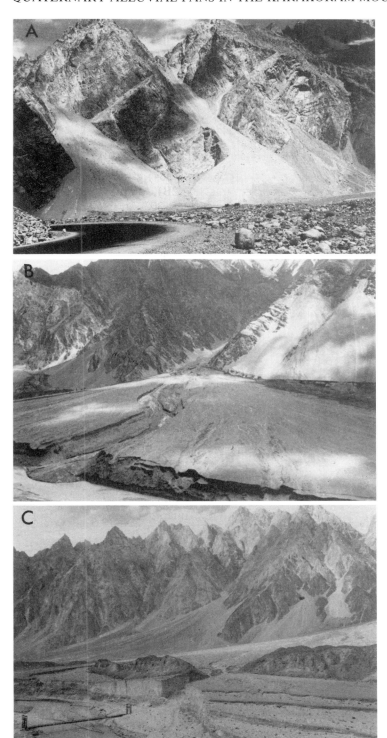

Figure 2.2. Sediment fans in the Karakoram Mountains: (A) Scree fans near Pasu; (B) Debris-flow dominated fans near Pasu. Note the fan-toe truncation and fanhead entrenchment; (C) Sediment fan near the Batura Glacier. The dark ridges are inliers of till within the fan. Individual debris-flows can be identified by different rock varnishes on the fan surface. Large debris-flows can easily be seen where the fan has been truncated at its toe. Note the high-angled scree slopes and cones that surmount the adjacent slopes

also provide some scale correction. Barometric levelling has been employed to establish heights for the maps using a 'leap frog' method (Pugh, 1975). Smaller scale features (*cf.* Figure 15) have been mapped on suitable scales using plane tabling, chaining, taping, and levelling (Pugh, 1975). Fan surface gradients were also recorded using a hand-held clinometer.

LOGGING OF SEDIMENTARY SECTIONS

Standard field logging techniques were used (Tucker, 1982) on many of the sections. However, some sections are many hundreds of metres high, often with vertical faces of several tens of metres developed in bouldery diamictons quite impossible to measure by direct methods. Photogrammetric techniques have been used in most cases to determine vertical distances, aided by field sketches. It is also important to construct logs or diagrams showing lateral variation and geometry: this is critical in the interpretation of the facies and in establishing relations between different sedimentary units (Eyles *et al.*, 1983; Shaw, 1986).

SEDIMENTOLOGICAL ANALYSIS

Sediments were classified in the field on the basis of particle size and shape, together with the evidence of bedding, jointing, bedforms, faulting, fabric, and superposition in terms of facies associations. Clast fabrics were measured in the field using standard methods. Oriented kilogram samples were collected for laboratory analysis. Particle size distribution was determined by a combination of wet sieving (B.S.1377, British Standards Institution, 1967) and X-ray size sorting for the $-70\,\mu m$ fraction using a Micromeritics 5000 ET SediGraph. Particle angularity and sphericity were estimated using the charts of Rittenhouse (1943) and Shepard and Young (1961). Petrographic pebble counts were made for particle sizes greater than 2 mm. X-ray diffractometry was used to establish the mineralogy of the silt and clay content, interpretation following Carroll (1974) and Brindley and Brown (1980). Optical microscopy of oriented thin-sections and scanning electron microscopy were used to examine the microfabrics of the diamictons.

DATING

A number of techniques have been used in this region to establish a chronological sequence: these include morphostratigraphy, superposition, degree of surface weathering, point compressive strength of boulder surfaces, rock varnish development, per cent lichen cover, and absolute dating (radiocarbon, thermoluminescence (TL), and palaeomagnetism). In our studies, relative dating techniques were quite successful. Two exceptions were point compressive strength of boulder surfaces using a Schmidt hammer and percentage cover of lichens. The Schmidt hammer did not yield reproducible results, and lichen cover percentage also proved impracticable because it appears to be controlled more by altitude than by the age of boulder emplacement, lichen being widespread only above *c.* 2800 m. Similar results were obtained in the upper Hunza River valley by Derbyshire *et al.* (1984).

The rock varnishes on surface boulders were found to be particularly useful in these studies. Rock varnish consists of 5 to 50 μm thick layers of manganese and iron oxide, plus hydroxides, clay minerals, and trace elements, considered to accrete on rock surfaces during summer thunderstorms. Biological processes are thought to assist the concentration of manganese and iron oxide under alkaline conditions (Dorn, 1983), scanning electron microscopy suggesting that small bacterial colonies of unidentified origin found in voids which match the morphology of the colonies are intimately associated with the varnished surfaces (Figure 2.3A). These varnishes become darker and thicker with age and Derbyshire *et al.* (1984) have used a rock varnish index as a means of relative dating of morainic boulders. Dorn (1983) used the cation ratio of $Ca+K:Ti$ to quantify the age of varnishes in the southwestern United States. However, these ratios are about ten times higher in the Karakoram than they are in the southwestern United States, and exhibit a very poor correlation with apparent age (Figure 2.4A).

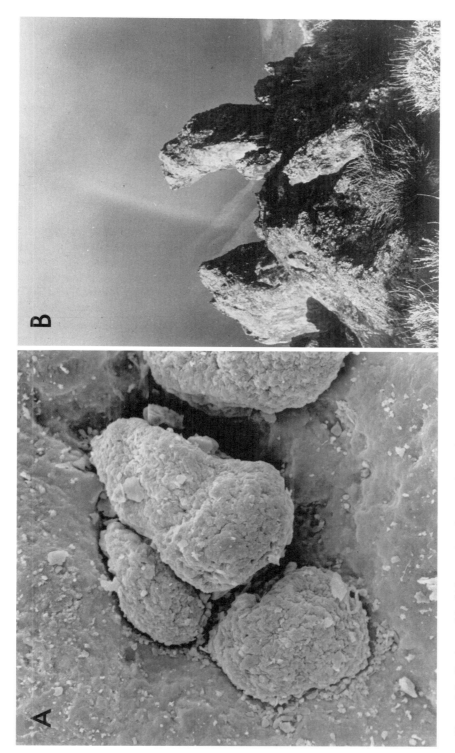

Figure 2.3. Rock varnish. (A) scanning electron photomicrograph of a rock varnish surface with several colonies of microbacteria corroding the rock surface and helping to produce the varnish; (B) Highly-weathered and corroded granite with a post-maximal rock varnish. The varnish is beginning to weather off the rock and granular disintegration is the dominant process. The photograph shows intense cavernous weathering

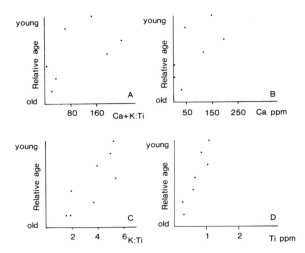

Figure 2.4. Geochemical variation within Karakoram rock varnishes, measured by laser spectrometer. See text for discussion

In the Karakoram, Ca overwhelms the ratio (cf. Figures 2.4B and 2.4D), its concentration being greater probably because of accretion of much windblown carbonate. The Karakoram varnish also differs in that, though often rich in Fe, little Mn is present. Nevertheless, there is a definite trend in cation ratios with age for the Karakoram varnish (e.g. Figure 2.4C) although, of course, absolute dating is required to qualify these ratios, and simple concentration of indicator cations (e.g. Figure 2.4D) may provide the best correlation of all. Pending the completion of absolute dating analyses, the varnish provides a relative scale of fan surface dating, although it is recognized that varnish colour is influenced by rock colour and that its range of variation even on a uniform lithology is not known in detail.

During the third and fourth centuries A.D., Buddists carved petroglyphs on some of the varnished surfaces of boulders and bedrock. Since that time only very weak varnishes (Munsell colour 2.57 8/5) have developed on the carved surfaces, suggesting that c. 1500 years is the minimum time required for a varnish to develop in this region. Many of the boulders have deeply weathered surfaces (Figure 2.3B) with second generation varnishes. Clearly, this complicates attempts at relative dating because there is no obvious way of establishing whether or not the varnish is a first generation feature.

Boulder weathering ratios, i.e. percentage of flaked and split boulders and percentage of pitted boulders, were tested in the Hunza valley by Derbyshire et al. (1984), but Porter (1970) showed in Swat that it was difficult to obtain satisfactory results using this method. However, it has value in cases where there are large contrasts in degrees of weathering between surfaces. Derbyshire et al. (1984) also showed that surface boulder frequency (dependent on sediment type) and slope angles were not a sound basis for relative dating of surfaces. Moreover, well-developed soils are extremely rare in this area (although thin calcretes and carbonate meniscus cements are occasionally found) so that dating using soil properties (Rockwell et al., 1985) is not possible.

Many of the sections contain thin lacustrine silts which are suitable for thermoluminescence dating, some sections containing as much as 20 m of glaciolacustrine silt. So far, however, only three thermoluminescence determinations have been published for the Hunza valley (Derbyshire et al., 1984) and all require independent laboratory reassessment. Moreover, there is little in the way of geomagnetic dating: very high sedimentation rates provide sections with high time resolution but with very limited age ranges, making it difficult to apply a dating method dependent on the recognition of secular magnetic variation events.

Finally, the prevailing arid climates of the past two to three million years have resulted in only very rare preservation of organic materials within the sediments, only three ^{14}C dates being available for this area (Woldstedt, 1965). In view of these difficulties, the ages of the sediments and landforms here are based essentially on relative dating controlled by a limited number of dated sites.

The Diamictons: Some Sedimentary Properties

The difficulty of distinguishing between in situ tills and tills reworked and resedimented both during and following initial glacial deposition is now generally recognized. In the Karakoram, the complexity of the relationship between the till deposits of large, steep glaciers and the widespread mobilization of diamictons of glacial and mass wasting (including earthquake triggered) origin by rapid melting of glaciers and snowpacks and by rainstorms makes the distinction between *in situ* tills and debris-flow deposits particularly difficult. However, a number of characteristics have been found useful in discriminating between glacial and non-glacial deposition. Debris may flow passively over underlying sediments without disturbing or eroding them. Some flow deposits have discrete shears and strong fabric anisotropy near their bases and show vertical grain-size differentiation, the surface of the flow units frequently being draped with fine silt as a result of late-stage dewatering and weak overland flow carrying fines. Glacial diamictons frequently have shear banding which is irregular and commonly steeply-dipping as a result of subglacial basal shear. Overconsolidation induced by ice overburden is also present in tills but is often associated with subhorizontal dilation joints with fissility (a property not recognized in the debris-flow deposits) produced by glacial unloading. Ice overburden also induces intrusion of subglacial till into joints in the underlying bedrock: this has not been seen in the debris-flow deposits, which rather tend to preserve subjacent surfaces and zones of shear owing to the weak vertical stresses imposed during debris-flow deposition.

Microfabric studies have shown that all subglacial tills in the Karakoram mountains contain coherent systems of microshears which can be interpreted in relation to ice movement direction. However, these are much clearer in some tills than in others, a variation which depends upon particle size and shape: these in turn are strongly influenced by mineralogy. Occasional microshears are found within the matrix of debris-flows but they are not as frequent or as directionally coherent (Figure 2.5). In deposits with similar petrography, mean grain anisotrophy and the degree of compaction (bulk density) is higher in the subglacial tills than in the flow debris.

Case Studies

THE LOWER GILGIT VALLEY

The lower Gilgit valley below the confluence of the Hunza and Gilgit rivers (Figure 2.6) is essentially an elongated basin about 40 km long and 3 to 5 km wide. In places, valley fill sediments exceed 700 m in thickness: the surface of these fills is dominated by low-angled (< 15°) sediment fan landforms and a series of terrace remnants along the valley sides (Figures 2.6 and 2.7A). Figure 2.8 shows the character of the sediments.

Derbyshire *et al.* (1984) showed that the Hunza valley has undergone at least three major glaciations, named the Shanoz, the Yunz (139 000 ± 12 500 TL yr B.P.), and the Borit Jheel (65 000 ± 3300 yr B.P.). The latter two glaciations produced major trunk valley glaciers that extended down the Hunza and Gilgit valleys into the middle Indus valley (Figures 2.9A and D). During the second of these glaciations (Yunz: Figure 2.9A), the Gilgit basin was enlarged by the confluence of Hunza and Gilgit ice streams. Ice was also supplied by the northern side valleys, the Manu, Bagrot, and Batkor. The diversion of the Manu valley ice by the main valley ice stream helped erode a palaeovalley east of Dainyor, producing a rock bar separating it from the main valley. The glacially-eroded valley slopes in the Gilgit area are difficult to recognize because of subsequent intense modification by slope processes, although good examples of stepped morphology typical of alpine glaciated slopes can be seen west of Gilgit. A well-defined break in slope about 300 m above the present river level marks the former level of Yunz ice. Rock benches are present at a similar height along the valley upon which rest polymictic diamictons containing striated clasts. There is little evidence of till on the steep slopes of the trunk valley: where tills are present they have been preserved under younger scree deposits which have subsequently been eroded. Indeed, most of the Karakoram valley sides are free of till, especially in the Lower Hunza, Indus, Skardu, and Shigar Valleys, the evidence of former glaciations in the form of till being fragmentary (Derbyshire

Figure 2.5. Scanning electron photomicrographs of vertical faces of greenschist dominated till (A) and debris-flow sediment (B), both from the Hindi area of the Hunza Valley. The till shows grain striation and low-angle shearing with associated matrix dilation

QUATERNARY ALLUVIAL FANS IN THE KARAKORAM MOUNTAINS

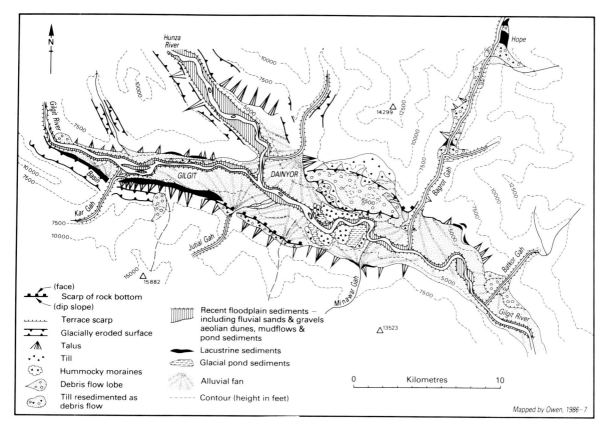

Figure 2.6. Geomorphological and sedimentological map of the Gilgit area

et al., 1984). However, where relative relief is lower and slopes are gentler, extensive till cover has been preserved, e.g. at Sost in the Upper Hunza valley (Figure 2.7B). The slopes in the Gilgit basin may have approximated this state very soon after deglaciation. We suggest that, soon after deglaciation, the combination of water-saturated tills, steep and long valley slopes, and the instability consequent upon the wasting of confining glacier ice, produced extensive debris-flows, substantially clearing the valley sides of till (Figures 2.9B and C). Such resedimented till forms a major component in the sediment fans, along with valley-floor tills preserved as inliers, and glaciofluvial outwash deposits (Figure 2.8) from the retreating glaciers. The Dainyor palaeovalley demonstrates the importance of the glaciofluvial component in the immediate post-glacier sedimentation, for it is infilled to a depth of several hundred metres by glaciofluvial sediments derived from the Manu and Hunza valley ice (Figure 2.9B).

The third glaciation (the Borit Jheel: c. 65 000 years B.P.) produced a large stream of ice that extended down the Hunza valley, but ice was less extensive in the Gilgit vlaley and probably did not extend as far as the Gilgit basin (Figure 2.9D). The Borit Jheel ice eroded valley sediments deposited by the Yunz glaciation except for a few isolated till remnants on rock benches east of Gilgit and in the base of the Dainyor palaeo-valley. Hunza ice blocked the Gilgit valley and produced a lake which extended at least 20 km up the Gilgit valley. As much as 30 m of lacustrine

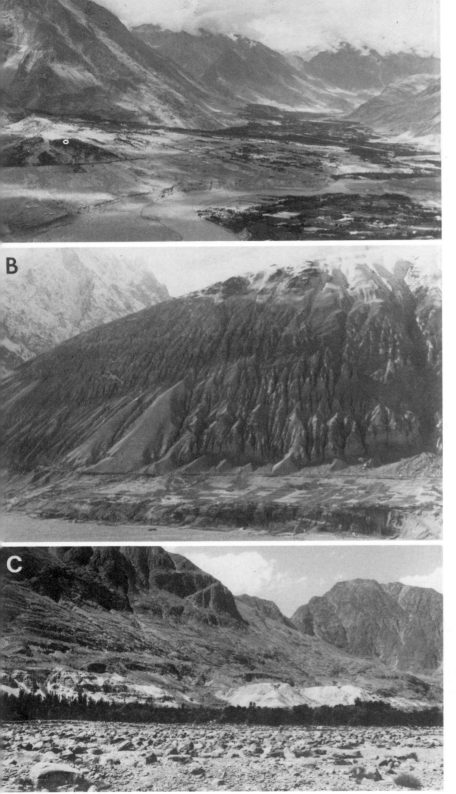

Figure 2.7. Sedimentary features near Gilgit. (A) Sediment fans, view north toward Gilgit. Lacustrine terraces in the far distance are evident by their light colour; (B) Till deposited on the valley walls near Sost in the upper Hunza Valley. Most of the valley walls in the lower Hunza and Gilgit area were covered with deposits like this before they were reworked into sediment fan deposits; (C) The Gilgit sediment fan with the Basin lacustrine terrace in the background. Note the glacially-eroded valley sides

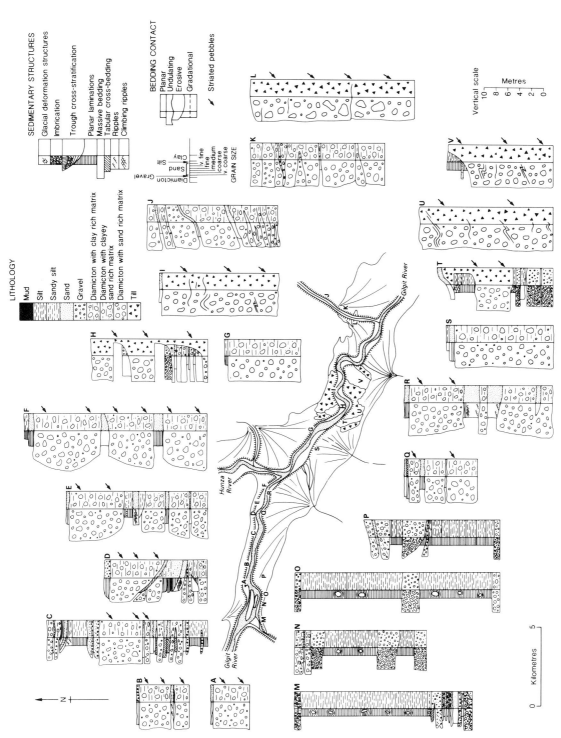

Figure 2.8. Simplified sedimentary logs for the sediments in the Gilgit Basin. The width of the sedimentary log is proportional to the dominant grain size of the sediment: a scale is included in the diagram

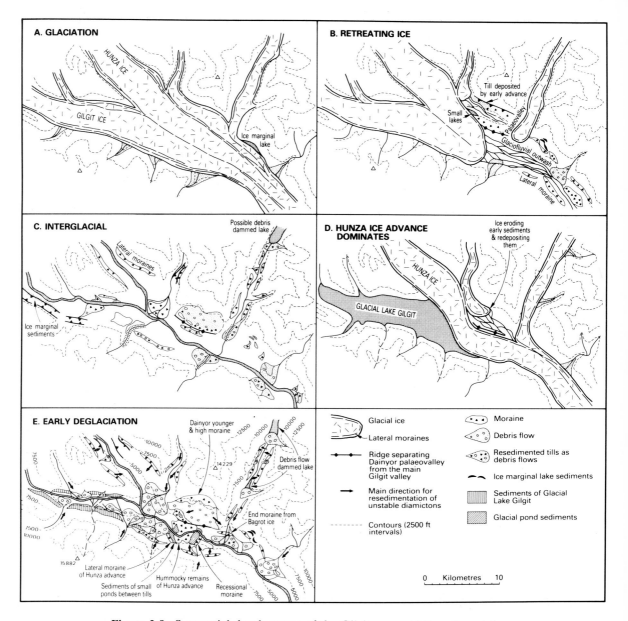

Figure 2.9. Sequential development of the Gilgit area and its sediment fans

sediments are present as terrace inliers within the fans west of Gilgit and consist of two units separated by a well-sorted 2 m-thick imbricated pebble sheet that formed during the breaching of the ice dam (Figures 2.7C and 2.8) on two occasions. Tongues of ice and meltwater streams from the Manu valley eroded the Dainyor palaeovalley fill, but added further to the sediment pile, increasing its thickness to over 700 m. Marginal lakes also developed along the sides of the valleys: these are now represented by planar-bedded silts interbedded with tills.

On retreat of the Hunza ice, till was exposed at locations where it had been deposited beneath or immediately adjacent to the ice, often on steep valley sides. A ridge of till south of Dainyor trending NNW–SSE marks the former lateral margin of the Hunza glacier and a N–S ridge immediately west of Minawar Gah is an example of a recessional moraine (Figures 2.6 and 2.8). Bagrot valley ice produced an end moraine at the junction of the Bagrot and the Gilgit valleys.

Unstable tills along the valley sides were resedimented as debris-flows, leaving the more compact till and lacustrine sedimentary outcrops as inliers. Some of the fans in this area are made up of as few as four debris-flows interbedded with glaciofluvial and fluvial sediments (Figure 2.8). It is evident that debris-flow and fan development occurred quite soon after glacial retreat and, in many cases, glacier ice may have been in close proximity to the developing fans. There is little evidence indicating the position of glacier margins during fan formation in the Gilgit area but at Sost and elsewhere in the upper Hunza valley, there is a very clear relationship between ice margins, marginal till deposits and debris-flow fan formation.

The main type of debris-flow deposits are low angled ($< 10°$) homogenous sheets of debris 5 to 20 m thick covering areas of up to 30 km^2 (Figure 2.10). This implies very wet, fluidized debris and rapid flow and sedimentation rates. Similar types of debris-flows have been described and observed in China (e.g. Deng, 1983; Lü and Li, 1986). Other debris-flows are more lobate in form, occurring as inliers and usually consisting of a series of superimposed debris-flow units several tens of metres thick (Figure 2.11C). Poorly-defined downslope stratification and vertical grain-size differentiation can be recognized, but the majority of these flows cannot be differentiated and each represents a single depositional event. The most impressive debris-flow is seen at the mouth of the Batkor valley. It consists of a 150 m-thick diamicton deposited in a single flow event (Figure 2.11B).

Debris-flow deposition across the more confined valleys such as the Bagrot produced blockages behind which lakes were impounded and in which tens of metres of lacustrine silts were deposited. At Hope in the Bagrot valley two adjacent debris-flows of considerable size produced two large lacustrine sequences which now form a major component of the fan terraces (Figure 2.11C). Figure 2.12 summarizes the complexities of valley diamictons resulting from the resedimentation of glacial silts. The surfaces of the debris-flow fans comprise thin caps of fluvial and debris-flow sediments. Fluvial erosion also modifies the surface forms producing small gullies. However, more dramatic fluvial erosion by the main trunk rivers has produced widespread fan-toe truncation and dissection (Figure 2.2B). Glacial meltwater flowing to the fanheads forms single active channels concentrating erosion and producing fanhead entrenchment (Figure 2.11A).

Debris-flows considered to be quite recent in origin, to judge from the presence of little or no varnish on their surface boulders, form a minor part of the fans (Figure 2.13A). Frequently they occur as small linear bodies of debris associated with the modification of adjacent steeply-sloping screes. Fewer than ten recent debris-flows greater in area than 5000 m^2 have been recognized in the Gilgit region (> 200 km^2). Larger debris-flow events occur more frequently along the main fanhead entrenched channel. Much of this material is carried directly to the trunk river and, as a result, this sediment makes no contribution to the growth of the fan (Figure 2.13B).

The fan surfaces are covered with surface boulders. These are frequently isolated and over 3 m in diameter, being the product of removal of the fine sands and silts from the diamictons during dust storms or by overland flow. Fine aeolian silts have accumulated during recent times and may be as much as 1 m thick. Such aeolian silts are not loess in the strict sense because they often contain small pebbles, indicating reworking by colluvial and overland flow processes (Figure 2.10).

Taken together, all the characteristics described above are consistent with the view that these sediment fans are relict features formed during a relatively short period of time following the last major glaciation of the Karakoram Mountains (Borit Jheel, $c.$ 60 000 yr B.P.). Resedimentation of till adjacent to other sedimentary

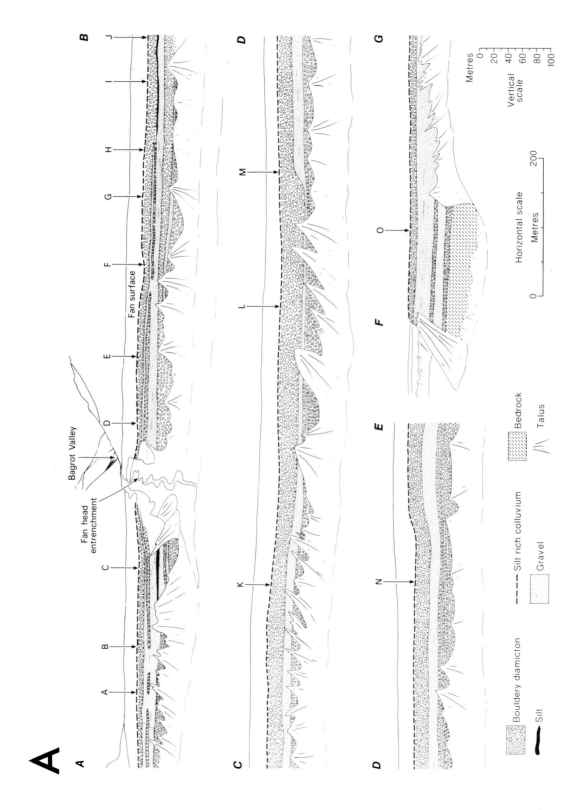

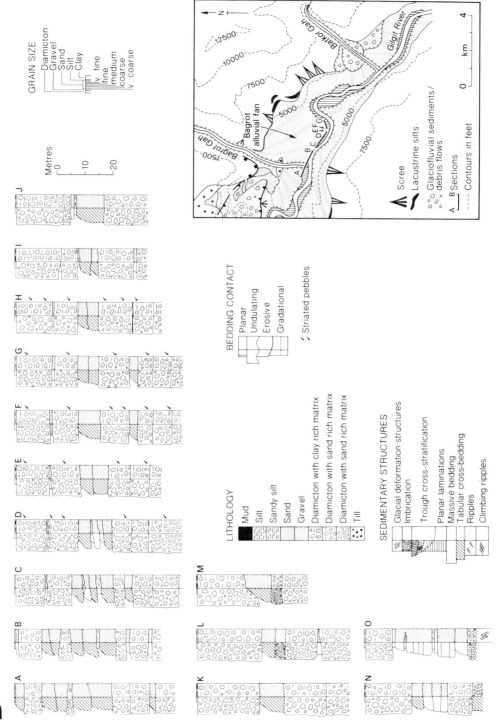

Figure 2.10. (A) Field sketch of the sediment fan near the Bagrot Valley. The diamictons represent debris-flow deposits while the gravels were deposited by fluvial deposits. See inset map for locations A to G. (B) Sedimentary logs (A to O) for the Bagrot–Dainyor alluvial fan

Figure 2.11. Debris-flows in the Gilgit Basin. (A) The Batkor debris-flow. The small houses in the foreground indicate the scale; (B) Lateral view of the Batkor debris-flow. This was one depositional event. Note the crude, subhorizontal stratification. The bouldery surface is due to deflation of fines; (C) Large debris-flows at the southern end of the Bagrot Valley. These debris-flows blocked the valley and produced large lakes that deposited silts which onlapped the debris-flows. One set of lacustrine silts can be seen in the foreground

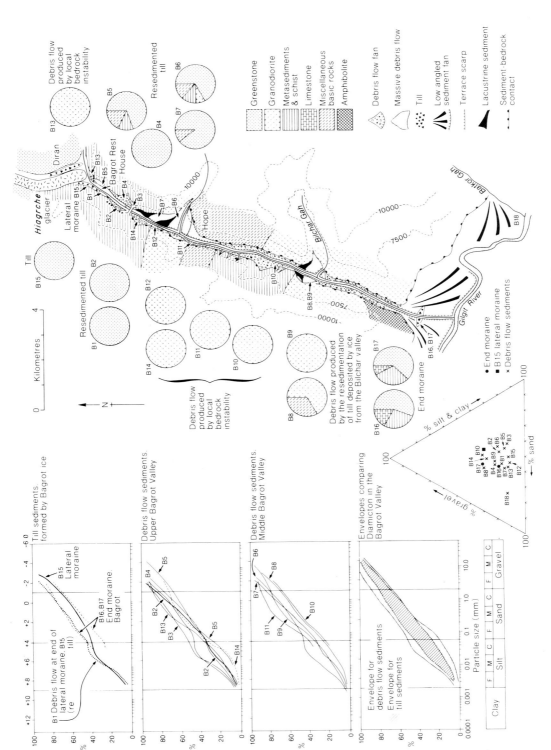

Figure 2.12. Diamicton deposits along the Bagrot Valley

Figure 2.13. Recent debris-flow. (A) Small flow easily identified by lack of rock varnish; (B) Recent debris-flow in a fanhead entrenched channel

units, plus previously-deposited glacial and paraglacial sediments, produced rapid infilling of the valleys in the form of fans which were then incised by the fluvial system and further modified by minor fluvial, aeolian, and debris-flow processes. A detailed example of sediment fans in and adjacent to glacial and paraglacial sediment associations is provided by the Pasu area of the upper Hunza valley.

THE PASU AREA

The upper Hunza area between Pasu and the snout of the Batura Glacier provides a good example of the complexity of the paraglacial and proglacial sediment fan environment (Figure 2.14). The geomorphology, sedimentology, and glacial history of this area have been described by Brunsden *et al.* (1984), Goudie *et al.* (1984), and Li Jijun *et al.* (1984). The Batura Glacier Investigation Group (1976, 1979), Derbyshire *et al.* (1984) and Li Jijun *et al.* (1984) recognized two major advances of the Batura Glacier during Quaternary time (the Yunz and Borit Jheel) and, during the Holocene, an expanded foot stage (Ghulkin I), four minor advances (Ghulkin II, Batura, Pasu I, Pasu II), and recent fluctuations of the ice fronts. The oldest glacial sediments present near the valley derive from the Borit Jheel stage and are overlain by glaciofluvial and glaciolacustrine sediments which, in turn, are overlain by Ghulkin I tills. The Ghulkin I tills have a dominantly granitic component with slates and limestone clasts in a silty clay matrix, and are referred to in Figure 2.14 as older moraine. They form a series of inlier ridges within the sediment fans on the north side of the Hunza (Figure 2.2C). The younger tills include those with strong dark varnish (Ghulkin II stage) on surface boulders, brown varnish (Batura stage) and the yellow varnish of the Pasu stage. These form distinct arcuate ridges that are readily correlated using morphostratigraphy and weathering criteria.

Figure 2.14 shows the distribution of the glacial and paraglacial sediments. The sediment fan in

QUATERNARY ALLUVIAL FANS IN THE KARAKORAM MOUNTAINS

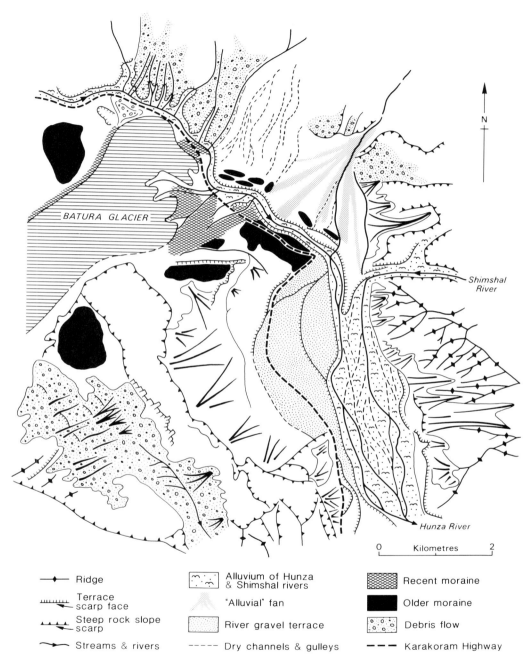

Figure 2.14. Geomorphological and sedimentological map of the Pasu area

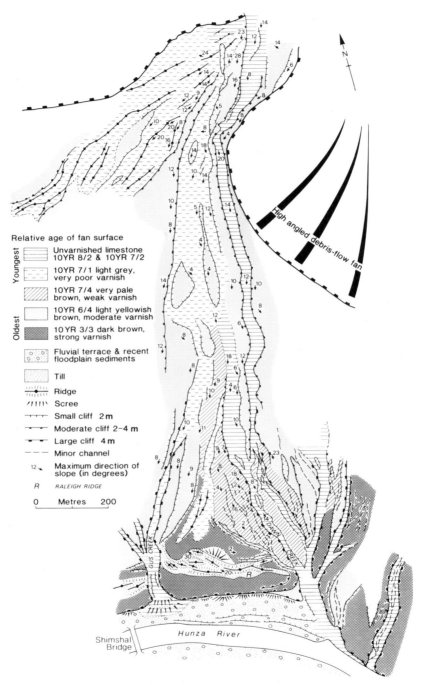

Figure 2.15. Geomorphological and sedimentological map of a sediment fan near the snout of the Batura Glacier

the northeast corner of the map was surveyed using plane tabling techniques at a scale of 1:10 000 and its surface was mapped on the basis of terrace levels, weathering, and varnish criteria (Figures 2.15 and 2.2C). Deep fanhead entrenchment and fan-toe truncation facilitated the study of the sedimentology of this feature. Five terrace levels were recognized on the fan's surface, the Ghulkin I till forming on E–W inlier ridge near the toe. Figure 2.16 shows logs and sketches of the sedimentology of the fan. The deposits consist of glaciofluvial sands and gravels which are present almost down to river-level having been tilted to the vertical (Figure 2.17A). These are unconformably overlain by a 2 to 4 m thick bouldery conglomerate. Above the conglomerate are two 5 to 10 m thick limestone-rich debris-flow units which can be traced laterally for several hundred metres. Each of these is capped by a thin, calcareous, silty sand. These debris-flows have a strong dark-brown (10YR 3/3) varnish and form the oldest surface on the fan, the surfaces being bouldery partly as a result of loss of fines by deflation. In addition, the fans are cemented in places so that the surface is calcreted (Figure 2.17B). Limestone-rich debris-flow sediments with a light yellow–brown (10YR 6/4) moderate varnish complete the sequence. Topographically, they make up the highest fan surface, their sheet-like form extending over large expanses. Fluvial channels have cut into the sediment fans and lenticular bodies of debris have been deposited in them. Three distinct series of lenticular flow debris can be recognized on the basis of surface boulder varnishes (oldest to youngest 10YR 7/4, 10YR 7/1, and unvarnished) and morphostratigraphic criteria. Each pulse of sedimentation was separated by a phase of fan channel incision (Figure 2.18). The channels often contain large boulders which are several

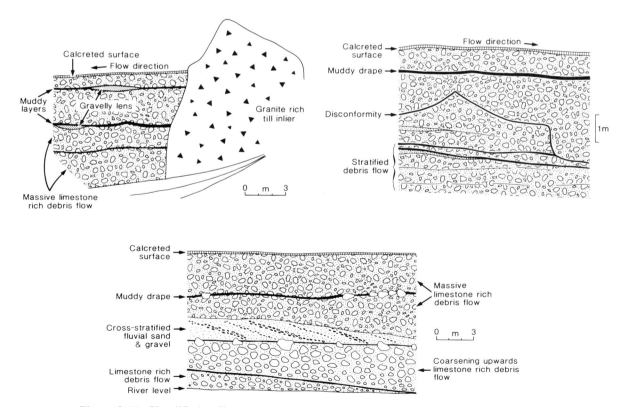

Figure 2.16. Simplified sedimentary logs for selected sections of the fan in Figure 2.15

metres in diameter derived from the larger debris-flows left behind by the incision process. Surrounding the main fan are steep (> 30°) scree cones and incised, very steep (< 30°) debris-flow deposits (Figures 2.2A and B). There is little evidence of sediments being supplied to the fan surface in recent times from these sources. However, the scree cones are undergoing modification by linear debris-flow activity on a small scale. Figure 2.18 shows the sequential development of the fan. Figure 2.18A represents the expanded foot stage of the Ghulkin I glaciation which overrode previously deposited glaciofluvial sands, tilting them to the vertical and depositing the very dark-brown weathered and varnished till ridges now forming inliers in the fan. Derbyshire *et al.* (1984) suggest that the Hunza River was pushed northwards by the Batura ice. Upon ice retreat, large masses of debris flowed down the main tributary valley (Figure 2.18B) and possibly from the side walls as resedimented till, but no true resedimented till is present in the fan since the debris-flows differ from the tills in having a single, uniform lithology. There then followed a stage of stagnation before the emplacement of the next series of large debris-flows (Figure 2.18C), as suggested by the marked difference in the colour of the varnishes (10YR 3/3) compared to the older ones (10YR 6/4). These two major depositional phases account for most of the fan volume, as the subsequent phase was one of fan incision with deposition of lenticular masses of debris within the channels (Figure 2.18D). The ultimate form of the fan was completed by three pulses of debris-flow as the main fan channel shifted position. At the same time, scree cones accumulated above the fan, their surfaces, in turn, being modified by the emplacement of lenses of flow debris (Figures 2.18E and F). There is little sign of present-day debris-flow processes acting on the surface of the fan, although some recent, small debris-flows can be recognized in the active channels. Deflation of fines appears to be the dominant current process on the fan surface. The single modern channel removes some sediment, but there is little evidence of any substantial deposition within the channel itself.

Figure 2.17. Features of the sediment fan near Batura Glacier. (A) Two limestone-rich debris-flows deposited over fluvial conglomerates and glacially tilted sands and gravels, exposed in the truncated toe of the fan; (B) Calcreted surface on the fan

Conclusion

The sediment fans of the Karakoram valleys constitute a distinctive geomorphological and sedimentological subtype within the alluvial fan environment, a distinctiveness which owes much to their proglacial situation. The Hunza and Gilgit valleys provide examples of the way such sediment fans derive essentially from a few, large, perhaps catastrophic events followed by modification during a greater number of smaller-scale

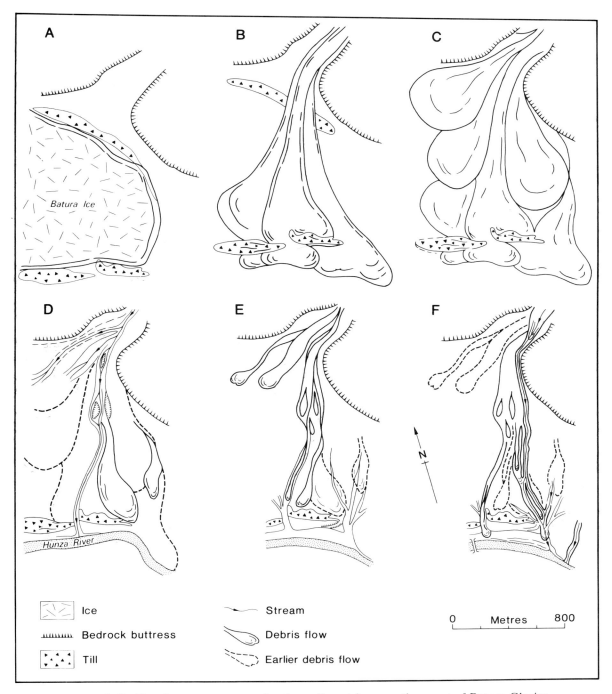

Figure 2.18. Development sequence for the sediment fan near the snout of Batura Glacier

events. Similar conclusions were reached in a study of the fans in Owens Valley, California by Beaty (1974) who used the case to defend the principle of catastrophism. Certainly, sediment fans provide evidence that large-scale deposition may be rapid in response to shifts in geomorphic thresholds reflecting local and regional tectonic and climatic events. There are several documented cases of tectonically-induced catastrophic events in the Karakoram Mountains (Henderson, 1959; Obbard, 1860; Shaw, 1871; Cunningham, 1854). However, although it is well known that the Karakoram Mountains are amongst the most active tectonic regions on Earth, tectonic activity appears to play only a minor role in sedimentation, climatically-driven catastrophic events such as rapid deglaciation being much more important. It appears probable that paraglacial sediment fans such as those exemplified here will prove to be widespread in glaciated mountains with high relative relief and severe climatic gradients.

References

Batura Glacier Investigation Group 1976. Investigation report on the Batura Glacier in the Karakoram Mountains, the Islamic Republic of Pakistan, 1974–1975. Batura Investigation Group, Engineering Headquarters, Peking. 123 pp.

Batura Glacier Investigation Group 1979. The Batura Glacier in the Karakoram Mountains. *Scientia Sinica*, **2**, 958–984.

Beaty, C. B. 1974. Debris flows, alluvial fans and revitalised catastrophism. *Zeitschrift für Geomorphologie N.F. Suppl.*, **21**, 39–51.

Brindley, G. W. and Brown, G. 1980. *Crystal structures of clay minerals and their X-ray identification*, Mineralogical Society, Spotiswoode Ballantyne Ltd., London.

Brunsden, D. and Jones, D. K. C. 1984. The geomorphology of high magnitude–low frequency events in the Karakoram mountains. In Miller, K. (Ed.), *International Karakoram Project Volume Two*. Cambridge University Press, Cambridge. 343–388.

Brunsden, D. *et al.* 1984. Particle size distribution on debris slopes of the Hunza Valley. In Miller, K. (Ed.), *International Karakoram Project Volume Two* Cambridge University Press, Cambridge. 536–579.

Bull, W. B. 1968. Alluvial fan cone. In Fairbridge, R. W. (Ed.), *The Encyclopedia of Geomorphology*. Reinhold, London. 7–10.

Bull, W. B. 1977. The alluvial fan environment. *Progress in Physical Geography*, **1**, 222–270.

Carroll, D. 1974. Clay minerals: a guide to their X-ray identification. *Geological Society of America spec. paper*, **126**, 80 pp.

Cunningham, J. D. 1854. *Ladak, Physical, Statistical and Historical with Notices of the Surrounding Countries*. London. Reprinted 1970, Soger Pub., New Delhi.

Deng Yangxin 1983. An explanation of the boulder clay formation near Yangjiaoling ridge of the Lushan by Debris–flow. *Journal of Glaciology and Cryopedology*, **5**(1), 33–46. (In Chinese)

Denny, C. S. 1967. Fans and pediments, *American Journal of Science*, **265**, 81–105.

Derbyshire, E. 1982. Glacier regime and glacial sediment facies: a hypothetical framework for the Qinghai Xizang Plateau, *Symposium on Qinghai Xizang Plateau (Tibet), Beijing, China. Proceedings Vol. II—Geological and ecological studies of Qinghai Xizang Plateau*, 1649–1656.

Derbyshire, E. *et al.* 1984. Quaternary glacial history of the Hunza Valley, Karakoram mountains, Pakistan. In Miller, K. (Ed.), *International Karakoram Project Volume Two*. Cambridge University Press, Cambridge. 456–495.

Derbyshire, E. and Miller, K. J. 1981. Highway beneath the Ghulkin. *Geographical Magazine*, **53**, 626–635.

Dewey, J. F. and Burke, C. A. 1973. Tibetan, Variscan and Precambrian basement reactivation products of continental collision. *Journal of Geology*, **81**, 683–692.

Dorn, R. I. 1983. Cation-ratio datings: a new rock varnish age-determination technique. *Quaternary Research*, **20**, 49–73.

Drew, F. 1873. Alluvial and lacustrine deposits and glacial records of the Upper Indus basin; Part I. Alluvial deposits. *Quaternary Journal of the Geological Society of London*, **29**, 441–471.

Eyles, N. *et al.* 1983. Lithofacies types and vertical profile models; an alternative approach to the description and environmental interpretation of glacial diamict and diamictite sequences. *Sedimentology*, **30**, 393–410.

Ferguson, R. J. 1984. Sediment load of the Hunza River. In Miller, K. (Ed.), *International Karakoram Project Volume Two*. Cambridge University Press, Cambridge. 581–598.

Ferguson, R. I. *et al.* 1984. Techniques for investigating meltwater runoff and erosion. In Miller, K. (Ed.), *International Karakoram Project Volume Two*. Cambridge University Press, Cambridge. 374–382.

Gansser, A. 1964. *Geology of the Himalayas*. Wiley, New York.

Gornitz, V. and Seeber, L. 1981. Morphotectonic analysis of the Hazara arc region of the Himalayas,

North Pakistan and Northwest India. *Tectonophysics*, **74**, 263–282.
Goudie, A. S. 1984. Salt efflorescence and salt weathering in the Hunza Valley, Karakorum Mountains. In Miller, K. (Ed.), *International Karakoram Project Volume Two*. Cambridge University Press, Cambridge. 607–615.
Goudie, A. S. *et al.* 1984. The geomorphology of the Hunza Valley, Karakoram Mountains, Pakistan. In Miller, K. (Ed.), *International Karakoram Project Volume Two*. Cambridge University Press, Cambridge. 359–411.
Haast, J. 1864. Report on the formation of the Canterbury Plains. *Provincial Council of Canterbury, Sess 22. Proceedings*, **63**.
Haast, J. 1879. *Geology of the Provinces of Canterbury and Westland, New Zealand*. Christchurch.
Henderson, W. 1859. Memorandum on the nature and effects of the flooding of the Indus on 10th August 1858 as ascertained at Attock and its neighbourhood. *Journal of the Asiatic Society of Bengal*, **28**, 199–219.
Hewitt, K. 1968. The freeze–thaw environment of the Karakoram Himalaya. *Canadian Geographer*, **12**, 85–98.
Le Fort, P. 1975. Himalayas—the collided range. *American Journal of Science*, **275A**, 1.
Li Jijun *et al.* 1984. Glacial and paraglacial sediments of the Huzna Valley North-West Karakoram, Pakistan: a preliminary analysis. In Miller, K. (Ed.), *International Karakoram Project Volume Two*. Cambridge University Press, Cambridge. 496–535.
Lü Ruren and Li Deji 1986. Debris flow induced by ice lake burst in the Tangbulang Gully, Gongbujiangda, Xizang (Tibet). *Journal of Glaciology and Cryopedology*, **8**(1), 61–71. (In Chinese)
Lyon-Caen, H. and Molnar, P. 1983. Constraints on the structure of the Himalaya from an analysis of gravity anomalies and a flexural model of the lithosphere. *Journal of Geophysical Research*, **88**, 8171–8191.
Mason, K. 1929. Indus floods and Shyok Glaciers. *Himalayan Journal*, **1**, 10–29.
Mason, K. 1935. The study of threatening glaciers. *Geographical Journal*, **85**, 24–41.
Mehta, P. I. C. 1980. Tectonic significance of the young mineral dates and rates of cooling and uplift in the Himalaya. *Tectonophysics*, **62**, 205–212.
Obbard, J. 1860. On the translation of waves of water in relation to the great flood of Indus in 1858. *Journal of the Asiatic Society of Bengal*, **39**, 266–274.
Owen, L. A. 1989. Neotectonics and glacial deformation in the Karakoram Mountains, Western Himalaya. *Tectonophysics*, **163**, 227–265.

Paffen, K. H. *et al.* 1956. Forschungen in Hunza–Karakorum. *Erdkunde*, **10**, 1–33.
Patton, C. P. *et al.* 1970. *Physical Geography*, Wandsworth, Belmont.
Porter, S. C. 1970. Quaternary glacial record in Surat Kohistan, West Paksitan. *Geological Society of America Bulletin*, **81**, 1421–1446.
Pugh, J. C. 1975. *Surveying for Field Scientists*, Methuen, London.
Rachocki, A. 1981. *Alluvial Fans: An Attempt at an Empirical Approach*, Wiley, Chichester.
Rhind, D. W. 1968. *The Terraces of the Tweed Valley*. Unpublished Ph.D. thesis. University of Edinburgh.
Rittenhouse, E. G. 1943. A visual method of establishing two dimensional sphericity. *Journal of Sedimentary Petrology*, **13**, 79–81.
Rockwell, T. K. *et al.* 1985. A late Pleistocene–Holocene soil chronosequence in the Ventura basin, Southern California, U.S.A. In Richards, K. S. Arnett, R. R., and Ellis, S. (Eds), *Geomorphology and Soils*. Allen and Unwin, London. 309–327.
Seeber, L. and Gornitz, V. 1983. River profiles along the Himalayas arc as indicator of active tectonics. *Tectonophysics*, **92**, 335–367.
Shaw, J. 1986. Discussion. Lithofacies types and vertical profile models: an alternative approach to the description and environmental interpretation of glacial diamict and diamictite sequences. *Sedimentology*, **33**, 151–155.
Shaw, R. B. 1871. *High Tartary, Yarkand and Kashgar and Return Journey over the Karakoram Pass*, London. Reprinted 1984 as Shaw, R. B. *Visits to High Tartary, Yarkand and Kashgar*. Hong Kong, Oxford University Press.
Shepard, F. P. and Young, R. 1961. Distinguishing between beach-dune sands, *Journal of Sedimentary Petrology*, **31**, 196–214.
Surell, A. 1851. *Etude sur les Torrents des Hautes Alpes*, Paris.
Tucker, M. E. 1982. *The Field Description of Sedimentary Rocks*, Geological Society of London Handbook Series. Milton Keynes, The Open University Press.
Whalley, W. B. *et al.* 1984. Rock temperature observation and chemical weathering in the Hunza region, Karakoram: Preliminary data. In Miller, K. (Ed.), *International Karakoram Project Volume Two*. Cambridge University Press, Cambridge. 616–633.
Woldstedt, P. 1965. *Das Eiszeitalter*. Ferdinand Enke Verlag, Stuttgart.
Zeitler, P. K. 1985. Cooling history of the NW Himalaya, Pakistan. *Tectonics*, **4**, 127–151.

CHAPTER 3

Ice Marginal Ramps and Alluvial Fans in Semiarid Mountains: Convergence and Difference

Matthias Kuhle
Universität Göttingen, Göttingen

Abstract

Ice marginal ramps are depositional landforms that develop over tens of kilometres along the margins of large piedmont glaciers in semiarid environments. The ramps extend with a 7–15° gradient several kilometres or even tens of kilometres into the forelands. They have proximal slopes that are tens of metres to several hundred metres high and frame the former terminal ice basins. The proximal slopes consist of till, which thins out towards the periphery. The overlying beds contain ice contact stratified drift, which, with increasing distance from the former ice margin, is succeeded by clearly-sorted glaciofluvial gravel. Under semiarid conditions the syngenetic contribution of two agents of transport (glacial, glaciofluvial) in forming one accumulation complex produces stratigraphic and phenotypical features that are rare in the glacial morphology of the temperate humid zones. For this reason, they are often misinterpreted. Being indicators of the ice margins, ice marginal ramps permit the accurate reconstruction of extensive piedmont glaciations in the semiarid highlands of subtropical latitudes.

Introduction

The genetic relationship between glacial systems and fluviatile (glaciofluvial) deposits was first recognized by Charles Martins (1841–1842). Picking up this idea A. Penck (1882) worked out a glacial sequence using the foreland of the Alps in southern Germany as an example. With this sequence, the subrecent extension and temporal variation of glaciers can be determined based on the spatial relations and stratigraphy of drumlin and esker zones, till sheets, moraine ridges, and glaciofluvial deposits. In areas with poorly-preserved or nonexistent end moraines (e.g. southern New England), this method has been extended so that the retreat positions of the last ice sheet can usually be reconstructed by using meltwater deposits alone, according to the 'morphologic sequence' concept (*cf.* Jahns, 1941, 1953; Koteff, 1974). Here it is essential that the criteria for differentiating between glaciofluvial and purely fluviatile deposits are not the sedimentary fabric e.g. grain-size distributions, but the morphologic features of the deposits and their spatial relationships with other landforms. During studies on Pleistocene glaciation in subtropical mountain re-

Alluvial Fans: A Field Approach edited by A. H. Rachocki and M. Church
Copyright © 1990 John Wiley & Sons Ltd.

gions, the author was able to work out a new type of glacial sequence: the so-called ice marginal ramps (IMRs), a specific combination of end moraines and outwash (glaciofluvial) deposits. Due to the great amount of glaciofluvial material contained in them, IMRs were previously misinterpreted as incised pediments by other authors (Czajka, 1957; Geological Map of Iran, 1972; Weise, 1974; Brunotte, 1986). Based on an expanded morphologic sequence in which large-scale glacial features (erosional forms, erratics) are also used as evidence, the glacial–glaciofluvial origin of IMRs has been proven (Kuhle, 1974, 1976, 1984a,b, 1985, 1987a, 1989).

Ice Marginal Ramps

The basic configuration of IMRs, which display a typical sequence in the field, can be described as follows. The moraine ridge surrounding a subrecent piedmont glacier was syngenetically covered by glaciofluvial debris from the former surface of the glacier ice. This debris cover extends from the culmination of the moraines down into the mountain foreland as ramps. Two types of glaciofluvial cover on the frontal moraines can be distinguished (Figure 3.1): in type A the glaciofluvial deposits start immediately after the culmination of the frontal moraine; in type B the moraine rampart and glaciofluvial material are truncated and unconformably covered by fresh deposits that dip in the direction of the terminal basin. Whereas in the initial stages the surface of the glaciofluvial debris on the outer slope of the moraine is steep (i.e. 25°), in the final stage the debris ramp dip at angles of 7–15°. These internal structures show that the IMRs are the result of the long-term processes of accumulation along stable ice margins (type A) in foreland regions. Minor fluctuations in the ice margins are reflected in the stratigraphy of the type B.

Contrary to the glaciofluvial debris of the IMRs, the traditional fan-shaped outwash plains are concentrated at the former glacier snout, the point of exit of channelized subglacial meltwater (*cf.* Figure 3.13). In the mountain forelands studied, the dip angle of these outwash plains was 2–4° and thus clearly shallower than the IMRs. IMRs have as yet been observed only in areas of piedmont glaciation in semiarid mountain regions. Since such glaciations do not exist under the present climatic conditions, the processes and dynamics of the formation of IMRs can be studied

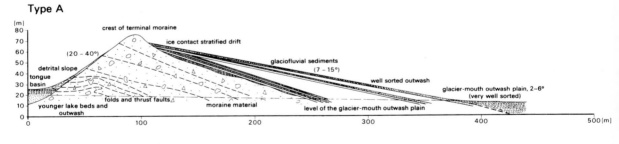

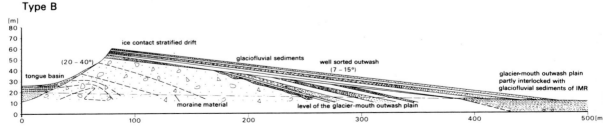

Figure 3.1. Types of the ice marginal ramps representing two stages of development. Type B evolves from Type A and, being the last stage, occurs more frequently

SEMIARID IMRS AND ALLUVIAL FANS

Figure 3.2. Characteristic ice-margin deposit in the semiarid Nanga Parbat Massif (200–300 mm per year precipitation at the level of the glacier terminus at 3000 m a.s.l.). This left lateral moraine of the Bazhin glacier ($\diamondsuit$; 35°13′N; 74°38′E; 3780 m a.s.l.) has formed since 125 ± 60 yr B.P. (^{14}C, 98.4 ± 0.8:modern). The meltwater, flowing over the surface of (dashed line) and seeping through (arrows) the moraine ridge, deposits material of the surface and lateral moraines in the manner of glaciofluvial beds. This example may serve as phase 1 of IMR development. (Photo M. Kuhle, 2.9.87); (B) The Tshaigiri glacier in the Nanga Parbat group photographed from 4100 m a.s.l. (35°11′N; 74°35′E). Meltwater discharge coming down from the moraine-covered glacier surface (squares) leads to the aggradation of stratified drift (arrows) covering the outer slope of the lateral moraine. These glaciofluvial ramps dip at angles of 19–28° and have relative heights of up to 250 m. With respect to their internal structure and processes they correspond to IMRs: they have a moraine core at the proximal parts which is superposed by glaciofluvial beds. Hence this example illustrates phase 2 of IMR formation. (Photo M. Kuhle, 5.9.87)

only using analogies. The processes occurring on the dam moraines of the Nanga Parbat massif (Figure 3.2) are analogous to those of the juvenile stage of IMRs. Steep, ramplike debris slopes (19–28°) are deposited here on the outer slope of the moraines by meltwater flowing out at the sides (Figure 3.2A). The picture near the glacier tongue 700 m below the equilibrium line shows that not only fluvioglacial activity, but also mass movement from the upper moraine can participate in the formation of the ramps. Whereas the fluviatile processes (rounding, sorting) are insignificant here, the mature forms of the subrecent IMRs show increasing sorting and rounding of the debris with distance from the source. The first comprehensive analysis of IMRs was made with reference to the Pleistocene glaciation of Kuh-i-Jupar (Kuhle 1974, 1976), a massif in the Kuh-Rud Mountains, which run parallel with the Zagros chains to the north (southeastern Iran). A further example is located in the Uspallata basin in the eastern part of the Aconcagua massif (Argentinian Andes, Kuhle 1984b). The biggest known IMR site is the Qinghai–Xizang Plateau (northeastern Tibet) where 18 localities were investigated in detail (Kuhle 1982b, 1987a,b). Late glacial IMRs were found in the forelands of Shisha Pangma and Gang Benchen (southern Tibet). Their proximity to present-day glaciers there renders the genetic connection immediately verifiable.

ICE MARGINAL RAMPS OF THE KUH-I-JUPAR (29°40′–30°15′N; 56°50′–57°35′E)

On the evidence of glacial deposits in the Kuh-i-Jupar massif (maximum summit height 4135 m a.s.l.), a two-phased mountain glaciation was reconstructed with a length of 17 km in the direction of flow and a lobe width of 20 km in the foreland at the time of the earlier glaciation (Kuhle, 1974, 1976). Valley glaciers flowed into the mountain foreland and coalesced to form wide piedmont ice lobes during the earlier glaciation. At the time of the later, less severe glaciation, the glacier termini reaching the foreland were generally separated from each other by intervening rock spurs. The

Figure 3.3. Late Pleistocene ice marginal ramp (arrow) in the northern Kuh-i-Jupar foreland. The ridge-like lateral moraine (X) delimiting the terminal basin on the right is a linear landform contrasting with the three-dimensional IMR. Arrow indicates the location from which Figure 3.4 was photographed. In the foreland (middle background) the IMRs of an older piedmont glaciation are visible. (Photo M. Kuhle, 9.7.73)

outlines of the respective terminal basins are marked by IMRs (Figure 3.3 and 3.4). In the case of the earlier piedmont glaciation these are long lines (Figures 3.3 background) whereas the IMRs of the later glaciation were often built up in a wedge shape by still separate, tapering glacier tongues (Figures 3.3 and 3.4). Fluvial dissection which continued to be active after their formation, created a characteristic converging system of small floodplain valleys (Figures 3.3 and 3.4, *cf.* also Figures 3.7 and 3.8).

In the mountains, the former catchment, this glaciation is recognizable over wide areas by landforms of glacial erosion; U-shaped valleys, overdeepened basins, roches moutonnées, glaciated valley flanks, transfluence cols (Kuhle, 1976: 127–179). Rock surfaces with still visible striae and glacial polishing were found (Kuhle, 1976, Figures 70 and 72) as well as remains of lateral moraines with proven erratic boulders (Kerman conglomerates on underlying massive biomicrite limestone: Kuhle 1976, Figure 76).

The characteristic glacial sequence of the IMRs begins with a ridge of typical diamictic (non-stratified) morainal material. Facetted, subrounded, rounded, and well-rounded erratic boulders of the Jupar limestone (biomicrite) and Kerman conglomerate (polymict), some of which are $6 \times 6 \times 5$ m and $2 \times 1.5 \times 1.2$ m in size, are embedded in a matrix of fine-grained non-stratified material comprising all fractions down to silt and clay (Figure 3.5A). Ice-contact stratified drift is deposited directly on the outer slope of the frontal moraine (type-A IMR) as shown in Figure 3.5B. Here the material of the frontal moraine has already been partially washed away. Grading does occur in places, but is only poorly developed. It is important to note the obvious, strong influence of the meltwater.

In the case of type-B IMRs the culmination of

Figure 3.5. (A) Moraine exposure in the Kuh-i-Jupar foreland (southern Iran) with coarse polymict boulders (light-coloured, limestone; dark, Kerman conglomerate) in a dense till matrix. The viewpoint is located 10 m below the highest moraine ridge surrounding the terminal basin of Figure 3.5D. The base height of the inner moraine slope is 2240 m. (Photo M. Kuhle, 8.6.73); (B) Transition zone between till and ice-contact stratified drift. Neither the grain-size distribution (55–60% silt, 45% of which has 20–30 μm) nor the rounding and the orientation (30% orientation in direction of transport) have changed. However, graded bedding and sizing begin to develop. The polymict, erratic material contains limestone (biomicrite ck and k 3/2, pale blocks) and the Kerman conglomerate (right, beside the scale). Kuh-i-Jupar north slope (29°55′N; 57°10′E) in the Darne Kanatkestan Valley. (Photo M. Kuhle, 19.8.73); (C) Outcrop of the glaciofluvial layers on the periphery of an IMR 1200 m away from the former ice margin. Graded bedding (dipping at 9–11°) and sorting have clearly developed. Coarse blocks are lacking, and the particles are subrounded, rounded, and well rounded. Compared with the location in Figure 3.5B the grain-size distribution has moved from mainly silt to mainly sand size (70% 63–200 μm). The glaciofluvial beds are deposited on Neogene sandstones (below ice axe). Kuh-i-Jupar north slope (30°07′N; 57°10′E) piedmont in the elongation of Darne Kanatkestan. (Photo M. Kuhle, 2.4.74); (D) Outwash fan with gently sloping beds (2–3°). The characteristics of the material correspond to those of Figure 3.5C. The orientation of particles (40–50% transverse to the direction of transport) is clearly fluviatile. Kuh-i-Jupar north slope (30°01′N; 57°12′E) piedmont in the elongation of Darne Kanatkestan. (Photo M. Kuhle, 12.9.73)

the moraine ridge is discordantly overlain by well-sorted, rounded, and well-rounded debris of finer particle sizes, with bimodal grain-size distribution (sand predominates in the fine-grained range). In the distal portions of IMRs, some 1 to 1.5 km away from the former ice margin there is very good sorting with a more confined grain-size range, but without any significant change in rounding (35–45% of the cobble components are rounded and well-rounded, only 10% are sub-rounded and *ca.* 2% are very well-rounded; Figure 3.5C). The ice marginal ramps are 500–1500 m long in the direction of deposition, and their surfaces are inclined between 7° and 15° (usually 12°). Typical outwash plains extend out from former glacier outlets as shallowly inclined (2–4°)

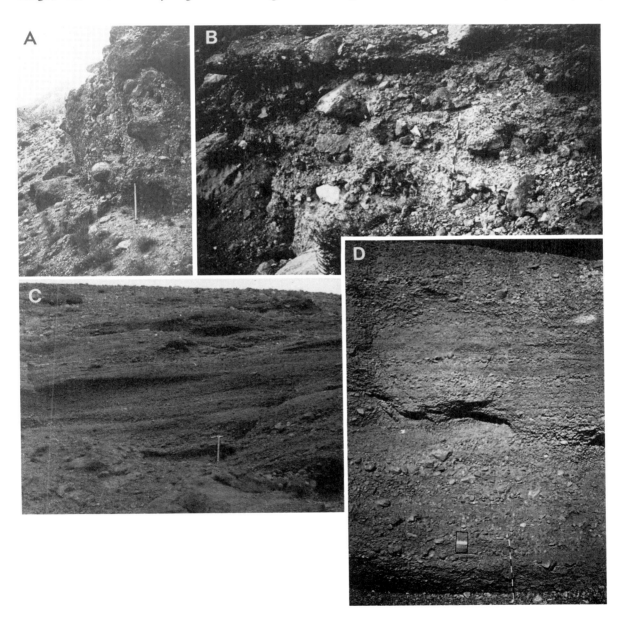

Figure 3.6. Pushed glaciolacustrine sediments deformed by the overriding Riss-age piedmont glacier of Kuh-i-Jupar. This compression at 2040 m a.s.l. is unconformably overlain by more or less stratified drift from a younger stadial moraine. (Photo M. Kuhle, 17.3.74)

fans and are situated between IMRs. They are easily recognizable, distinct morphologic features (Figures 3.3 and 3.9). Their sediment fabric, however, cannot be distinguished from that in the distal portions of IMRs, presuming the same transport distance. Figure 3.5D shows an outcrop in a younger outwash plain containing cobble components (limestones) of which 40% was subrounded, 40% rounded, and 20% well-rounded. The lower degree of rounding is a function of the shorter transport distance (800 m) from the ice margin.

The glacigene interpretation of IMRs is further confirmed by evidence of classical ice thrust features (*cf.* Flint 1971:121–124). For example, 4 km away from the mountain foot there are heavily compressed and folded lake sediments (Figure 3.6) exposed. They cover a distance of 200 m and are up to 14 m thick. The overlying layers consist of up to 8 m of till, covered by 3 m of glaciofluvial material belonging to an IMR (Kuhle, 1976:90–97). The lacustrine sediments were formed in an interstadial terminal lake that was impounded by the IMRs of the earlier stage of glaciation.

ICE MARGINAL RAMPS IN THE USPALLATA BASIN (CENTRAL ANDES)

Full glacial glaciers flowing down from the 5000 m high Cosinero del Chacay (adjoining the

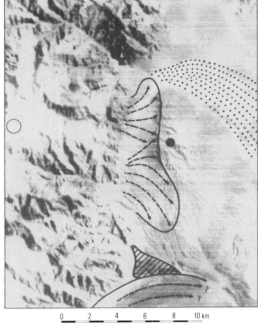

○ culmination of Cos. del Chacay   ⫽ direction of ice flow
● ice marginal ramps   //// Morainic material with striated clasts

Figure 3.7. Satellite image (NASA ERTS E–2022–13452–701, 13.2.1975) of the Cosinero del Chacay E slope and the western Uspallata basin, where the Ice Age ice margin forked into two lobes and deposited an ice marginal ramp (Figure 3.8). At the bottom, the lateral or end moraine deposited from the south by Mendoza glacier is visible. Dots indicate conventional outwash

Aconcagua Massif in the northeast) reached the Uspallata Basin and were diverted northward and southward by a steep, Tertiary rock bar (Figure 3.7). The bedrock here is covered by frontal moraines up to 420 m high. The rounded and facetted boulders embedded in a fine matrix exclude an origin due to landslides. Furthermore, directly on the distal side of the moraine culmination there are typical steep, glaciofluvial gravels that underwent postgenetic dissection by small V-shaped valleys (Figure 3.8). Their internal structure, external appearance, and situation (at the margin of a terminal basin) correspond in all details with the IMRs occurring in Kuh-i-Jupar (Type A; *cf.* Figures 3.3 and 3.4).

The glacial origin of these forms was clearly confirmed by the extensive evidence of the glacial sequence (i.e. mapping of the spatial distribution of striated clasts and erratic boulders, Kuhle, 1984b).

ICE MARIGINAL RAMPS ON THE TIBETAN PLATEAU (QUINGHAI–XIZANG PLATEAU)

In a study area measuring 450 × 820 km in northeast Tibet (38°–40°N and 34°50′–36°N; 95°–102°E) full glacial IMRs were investigated in detail at 18 locations in the forelands of the Kuen Lun and Quilian Mountains (Kuhle, 1982b, 1987a,b). An ideal sequence of terminal basins, IMRs, and glacier-mouth outwash plains occurs at 35°13′N; 97°50′E on the south slope of the Kuen Lun Mountains. With a frontal height of more than 100 m, the IMRs stretch over 10 km down into the foreland, where they interlock with the glacier-mouth outwash plains (Figure 3.9). The detailed evidence of these forms is given in Kuhle (1982b, 1984a, 1987a,b) and the evidence of a full glacial ice sheet on the Tibetan Plateau and surrounding mountain ranges is given in Kuhle (1980–1988). One of the rare cases where IMRs occur immediately in front of recent glaciers is found on the north slope of Shisha Pangma (8046 m) and Gang Benchen (Figures 3.10, 3.11, and 3.12). These relict forms of Late glacial glaciers attain maximum thicknesses of 500–600 m and extend as much as 16 to 18 km down into the foreland, as far as 5050 m elevation. The typical postgenetic dissection of the IMR surfaces is missing here (compare Figure 3.11 and Figures 3.3 and 3.7), since these Late glacial landforms are still situated in a periglacial environment where the movement of several metres thick solifluction sheets over wide areas impedes the development of linear fluvial erosion.

Affinities and Differences Between IMRs and Alluvial Fans of Other Origin

Due to the great amount of fluviatile material in IMRs, especially near their surfaces, they have been interpreted frequently as dissected alluvial pediment covers (Czajka 1957; Weise 1974; Brunotte 1986). The dissection of alluvial covers, however, does not generally produce the tongue-shaped basins typical for the morphologic sequence of IMRs (Figures 3.3, 3.4, 3.8, and 3.9) but linear erosional channels separated by stepped residual surfaces which have maintained their greatest elevations at the direct contact to mountain spurs, where they are protected from erosion by the mountain spurs (Figure 3.13).

The source area for dissected alluvial fans can be determined by projecting the inclination of their bedding. Using this method of the 'extrapolated fanhead' the source area of the steeply dipping beds (7–15°) of the IMRs would have been situated far above the mountain summits. The possibility that these features represent tilted Neogene alluvia (valley fill) also has been clearly

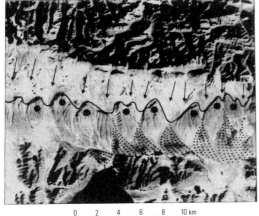

● ice marginal ramps (IMR) ⟵ direction of ice flow
∼ glacier margin ⋮ glacier mouth outwash plains

Figure 3.9. Ice marginal ramps on the Kuen Lun south slope (35°13′N; 97°50′E) mark the margin of a Late glacial piedmont glacier. The position of the small IMR valleys at the right angles to the frontal moraines is characteristic. Unlike relict alluvial fans, which are convex in the centre, these landforms have a concave cross-profile. There the small valleys converge into a central depression. Field investigations in 1981 established the morainic nature of the substratum (Kuhle, 1987a). (Photo: NASA ERTS E–2691–03112–701, 13.12.1976

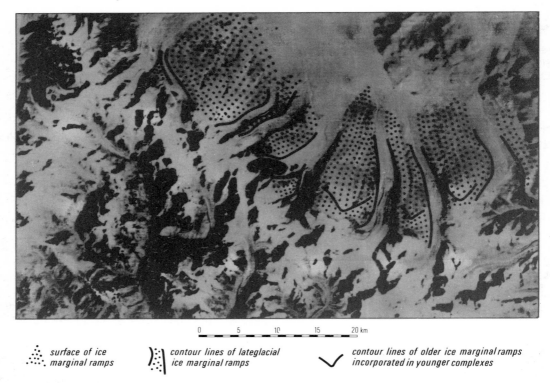

Figure 3.12. Late glacial IMRs on the Shisha Pangma north slope (north side of the Himalayas between 8046 and 5020 m a.s.l., 28°25′N; 85°48′E). Glaciers still reach their base today (see Figures 3.10 and 3.11). Owing to Neoglacial and historical glacier oscillations, the original ground-plan form of the IMRs was distorted by moraine deposits. The till and gravel ramps (*cf.* Figure 3.9) once stood isolated in the foreland and were subsequently directly joined to the mountain valley mouths by means of the lateral moraines deposited by subsequent, increasingly channelled glacier tongues. (Photo NASA ERTS E–2662–03542–702, 14.11.1976)

ruled out (Kuhle, 1976:97–102). Neogene dislocation zones and the IMRs are not related. In zones both affected by Neogene tectonics and having IMRs the morainic material and adjoining fluvioglacial deposits lie undisturbed upon an erosional surface that cuts through the Neogene layers, so the tectonic processes had ended before the time of glacial accumulation. In addition, the small-scale variations in the dip direction of IMRs contrast fundamentally with large-scale mountain building.

There is a direct, regular relationship between the distribution pattern of IMRs (i.e. their distance from the catchment area) and the altitude and exposure of their catchment (see morphological maps in Kuhle, 1976, Figures 162–164, and Figure 3). This means that IMRs which have similar exposures and ages but have differing distances from catchment areas of differing altitude will produce a uniform altitude of the former glacier equilibrium line. This is based on a calculation method developed for alpine moraines (v. Höfer, 1879; Kuhle, 1986b). A direct relationship between the basal surface and height (volume) of IMRs and the size of the catchment area does not exist. For alluvial fans, however, there is a direct relationship between their size and extension and the drainage basin area, but not altitude or exposure (Bull, 1964). The characteristic morphologic sequence in which the glaciofluvial deposits of IMRs occur allows only a glacial–morphological interpretation. Mudflow

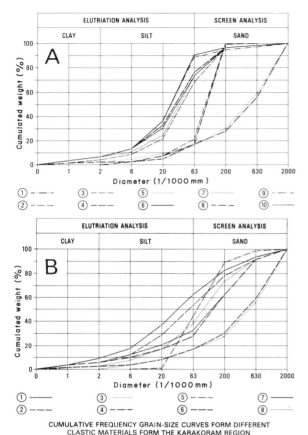

Figure 3.14. (A) Grain-size curves of Quaternary sediments in the arid high-mountain environment between the Aghil main chain, over the western Kuen Lun to the northern piedmont areas leading down into the Tarim Basin (36°20′N; 38°50′E; altitudes between 4000 and 1400 m a.s.l.), sub-2 mm fractions. Curves 2 and 8 show grain-size distribution from widespreaded, 1 to 2° inclined alluvial fans on the piedmonts at 1500 m, 80 km away from the solid rock of the mountains. Curve 6 (till; 3740 m) shows the same characteristics as curve 9 (mudflow). In both cases the source rock is the Kuen Lun granite, and the distance of transport amounts to 15 km. Curves 1 and 5 are characteristic for alluvial terraces at the bottom of the Yarkand Valley (3800 m a.s.l.); the material contains granites from the Kuen Lun slope and metamorphic rock (clay, silt, and sandstones) from the Aghil slope; (B) Grain-size curves of alluvial sediments. Samples 5 and 8 are from very large, gently sloping (1°) alluvial fans on the piedmont. The material is polymict. The distance of transport amounts to more than 100 km (3850 m a.s.l.). Curves 1, 2, 4, and 7 show gravel deposits with differing admixtures of talus cone material on the bottom of a narrow

deposits (diamicton) in foreland areas which have been fluvially covered over (Brunotte, 1986) must be excluded as well because the fluviatile deposits which lie steeply on and beyond the culmination of the diamictite must have been directly connected to a glacier surface as a stable level of meltwater discharge.

There have occasionally been demands that the genetic interpretation of IMRs be based on the quantitative analysis of their sediment fabric. Quantitative analysis of moraine material originally was developed for the Scandinavian ice sheets, but only long after the glacial genesis of the deposits had already been explained (e.g. Esmark, 1827; Bernhardi, 1832; Torell, 1875). The oldest method of quantitatively analysing glacial drift originated from Ussing and Madsen (1897). Dreimanis (1939) described the method of elutriation analysis, Raistrick (1929) that of heavy-mineral analysis, White (1934) that of mineral analysis, and Münnich (1936) that of grain-size distributions. All of these methods are founded on one principle: material known to be of moraine origin is described quantitatively, and the changes within the area of occurrence are then interpreted to provide information on ice movements and sources. Quantitative analysis does not give evidence of the genetic origin (e.g. differentiation of mudflow and till materials) but helps in discovering small-scale variations within specific depositional features. The studies of Dreimanis and Vagners (1971) showed a typical dependency of the particle sizes in the till matrix on transport distance and initial material. However, there is the requirement that the initial rock material has a defined source area. These conditions are generally given in flat land areas. In mountainous areas local interference of genetically different processes is produced by the relief energy. Thereby, there is a great mixing of primary and secondary debris masses which is controlled by chance.

Kuen Lun valley. The granite and metamorphic detritus has been transported 10 km at the most. Curves 3 and 6 show also well-graded gravel deposits consisting of granitic and metamorphic detritus after transport of up to 100 km (middle Yarkand Valley; 3800 m a.s.l)

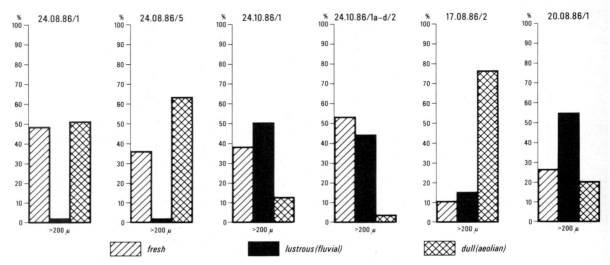

Figure 3.15. Surface texture of sediment grains greater than 200 μm. Samples 24.08.86/1 and /5 (*cf.* Figure 3.14B, Nos. 1, 7) show the characteristics of very briefly transported (less than 10 km) gravel and talus material from a canyon in the Kuen Lun: the fresh, angular grains (I) have the same portion as the aeolian grains (III) while the fluviatile-polished material (II) is almost completely lacking. Samples 24.10.86/1 and 24.10.86/1a–d/2 were all taken from outwash terraces in semiarid Yarkand Valley; however, the distance to the source area varies. With a greater transport distance, sample 24.10.86/1 shows a predominance of group II (fluvial grains), followed by the aeolian material (III) and has only a very small portion of group I. Due to the comparably long fluviatile transport of sediment 17.08.86/2, (*cf.* Figure 3.14A, No. 8) group I (angular, fresh) takes almost no part in it. Group III, aeolian transport, predominates, which is typical for sediments 200 μm in full-arid environments. Sample 20.08.86/1, (*cf.* Figure 3.14A, No. 6) shows the characteristics of a till at a distance of up to 15 km from the source area. The fluviatile-polished grains (II) predominant; it cannot be determined whether the group I grains must be attributed to normal weathering or to glacial erosion and transport. The aeolian material (III) still has a portion of 20% and seems to have been blown from a distance

Detailed quantitative analyses were carried out on sediments from the north slope of the Karakoram. The grain-size curves (Figure 3.14) show nearly homogeneous sediments of genetically differing origins. Microscope analysis of surface feature (Figure 3.15) indicates a random mixture of unworked (angular grains) aeolian (dull grains) and fluvially-polished grains within till, glaciofluvial, and purely fluvial deposits. However, materials of identical grain-size distributions and surface textures can be clearly distinguished visually using their stratification and position (Figure 3.16). Glaciofluvial deposits cannot be differentiated from till by means of particle size distributions (Sugden and John, 1976; compare their Figures 11.11 and 16.7). Moraine deposits can be more different from each other (*cf.* Figures 3.5A and 3.16B) than from fluvioglacial deposits (Figure 3.16A). In such cases quantitiative analysis may show at best the chaotic depositional conditions, but has no genetic evidence. On a higher level of integration, such chance variations can be disregarded. The differentiation of IMRs from alluvial deposits of the origin clearly can be based on the morphologic sequence. Quantitative analysis of the sediment fabric can provide a more detailed description of individual cases, but cannot give a primary proof of the genetic origin.

Characteristics of Ice Marginal Ramp Formation

On the basis of the evidence to date, the following environmental characteristics of IMRs may be inferred:

Figure 3.16. (A) Till outcrop (arrow) in the lateral moraine of the Neoglacial (younger than 12 870 ± 180 ^{14}C yr B.P.) K2 glacier (Karakoram north slope, 36°03′N; 76°28′E; 4400 m a.s.l.) overlain by the glaciofluvial (circle) beds of lateral outwash cones with very similar grain-size distribution. Both types of sediment contain up to 70% silt. In the outwash the portion of sand is slightly higher than in the till, where clay dominates over sand. The roundness of particles also coincides. The most distinctive feature is the graded bedding in one case and the lack of grading in the other. (Photo M. Kuhle, 2.9.86); (B) Late-glacial lateral moraine (younger than 12 870 ± ^{14}C yr B.P.; 20.1 ± 0.1% modern) at the Skamri glacier (background) in the Muztagh Valley (Karakoram north slope, 36°02′N; 76°20′E; 4500 m a.s.l.). The only difference to the mudflow deposits (Figure 3.5C) having the same grain-size distribution (65–70% in the range 6–63 μm) is the completely diamict character of these sediments (here also with polymict crystalline and clastic blocks). (Photo M. Kuhle, 14.10.86); (C) Mudflow fans in the Shaksgam Valley (Karakoram north slope, 36°07′N; 76°38′E; at 4050 m a.s.l.). They are built up of resedimented till, containing granite and limestone (dolomite:Do 90%, Ca 10% and calcite: Ca 90%, Do 10%) blocks. The glacial diamicton was transported 4–6 km by mudflows. The grain-size distribution remained the same, however, in contrast to tills the mudflow fan shows distinct bedding. Each mudflow event established a new layer (arrows). (Photo M. Kuhle, 20.10.86)

1. IMRs are formed only on the margins of large, unconfined piedmont glacier lobes; barriers, such as valley flanks, along the termini prevent their formation.
2. IMRs are formed only in association with large bodies of ice, whose reaction to short-term climatic fluctuations is slow enough for their ice margins to be sufficiently stable to build up deposits with the dimensions of IMRs.
3. These stationary glaciers react to climatic fluctuations with an increase or reduction in size of the lobe front; the resulting change in the angle of deposition in the tongue marginal area is expressed in the typical erosion and depositional stratigraphy of the IMR.
4. Retreating glacier snouts not only decrease in thickness, but their edges are also wasted and buried. The original lobe sharpens into a wedge shape; this is often reflected in the preserved, cone-shaped configuration of the terminal basin (Figure 3.9).
5. The situation of the IMRs in relation to the permafrost line has phenotypical significance: full glacial IMRs now located far below the permafrost line have undergone intensive fluvial dissection (their characteristically converging V-shaped valley system provides a useful criterion to distinguish them from the discharge pattern on dissected alluvial cones when interpreting satellite images), whilst Late or Neoglacial IMRs that are presently still situated within the permafrost zone have a level surface due to the movement of several metres thick sheets of creeping waste (*cf.* Figures 3.11 and 3.12).
6. The IMRs known at the present time exist in semiarid environments. From this the hypothesis may be drawn that, because of the relative aridity, the glaciers could not have expanded into the temperate altitudinal zone with year-round subglacial meltwater flow even at the time of the maximum, full glacial fall in temperature; the glaciers remained within the permafrost limit and were subject to the seasonal alternation between subglacial and supraglacial meltwater debris release.

Results and Consequences

Ice Marginal Ramps are the result of a complex of factors in which climatic parameters (cold/semiarid) together with spatial structures (extensive piedmont glaciation near the permafrost line) and temporal factors (persisting ice margins) attain a specific effectiveness. The basic pattern comprises foreland terminal basins which are delimited by the proximal slopes of ramp-shaped deposits with 7–15° distal gradient. Their long profiles show a continuous transition from clearly morainic deposits to clearly fluvially bedded and sorted gravels. Two agents (glacial/glaciofluvial) combine under specific climatic, spatial, and temporal conditions to build up one complex of deposition. This leads to emergence of novel phenotypical characteristics unknown in the classical glacial morphology of the temperate humid zones.

Within the reference limits of this formation pattern each of the three originally independent dimensions possesses a variability that is individually expressed in the phenotype of the IMRs: dissected/planated surface; wedge-shaped/elongated pattern; varying proportion of glaciofluvial to morainic substratum in thickness and extent. Their magnitude, characteristic three-dimensional configuration and link with the sparsely-vegetated semiarid environment mean that IMRs are indicators of former ice margins that are also identifiable on satellite images (Figures 3.7, 3.9, and 3.12). This could be a promising approach especially for research on inaccessible areas in High Asia or the Andes of South America. It should be emphasized, however, that IMRs have up to now, been identified only by means of direct fieldwork. Identification on the basis of satellite imagery should be confirmed by ground checks.

Up to now, IMRs have proved to be characteristic landforms of semiarid piedmont glaciations in subtropical latitudes. For this reason they should be accorded key of importance in Ice Age research.

Identifiable by satellite imagery interpretation combined with selective field work, ice marginal ramps provide the possibility of reconstructing the exact limits of the extensive Ice Age subtropical highland glaciation of Asia and South America and, thus, decisively advancing knowledge of the energy balance and development of the Ice Ages (Kuhle, 1987c, 1988b).

Acknowledgements

I wish to thank the University of Göttingen for financial support of two expeditions to Iran (1973, 1974) and the German Research Society (DFG) which funded field work in the Dhaulagiri and Annapurna Himalaya (1976/1977), in the Andes (1980), and in the Khumbu Himalaya on Everest south slope (1982). The expeditions to northern Tibet (1981), to southern Tibet on the north slope of the Himalayas (1984), and to the north slope of K2 (1986) were financed by the Academia Sinica, the German Research Society, and the Max Planck Society (MPG).

References

Bernhardi, A. 1832. Wie kamen die aus dem Norden stammenden Felsbruchstücke und Geschiebe, welche man in Norddeutschland und den benachbarten Ländern findet, an ihre gegenwärtigen Fundorte? *Jahrbuch für Mineralogie, Geognosie und Petrefaktenkunde (Heidelberg)*, **3**, 257–267.

Brunotte, E. 1986. Zur Landschaftsgenese des Piedmont an Beispielen von Bolsonen der Mendociner Kordilleren (Argentina). *Göttinger Geographische Abhandlungen*, **H.82**.

Bull, W. B. 1964. Geomorphology of segmented alluvial fans in western Fresno County, California, *U.S.G.S. Professional Paper*, **352-E**, 89–128.

Czajka, W. 1957. Das innerste Längstal am Ostrand der Argentinischen Puna. El Cajón de San Antonio, *Jahrbuch der Geographischer Gesellschaft Hannover*, **1956/57**, 153–177.

Dreimanis, A. 1939. Eine neue Methode der quantitativen Geschiebeforschung, *Zeitschrift für Geschiebeforschung*, **15**.

Dreimanis, A. and Vagners, U. J. 1971. Bimodal distribution of rock and mineral fragments in basal tills. In Goldthwaite, R. P. (Ed.), *Till, A Symposium*, Ohio State University Press. 237–250.

Esmark, J. 1827. Remarks tending to explain the geological history of the Earth. *Edinburgh New Philosophical Journal*, **3**.

Flint, R. F. 1971. *Glacial and Quaternary Geology*. Wiley, London.

Geological Map of Iran 1:100,000 Sheet 7449 Rayen 1972. Geological Survey of Iran by Djoković, J., Dimitrijević, M. N., Dimitrijević, M. D., and Cretić, S.

Höfer, H. V. 1879. Gletscher – und Eiszeitstudien. *Sitzungsbericht der Akademie der Wissenschaften Wien, Math.-Nat. Kl. 1*, **79**, 331–367.

Hövermann, J. and Kuhle, M. 1985. Typen von Vorlandsvergletscherung in Nordost-Tibet, *Regensburger Geographische Schriften* **19/20**, 29–52.

Jahns, R. H. 1941. Outwash chronology in northeastern Massachussets abstract. *Geological Society America Bulletin*, **52**, 1910.

Jahns, R. H. 1953. Surficial geology of the Ayer Quadrangle, Massachussets. *U.S.G.S. Geological Quadrangle Map*, **GQ-21**.

Koteff, C. 1974. The morphologic sequence concept and deglaciation of southern New England. In Coates, D. R. (Ed.), *Glacial Geomorphology*. Allen and Unwin, London. 121–144.

Kuhle, M. 1974. Vorläufige Ausführungen morphologischer feldarbeitsergebnisse aus dem SE-Iranischen Hochgebirge am Beispiel des Kuh-i-Jupar. *Zeitschrift für Geomorphologie N.F.*, **18**, 472–483.

Kuhle, M. 1976. Beiträge zur Quartärmorphologie SE-Iranischer Hochgebirge. Die quartäre Vergletscherung des Kuh-i-Jupar. *Göttinger Geographische Abhandlungen*, **67**, 1, 209 pp., 2, 105 figs.

Kuhle, M. 1982a. Der Dhaulagiri und Annapurna Himalaya. Ein Beitrag zur Geomorphologie extremer Hochgebirge. *Zeitschrift für Geomorphologie Supplementband*, **41**, 1, 229 pp., 2, 184 figs.

Kuhle, M. 1982b. Was spricht für eine pleistozäne Inlandvereisung Hochtibets? *Sitzungsberichte der Braunschweiger Wissenschaftlichen Gesellschaft, Göttingen*, **6**, 68–77.

Kuhle, M. 1982c. *DFG-Abschlussbericht über die Ergebnisse der Expedition in die Südabdachung von Cho Oyu, Mt. Everest und Lhotse*, 1982.

Kuhle, M. 1984a. Zur Geomorphologie Tibets. Bortensander als Kennformen semi-arider Vorlandvergletscherung. *Berliner Geographische Abhandlungen*, **36**, 127–137.

Kuhle, M. 1984b. Spuren hocheiszeitlicher Gletscherbedeckung in der Aconcagua-Gruppe (32–33°S). *Zentralblatt für Geologie, Paläontologie, part I, Verhandlungen des Südamerika Symposiums 1984, Bamberg*, **11/12**, 1635–1646.

Kuhle, M. 1985. Ein subtropisches Inlandeis als Eiszeitauslöser. Süd-Tibet und Mt. Everest-Expedition 1984. *Georgia Augusta, Göttingen*, 35–51.

Kuhle, M. 1986a. Die Vergletscherung Tibets und die

Entstehung von Eiszeiten. *Spektrum der Wissenschaft*, **9**, 42–54.

Kuhle, M. 1986b. Schneegrenzberechnung und typologische Klassifikation von Gletschern anhand spezifischer Reliefparameter. *Petermanns Geographische Mitteilungen*, 41–51.

Kuhle, M. 1987a. *Glacial, nival and periglacial environments in northeastern Qinghai–Xizang Plateau*. Reports on the northeastern part of teh Qinghai–Xizang (Tibet) Plateau by Sino-West German Scientific Expedition, Science Press, Beijing. 176–244.

Kuhle, M. 1987b. *The Problem of a Pleistocene inland glaciation of the northeastern Qinghai–Xizang Plateau*. Reports on the northeastern part of the Qinghai–Xizang (Tibet) Plateau by Sino-West German scientific expedition. Science Press, Beijing, 250–315.

Kuhle, M. 1987c. Subtropical mountain and highland glaciation as Ice Age triggers and the waning of the glacial periods in the Pleistocene. *Geo Journal*, **14**, 4, 393–421.

Kuhle, M. 1988a. Zur Auslöserrolle Tibets bei der Entstehung von Eiszeiten. *Spektrum der Wissenschaften*, **1**, 16–20.

Kuhle, M. 1988b. Die eiszeitliche Vergletscherung West-Tibets zwischen Karakorum und Tarim-Becken und ihr Einfluss auf die globale Energiebilanz. *Geographische Zeitschrift*, **76**.

Kuhle, M. 1989. Ice marginal ramps: an indicator of semi-arid piedmont glaciations. *Geo Journal*, **16**.

Martins, Ch. 1841–42. Sur les formations régulières du terrain de transport des vallées du Rhin Antérieur et du Rhin Postérieur. *Bulletin de la Société Géologique de France*, **13**.

Münnich, G. 1936. Quantitative Geschiebeprofile aus Dänemark und Norddeutschland mit besonderer Berücksichtigung Vorpommerns, *Abhandlungen Geologisch–Paläontologisches Institut der Universität Greifswald*, **15**, *Beiheft zur Zeitschrift für Geschiebeforschung*, **12**.

Penck, A. 1882. *Die Vergletscherung der Deutschen Alpen*, Barth, Leipzig.

Penck, A. and Brückner, E. 1901–1909. *Die Alpen im Eiszeitalter*, Tauchnitz, Leipzig.

Raistrick, A. 1929. the petrology of some Yorkshire boulder clays. *Geological Magazine*, **66**, 337–344.

Sugden, D. E. and John, B. S. 1976. *Glaciers and Landscape*. Edward Arnold, London.

Torell, O. 1875. *Undersökningar öfver istiden*. Öfvers Vetensk. Akad. Förhendlingor Stockholm 1872, No. 10; 1873 no.1, Zeitschrift der Deutschen Geologischen Gesellschaft.

Ussing, N. V. and Madsen, V. 1897. Beskrivelse til det geologiske kortblad Hindsholm. *Danmarks Geologiske Undersökning*, **1.R.**, 2.

Weise, O. 1974. Zur Hangentwicklung und Flächenbildung im Trockengebiet des Iranischen Hochlandes. *Würzburger Geographische Arbeiten*, **H.42**.

White, G. W. 1934. Soil minerals as a check on the location of the Wisconsin–Illinoian drift boundary in North Central Ohio. *Science*, **79**, 549–558.

CHAPTER 4

Anatomy of a White Mountains Debris-Flow —The Making of an Alluvial Fan

Chester B. Beaty
University of Lethbridge, Lethbridge

Abstract

The debris-flow origin of alluvial fans of the White Mountains, California and Nevada, U.S.A., is described. White Mountains debris-flows are the logical consequence of a series of interconnected events within and adjacent to the range, starting with detachment of individual rock fragments by weathering and ending with deposition of lobes or tongues of coarse rubble on the fans. By means of a variety of transport mechanisms, weathered material moves down the slopes toward and ultimately onto trunk canyon floors. Here, it accumulates through time, eventually to be flushed from the mountains by a cloudburst-generated torrent. Each debris-flow episode adds measurable volume to the affected fan, and the fans appear to consist primarily of superimposed debris-flow deposits, with only minimal amounts of sediment provided by other fluvial processes.

Debris-flow activity is highly sporadic and irregular, dependent upon the availability of unconsolidated materials on canyon floors and, equally importantly, upon the chance occurrence of intense thunderstorm rain in a single catchment basin. Precise determination of frequency is impossible, but an 'average' White Mountains alluvial fan could be built in about 750 000 years by three 'average' debris flows per millenium.

Introduction

Alluvial fans are a prominent element in the landscapes of the interior American Southwest. Those flanking the White Mountains of California and Nevada, particularly the fans on the western side of the range, are especially well developed and conspicuous. It is the purpose of this chapter to describe in some detail processes of origin of these fans, with specific attention being given to the role of debris-flows in their construction.

LOCATION AND GENERAL CHARACTERISTICS OF STUDY AREA

The White Mountains are the westernmost of the American Great Basin ranges in southeastern California and southwestern Nevada (Figure 4.1). The northern end of the range is at Montgomery Pass, elevation 2180 m, and Westgard Pass, elevation 2230 m, is taken to mark its southern limit. The length of the mountains between the passes is about 70 km.

Alluvial Fans: A Field Approach edited by A. H. Rachocki and M. Church
Copyright © 1990 John Wiley & Sons Ltd.

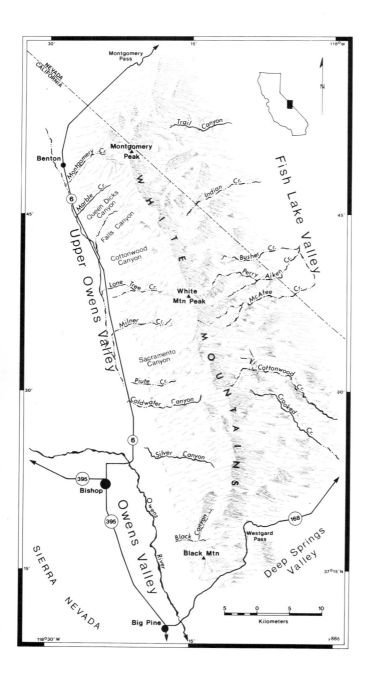

Figure 4.1. Index map showing location of the White Mountains and specific features mentioned in text

THE MAKING OF AN ALLUVIAL FAN

Most of the range is bounded on the west by the northern extension of Owens Valley, California, here designated 'Upper Owens Valley'. Fish Lake Valley, Nevada, lies immediately to the east. Between the two lowlands, the mountains have a width which increases from about 16 km in the north to 32 km in the south.

White Mountain Peak, the most elevated point in the range, attains an altitude of 4342 m, only 70 m lower than Mt. Whitney, highest summit of the 48 contiguous American states. Average height of the crest varies from 3000 to nearly 4000 m. Upper Owens Valley has elevations increasing from 1250 m near Bishop, California, to 1950 m at the western foot of Montgomery Pass. Fish Lake Valley has an average elevation of 1520 m. Local relief is thus of the order of 1800 to 2500 m in horizontal distances of 6.5 to 12.8 km, and slopes within the mountains and along the margins of the range—especially on its northwestern flank—are correspondingly steep.

GEOLOGIC FEATURES

In gross morphology, the White Mountains are thought to be a single, large, tilted fault block (Pakiser *et al.*, 1964; Bateman, 1965; Albers and Stewart, 1972). Relative tilting has been toward the east, and parts of the western front have many of the geomorphic features traditionally associated with recent faulting, including spectacular triangular facets in the northern sector (Figure 4.2). Along most parts of the range the transition from mountain slope to valley fill is abrupt, with a distinct boundary between the two. Owens Valley and its extension to the north are presumed to be structural grabens, downfaulted between the Sierra Nevada to the west and the White Mountains to the east within the last few million years (Rinehart and Ross, 1957; Pakiser *et al.*, 1964; Wahrhaftig and Birman, 1965). East of the range, Fish Lake Valley is also believed to be relatively downdropped, with a more-or-less unbroken line of youthful normal faults in alluvium extending along the margin of the mountains (Albers and Stewart, 1972). Normal faulting on both sides of the range has continued into the

Figure 4.2. West flank of White Mountains as seen from the floor of Upper Owens Valley

Holocene Epoch. As a discrete, tilted fault block, however, the mountains have probably been in existence for at least 2.5–3 million years (Beaty, 1970).

The White Mountains consist essentially of a granitic core (the Inyo Batholith) partially covered by masses of sedimentary, metamorphic, and volcanic rocks (Albers and Stewart, 1972; Strand, 1974). Marked differences exist in the size of debris on alluvial fans built by streams coming from areas of different bedrock. Fans below granitic sectors of the range consist in the main of sand, cobbles, and numerous larger boulders. Their surfaces are rough, with occasional blocks of extreme size far from canyon mouths. Fans derived from metamorphic and sedimentary segments of the mountains tend to contain a large proportion of silt, sand, pebbles, and cobbles; these depositional forms are characterized by comparatively smooth surfaces with few large boulders. Bedrock type of individual drainage basins thus appears to be of significance in determination of the *details* of fan morphology.

FAN FORMATION

The alluvial fan, as a distinctive element of the landscape, has long been recognized as a major geomorphic feature in mountainous arid and semiarid regions. More than 100 years ago, G. K. Gilbert (1882) indicated in general terms the processes by which fans are constructed:

> ... when water leaves the margin of the rocky mass [i.e. a desert range] it is always united into a comparatively small number of streams, and it is by these that the entire volume of detritus [from the mountains] is deposited. About the mouth of each gorge a symmetric heap of alluvium is produced—a conical mass of low slope, descending equally in all directions from the point of issue; and the base of each mountain exhibits a series of such alluvial cones, each with its apex at the mouth of a gorge and with its broad base resting upon the adjacent plain or valley. Rarely these cones stand so far apart as to be completely individual and distinct, but usually the parent gorges are so thickly set along the mountain front that the cones are more or less united and give to the contours of the mountain base a scalloped outline.

Since this brief description was written, the literature on alluvial fans has proliferated. As a modest contributor to more recent studies (Beaty, 1963; 1970), the present author has watched with interest the development of what has become a geomorphologic subspecialty: alluvial fan morphometric investigation. Intermittent observations over a 30-year period suggest strongly that in the White Mountains of California and Nevada, at least, the fans have been built mainly by debris-flow deposition.

Evidence of origin is of two kinds, that represented by surficial morphologic features and that revealed in stream cuts on the fans. Their internal structure, about which few solid facts are available, seems strongly to resemble that envisioned long ago by C. E. Dutton (1880):

> In the vicissitudes to which a stream ... is subject it occasionally happens that indirect causes have set it at work cutting into its cone; dissecting it, so to speak, by a deep cut and laying bare its anatomy. Our surprise is often great at finding the cone wonderfully well stratified, but in a peculiar way. The most perfect stratification is present when the dissecting cut is made radially. But when a cut transverse to the radius is made by excavation of another stream, the stratification, though still conspicuous, is much less uniform and harmonious. The cone appears to be built by long radial or sectorial slabs superposed like a series of shingles or thatches.

The structure of the White Mountains fans, so far as has been ascertained, appears closely to accord with Dutton's description, and it is the purpose of this report to provide an account of how the superposed 'radial or sectorial slabs', which in most cases are simply individual debris-flow lobes, reach their positions on the depositional forms fringing the range.

The Debris-flow Process in the White Mountains

EVIDENCE FROM HISTORICAL EVENTS

Examination of newspaper files (Kesseli and Beaty, 1959) has revealed a distinctive pattern of flooding and debris-flow in the White Mountains during the historical period. Two significant facts emerged from that investigation:
1. There has been a decided increase in the number of reported floods during the 20th century.
2. There has been a notable concentration of flooding in the summer months, particularly July and August.

As was mentioned in the earlier study, the increase in reported floods coincided with an increase of population in the area, an extension of the highway network, and a rise in the number of vehicles using the regional road system. Flooding has increasingly interfered with human activity and, not surprisingly, has more often been reported in the press. Similar trends in flooding occurrence, as reflected by newspaper accounts, have been noted by other investigators (for example, Woolley, 1946).

So far as seasonal distribution is concerned, the historical record clearly indicates that significant floods in the White Mountains occur predominantly in the warmer months, with about 60% of those reported by the news media or remembered by long-time residents having taken place in July and August. Since most of the reported and observed debris-flows have been triggered by intense thunderstorm rain, the importance of warm-season precipitation in this context is difficult to overestimate.

During the last 30 to 40 years, at least nine

floods with prominent debris-flows have occurred in the White Mountains, and most of these have been discussed in the literature (Beaty, 1963, 1968; Filipov, 1986). All but one were generated by thunderstorm rain; the lone exception apparently was the result of copious snowmelt, although firm evidence regarding that flood has been difficult to obtain (Beaty, 1968). More recently, on 18 July, 1984 a debris-flow developed in a small drainage basin on the east side of the White Mountains. The next morning the author photographed a remarkable collection of logs brought to the apex of the alluvial fan by that event (Beaty, 1985) but did not have an opportunity to investigate the flow comprehensively.

Although differing in detail, all of these flows shared a number of fundamental physical characteristics:

1. The bulk of the debris involved appears to have come from the floor of the trunk canyon within the mountains.
2. In each drainage system, scouring to bedrock was common along much of the lower main canyon.
3. The mobile debris tended to remain in active channels on the upper parts of the fans, spilling out in places but generally confined to the deeper cuts.
4. A crude sorting of the debris occurred on the fans, with an irregular but perceptible decrease in mean size in the downfan direction.
5. The lower ends of the debris-flows shaded imperceptibly into mudflows, with heavily charged muddy water constituting much of the volume.
6. While the evidence is somewhat less compelling, a surge of high-water flooding evidently *followed* deposition of most of the coarser debris, in the process deepening active channels on the fans and redistributing some of the recently-deposited finer sediments toward and onto the fan perimeter.
7. As is unfortunately so often the case, one can only speculate about the amount and intensity of precipitation responsible for the flows, since none of the weather stations in the sparse regional network recorded 'excessive' rainfall on the days in question.

EVIDENCE FROM FAN MORPHOLOGY

Many of the alluvial fans of the White Mountains exhibit convincing evidence of the primary role assumed by debris-flows in their construction. One such fan is that of Queen Dicks Canyon (formerly known as Falls Creek) on the western flank of the northern part of the range (Figure 4.1). Seven overlapping debris-flows can be distinguished on its upper surface (Figure 4.3). All are morphologically similar, each consisting of an elongated pair of ridges bisected (roughly) by a water-cut channel. The individual masses of rubble extend downslope in comparatively narrow heaps ending in blunt-nosed snouts. The flows are 100 to 275 m long, with widths of 10 to 15 m.

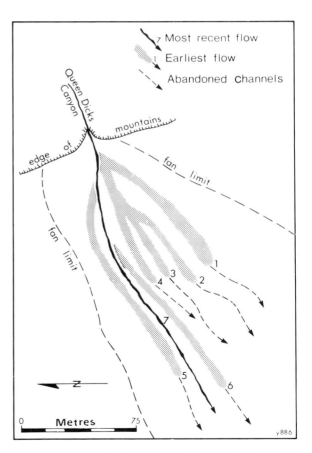

Figure 4.3. Sketch of apex of Queen Dicks Canyon fan showing seven successive debris-flows. Traced from an aerial photograph

Mean thickness is 1.5 m. The central channels have been cut through the deposits and into the older, underlying materials to depths of 0.5 to 1 m. The modern active channel—the channel leading directly from the lower canyon onto the fan surface—is in the middle of the most recent flow. Application of the principle of superposition allows determination of the chronological order in which each of the flows occurred.

What is apparent here is that successive debris-flows have added appreciably to the volume of the Queen Dicks Canyon fan. They have created a corrugated surface on its upper half, and the central channels have enhanced the ruggedness of that surface. With variations, most of the other young, steep fans along the western side of the White Mountains show comparable surface morphology (Beaty, 1960; 1970), with well-defined debris-flow lobes flanked by lateral ridges or mudflow levees (Sharp, 1942) that accentuate the microrelief. Clearly, debris-flow deposition has provided a significant proportion of the more recent volume additions sustained by these fans.

Further evidence of the importance of debris-flow deposition in fan formation can be gained from consideration of the significance of the characteristic *radial* channel pattern found on many of the fans. This pattern is in large part accounted for by the fact that channel changes produced by plugging and temporary damming of active channels during debris-flows have been frequent (Figure 4.4). The irregular surface of many fans is directly attributable to the presence of widespread abandoned channels as debris-flows have been diverted to other courses. Causes of channel abandonment are diverse; I have described some of them (Beaty, 1963), and others have commented on the process (Eckis, 1928; Hooke, 1967; Bull, 1968; Filipov, 1986). Two points are worth emphasizing in this connection:

1. Aggradation of the entire length of an active channel is *not* necessary for its abandonment; plugging may occur near the apex or close to the perimeter of a fan.
2. Numerous radial channels on a fan are *not* necessarily indicative of active dissection. Rather, in many cases the channels were cut during episodes of debris-flow *deposition* and subsequently abandoned (Hooke, 1967).

Figure 4.4. Classic example of a channel-plugging boulder on the Queen Dicks Canyon fan. Below the boulder, the abandoned channel has been dry for an undetermined period, probably for tens of thousands of years. (A) Low-altitude aerial view; (B) ground view of same boulder

Finally, the size distribution of materials at the surface and in channel walls on White Mountains fans strongly argues for debris-flow deposition. While there is an irregular decrease in average diameter downslope, the presence of large boulders far from canyon mouths (Figures 4.4 and 4.5)

THE MAKING OF AN ALLUVIAL FAN

suggests transportation by some agent other than running water (Buwalda, 1951; Beaty, 1963; Hooke, 1967; Rodine and Johnson, 1976; Hampton, 1979). Boulders as much as 2–3 m long are exposed in cuts at the lower margins of many of the west-side fans. It is difficult to imagine how they could have reached their present positions other than by debris-flow transportation and deposition (Rodine and Johnson, 1976; Hampton, 1979).

Figure 4.5. Granite boulders on Marble Creek alluvial fan 1.5 km from canyon mouth. Person at left of left-hand boulder gives scale

Morphologic characteristics of the White Mountains alluvial fans provide a reasonably consistent picture of the manner of their construction. Most of the fan surfaces are covered with identifiable, ancient and recent debris-flows. Frequent channel changes during floods and debris-flows and the abandoned channels produced thereby are readily observable and account for the uneven nature of the upper surfaces of many of the fans. Stream cuts reveal superposed debris-flows. And the size distribution of surficial materials on the fans indicates that deposition by running water has been of minor importance. The conclusion is therefore drawn that debris-flows have contributed by far the greater bulk of the material in the fans.

How do debris-flows occur?

The sequence of events leading inexorably to a debris-flow begins with weathering of bedrock, whereby individual particles are detached to start their intermittent journey toward ultimate deposition on an alluvial fan. Every piece of material in a debris-flow lobe has its origin in bedrock somewhere within the catchment basin of the responsible stream. Each discrete fragment has a unique history of movement, of course, but valid generalizations can be made by considering the limited number of factors that control the behaviour of debris in transit from outcrop to fan.

Bedrock type and structure and microclimate are the most significant determinants of weathering effectiveness. Availability of moisture, or lack of it, is probably the single most important climatic factor. Spacing of joints is of primary importance insofar as the *size* of weathering products is concerned, with the most massive pieces of rock coming from the granitic sectors of the northern White Mountains in which widely spaced joints are common. As noted, fans below these parts of the range have a rugged surficial microtopography, with many large boulders well out on their surfaces (Figure 4.5). Closely-jointed metamorphic and sedimentary bedrock, in contrast, tends

Figure 4.6. Talus from metamorphic bedrock on south wall of lower Milner Creek canyon

to produce debris of much smaller calibre, with joint-bounded fragments of only moderate size predominating (Figure 4.6). Fans built by streams draining these bedrock types have relatively smooth surfaces, with very few large boulders strewn about on them.

Description of the transport of material from the mountains to the fans is best undertaken through consideration of gradational activity in specific parts of the range. From the summit to the flanking alluvial deposits, detritus is in highly episodic motion, and the debris-flow is but the final stage of the overall process of translocation.

EVENTS ALONG THE CREST

Movement of debris from the crest into headwater basins is largely in the form of wet mass-movement streaming, principally as a variety of periglacial soil creep. Much of the summit, particularly the area north of White Mountain Peak, is characterized by rolling tracts of regolith-mantled terrain, at the margins of which streams and sheets of angular, weathered material are transporting debris into upper segments of adjacent drainage systems (Figure 4.7). Evidence of this sort of action is widespread throughout the mountains, and one must conclude that a large volume of detritus reaches positions in uppermost headwater basins directly as a result of mass movements, the details of which are obscure.

The narrowness of most of the crest of the White Mountains makes it improbable that glaciers of any significant size were active literally on the top of the range during the Pleistocene Epoch. Although snow occasionally collects to impressive depths along the summit, strong winds, exposure to the sun with limited melting, and sublimation (Beaty, 1975) act to limit its accumulated volume. At present, much of the snow that falls at higher elevations is blown by prevailing westerly winds into headwater basins and upper canyons of the east-side stream systems. Since the freshest glacial deposits are found in these same headwater localities, it seems likely that approximately the same situation prevailed during the Pleistocene. Glaciation, in short, appears not to have been an important agent of debris production and transport along the crest.

Running water seems also to be of minimal importance as a gradational agent in much of the summit area. The few rather minor surface streams in the higher parts of the range are flowing in broad, debris-mantled valleys and seem to be maintaining themselves in the face of great geomorphologic odds. Mass movements from adjacent slopes are delivering almost more debris to valley floors than the straggling streams are capable of removing. On rare occasions, parts of the crest are saturated by thunderstorm rain, but during most of the year, availability of surface water appears to be severely limited (Beaty, 1975).

In summary, it can be stated that gravity transfer is the primary agent of morphologic change along the crest of the White Mountains. By mass movement, uniformity of the gentle summit surface is maintained and comminuted debris is carried toward and into organized stream systems.

EVENTS ON CANYON WALLS

The thesis of this chapter is that materials of all sizes are constantly being moved downslope by a variety of processes, finally to be flushed completely out of the White Mountains and delivered

Figure 4.7. Weathered debris streaming into valley head on west side of summit, northern White Mountains

to a fan. Movement of any individual particle must necessarily be highly erratic, and most fragments probably have a fairly long residence time on the bottom of a trunk canyon. Their transportation to trunk canyon floors is largely an effect of retreat of the canyon walls, by which virtually all of the accumulating detritus in any stream system reaches positions of temporary stability.

Processes acting on canyon walls include: (1) direct gravity fall, by which piles of coarse talus accumulate below bedrock outcrops; (2) wet mass movements, active to some degree on most slopes; (3) erosion by running water, including gullying and rill cutting.

Many slopes in the White Mountains, particularly higher, steeper segments in headwater basins, consist of bedrock outcrops and extensive associated talus accumulations. From areas of exposed bedrock, much debris is transported canyonward as it is released by weathering. Individual large blocks may bound and roll all the way to the nearest canyon floor. Most loose material, however, is found in extremely steep accumulations of talus which head at the base of the outcrop supplying the fragments and widen below to form long cones of relatively coarse debris.

An additional factor in the delivery of large boulders directly to canyon floors is earthquake shaking. The White Mountains were vigorously jolted during the period 20–22 July, 1986, as a result of a series of nearby magnitude 5+ earthquakes (Cockerham *et al.*, 1987). Numerous bedrock blocks were dislodged during this episode, the largest of which yet discovered (Figure 4.8) came to rest on the floor of lower Marble Creek in the northern part of the range (see Figure 4.1 for location). It may well be the case that many of the most massive boulders on alluvial fans in the region were loosened and shaken free within the mountains during seismic events of the past. Such blocks would simply become part of the accumulating rubble on trunk canyon floors awaiting inevitable removal by debris-flow and/or flooding activity.

Steeper slopes in lower trunk canyons are often covered by discrete sheets of angular detritus, separated by bedrock outcrops and clumps of trees and shrubs. In the drier drainages of the southern part of the range, precipitous, coalescing talus cones often reach the edges of valley floors (Figure 4.9). Certainly a significant volume of fragmented material is delivered directly to the trunk-canyon floors by talus streaming.

Additionally, although volumetrically probably much less important, wet mass-movement forms are common on many slopes within the White Mountains. Because of the prevailing climatic aridity, most of the more gentle slopes in the range lack a true turf or sod veneer; their surfaces consist of loose debris partly covered by interspersed trees, shrubs, and bunch grass, with here and there a patch of bedrock. Depth of regolith varies widely, and it is in the thicker accumulations that evidence of mass movement is most conspicuous.

Small lobes or tongues of finer debris (Figure 4.10) are particularly common on north- and east-facing slopes in the sedimentary and metamorphic terrain of the southern part of the mountains. They are less frequently seen in the northern, mainly granitic sector. They are most often found on slopes of slight inclination, 5° or less, on which

Figure 4.8. Boulder in lower Marble Creek canyon dislodged during earthquakes of July 1986. Maximum dimensions: 11 × 10 × 8 m

Figure 4.9. Steep talus cones resting on the floor of lower Piute Creek. Buildings give scale

Figure 4.10. Small debris lobe on slope above mouth of Piute Creek. Hat, notebook, and pencil outline margin of flow

retention of soil moisture is favoured, leading, from time to time, to saturation of the upper regolith and consequent mobility.

The material involved in these small-scale flows is a mixture of particles varying in size from clay and silt to pebbles and pieces of rock up to a few cms long. Rarely do the lobes exceed 1 m in length and 40 to 50 cm in width. A thickness of 4 to 6 cm is average. Individually, these small masses of debris are not imposing, but collectively they must bring about the shifting downslope of a considerable volume of clay, silt, sand, and pebbles. For example, on the north side of a ridge in the southern part of the western flank of the range more than 100 minor flows of this sort were noted in a horizontal distance of about 600 m. Although actual counts were not made, other ridges in this sector of the mountains appear to have a comparable number of flows.

Numerous slopes in the White Mountains are furrowed by systems of parallel or semiparallel gullies and rills that are obviously water-cut features (Figure 4.11). Such slopes are retreating during hard rains (Beaty, 1959) by a process analogous to what was termed 'gully gravure' by Kirk Bryan (Bryan, 1940).

Briefly, at times of heavy precipitation, water collects in the gullies and washes out weathered debris accumulated from adjacent ridges. At the same time, gully floors are deepened by scour, and a certain amount of ridge-lowering is accomplished by minor rill cutting and washing. During periods between intense rains, weathering loosens material on the ridges and in the gullies, preparing it for transportation. During and following light precipitation, minor mass movements and rill cutting shift finer debris onto gully floors.

Figure 4.11. Gullied slopes on the lower west flank of the White Mountains north of Piute Creek. Copied from a colour slide

THE MAKING OF AN ALLUVIAL FAN

The ridges are lowered by these processes. Gullies are places of accumulation, ridges places of removal. The slope is smoothed, and the morphologic effect of a completed cycle is a lowering of the entire surface.

In time, another heavy rain falls; surface runoff collects in the partially filled gullies, again flushes them out, and intensifies local relief. The cycle is repeated; slice by slice the slope is literally peeled away. The process appears to be most active on metamorphic and sedimentary slopes covered with a thrifty growth of shrubs and grasses.

Immediately below many of the gullied slopes are areas of more-gentle inclination on which debris from above has collected. On these, evidence of small-scale mass movement is often conspicuous. Gullying is but one of a series of related activities by which weathered detritus is shifted toward and onto valley floors.

A wide suite of gradational processes is thus responsible for movement of debris from canyon walls to the bottoms of the trunk canyons. Steeper bedrock slopes give rise to impressive accumulations of talus. On occasion, larger blocks, perhaps loosened by seismic activity, fall directly to canyon floors. Mass movement forms are common, although individually the small lobes or tongues of finer materials are not impressive. Gullying on some slopes is striking, small-scale rill cutting on others much less so. All of the processes, whatever the details of their mechanics, function to transport material downslope, initially into tributary valleys, ultimately onto the floors of trunk canyons. Over short periods of time, movement appears to be highly irregular; in the long span, however, the interrelated events are essentially continuous.

THE TRUNK CANYONS

The slopes above trunk canyon floors in the White Mountains can be thought of as being the sites of a series of interconnected transportation systems, all operating to move material toward and ultimately onto the floors themselves. Here, alluvium and colluvium collect, sooner or later to be swept away by a major flood and/or debris-flow.

Figure 4.12. Coarse debris on floor of tributary, lower Piute Creek canyon. Truck gives scale

Debris comes to the trunk canyons by a number of processes. Most important appears to be delivery from tributary drainages (Figure 4.12). Thickness and width of unconsolidated material on main canyon floors increases at and below the mouths of tributary valleys. Profiles of the surface of alluvium–colluvium in trunk canyons demonstrate a reduction of gradient at and above the junctions of tributaries and a steeper-than-average gradient below these points (Figure 4.13). The common occurrence of steep fans or cones at the mouths of smaller tributaries (Figure 4.14) implies that the streams on the trunk canyon floors 'normally' do not have the ability to remove the excess load represented by the piles of unconsolidated materials. Under conditions of flood, however, such debris can be eroded and redistributed along the lower trunk canyon floor or perhaps transported entirely out of the mountains and onto the alluvial fans below.

In addition to delivery from tributary canyons, debris reaches trunk canyon floors through the functioning of other processes on canyon walls. Talus streaming, direct gravity fall of larger blocks, small-scale mass movements, and minor water-cut features testify to the effectiveness of

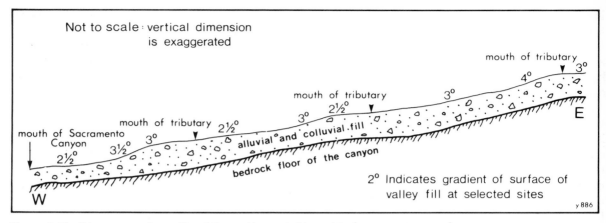

Figure 4.13. Diagrammetric profile of floor of lower 1.5 km of Sacramento Canyon showing variations in slope of surface at and below mouths of tributaries

slope retreat in shifting weathered material always downward. But it is believed that the amount of rubble thus moved is considerably less than that supplied by the tributary drainages, in which the transporting power of occasional running water is greatly increased by concentration into definite channels.

Only very modest valley deepening, if any, is effected by so-called 'normal' runoff (i.e. non-flood runoff) in trunk canyons of White Mountains drainage systems. Little evidence of bed and suspended loads is seen in the perennial streams, which exhibit clear flow most of the time. Under conditions of average discharge, such streams may theoretically have a large capacity for load, but the size of the available debris is such that they are incompetent to transport most of the coarse material. The load a stream can carry thus depends not solely upon its capacity and competence at a given discharge, but, just as importantly, upon the calibre, or size, of available debris in its channel.

During intervals between major torrents, stream beds in the White Mountains become rapidly adjusted to conditions of long-time average discharge. Fine debris is carried away following a flood, and the coarser rubble left behind as high water recedes accommodates itself to a distribution along the stream course consistent with the location of areas of pronounced turbulence. All White Mountains streams have relatively high gradients, but the sites of most vigorous turbulence during normal flow are (1) plunge pools below waterfalls and (2) zones of rapids and short stretches immediately downstream from these. The effects of turbulence evidently are subdued in the intervening reaches.

Thus, during periods between major floods and/or debris-flows the material that can be lifted and transported out of areas of maximum turbulence is rather quickly removed, to be deposited

Figure 4.14. Steep talus cone at mouth of minor tributary, lower Piute creek canyon. Inclination of the cone is about 25°

in positions of stability along reaches characterized by more gentle flow. Whether the stream is flowing in a bedrock channel or on a floor of alluvial and colluvial fill, percussion and abrasion by bed and suspended loads are minimal, and active deepening of the canyon does not occur.

Dramatically different conditions prevail in the valleys during floods. The most important change is a vast increase in discharge, endowing the creeks with much greater capacity and competence. Turbulent flow probably characterizes most of the stream course within the mountains during a large flood, and corrosion of the canyon floor will then take place most actively. Equally importantly in the present context, large volumes of debris will be shifted downstream and perhaps out of the highlands.

It is as a result of the infrequent but spectacular cloudburst floods that really significant changes on canyon floors are brought about, including removal of accumulated alluvium and colluvium. The most convincing evidence of the potency of cloudburst flooding is represented by the morphological effects of major debris-flows that have been associated with some of the more recent torrents. In the White Mountains, such flows are generated in trunk canyons within the range and move quickly through the systems and out onto alluvial fans. Their development depends upon the presence of unconsolidated debris in the channels, debris that becomes mobilized and incorporated into the fluid mass that surges through the canyon during a major flood. They move as wetted, coherent mixtures of material of all sizes, and very probably they resemble wet concrete in appearance and consistency. Their mobility is endowed by the contained water, and they are powered by the force of gravity. When they come to rest on a fan, they apparently 'set' in much the same way commercial concrete does, and very little free water drains from the masses (Fisher, 1971). Estimated and measured water content may vary from 15 to 60% by weight or volume (Sharp and Nobles, 1953; Pierson, 1981).

While not necessarily universally representative, the effects of major debris-flows in three drainages on the west side of the White Mountains in July 1952 are instructive in furnishing examples of canyon-floor modifications (see Kesseli and Beaty, 1959, and Beaty, 1963, for details). The affected systems (Cottonwood Canyon, Lone Tree Creek, Milner Creek) head in the highest part of the range near White Mountain Peak (Figure 4.1). All have steep trunk canyons, with gradients varying from 230 to 300 m/km^{-1}. All of the canyons are cut in metavolcanic, metasedimentary, and granitic bedrock, weathering of which has provided abundant debris of all sizes. All exhibit an extreme narrowing of the lower trunk canyon at or not far above its mouth. At their narrowest, these valleys are no more than 3 to 5 m wide on their floors, with near-perpendicular walls rising 25 to 35 m in the confined reaches. These constrictions exerted a pronounced influence on the behaviour of fluid debris in the canyons during the floods.

Some idea of the depth of mobile debris during passage of the floods was gained from examination of lower canyon walls. Patches of mud and smaller pieces of rock were found 13 to 20 m above present canyon floors in all three drainage systems, indicating that depths of water and debris of at least 10 to 16 m were reached at several locations. Only at particularly narrow parts of the canyons and at canyon mouths, however, did such depths of material pile up, and it is not supposed that at the height of the floods there were continuous masses of viscous rubble of this thickness flowing through the lower canyons.

More probably, temporary dams were formed at canyon constrictions, building up around piles of larger boulders and logs. Collapse of the dams from below or downcutting from the top released surges of water and debris intermittently, and the movement of material downcanyon and onto fan surfaces must have been far from uniform.

Consistency of the material in the canyons during flooding is problematical (see extended discussions in Fisher, 1971; Enos, 1977; Pierson, 1981; Costa, 1984), but debris remnants on the walls are now compactly solidified and resemble hardened concrete, with gravel, cobbles, and the occasional smaller boulder set irregularly in a matrix of sand and finer particles. The heterogeneous distribution of debris indicates lack of sorting by running water and suggests that the material moved was a well-wetted, coherent mixture of fines and larger fragments.

Source areas for the debris that left its mark many metres above canyon floors are somewhat conjectural. All three creeks are now flowing on or near bedrock throughout most of their trunk canyons. Moreover, in each drainage system one of the headwater branches is also on bedrock, while adjacent forks are alluvium-floored and support clumps of vegetation pre-dating the 1952 floods. The headwater branch of each creek lacking unconsolidated material on its floor is the one whose uppermost tributary is closest to White Mountain Peak; an extreme localization of the cell of heaviest precipitation in the general thunderstorm responsible for the floods is thus indicated.

In the headwater areas of all three drainages there is no unequivocal evidence of landsliding as a source of debris during these floods; there are no unmistakably fresh scars on slopes, nor are there remnants of fresh masses of debris on tributary or trunk canyon floors. There is no morphologic evidence at the junctions of tributary with trunk canyons to suggest that the tributaries contributed any notable amounts of material to the main canyons during the floods.

The conclusion is therefore drawn that the bulk of the rubble involved came predominantly from the floors of trunk canyons and from one of the upper branches of each system.

There is excellent evidence of removal of debris from the floor of Cottonwood Canyon. Here, a stand of dead cottonwood trees occupies matched alluvial terraces 4–6 m above the creek bottom in the lower 600 m of the canyon (Figure 4.15). Before 1952, these trees were alive and growing on a flat extending completely across the bottom of the canyon; the creek was then in a shallow cut in the middle of the flat (R. Springer, personal communication, 1957). During the flood of 1952, a body of material roughly 7 to 10 m wide, 5 to 7 m thick, and about 600 m long was removed from the canyon and transported to the alluvial fan below. Lowering of the stream bed caused the death of the trees on the terraces, as their root systems could no longer tap groundwater in the alluvial fill.

With variations, comparable morphologic changes were produced in Lone Tree and Milner creeks. Quite apparently, the major effect in the trunk canyons was removal of significant volumes of unconsolidated debris, which was carried out of the mountains and deposited on the alluvial fans along the range front. The affected valleys are now characterized by bedrock channels throughout much of their lengths.

The gradational history of a given drainage system should thus follow a definite pattern. A cloudburst flood and debris-flow will sweep clear much of the trunk canyon floor, leaving a relatively clean bedrock channel. Delivery of debris from canyon walls and tributary drainages will in time result in accumulation of another body of alluvium and colluvium in the trunk canyon. Sooner or later a heavy rain will fall again, and a flushing out of the canyon will occur anew. The pattern is one of debris transport on a discontinuously moving conveyer belt of variable velocity from outcrop to fan. The accumulation–removal cycle is repeated at irregular intervals, and the mountain mass slowly wastes away.

Figure 4.15. Looking downstream 150 m above mouth of Cottonwood Canyon. Depth of fresh cut at this point about 5 m. From Beaty, 1963

THE ALLUVIAL FANS

While very few eyewitness descriptions of debris-flow behaviour in trunk canyons are available, the movement and eventual cessation of flows on fans have been seen in a number of cases, including one excellent example from the White Mountains. In this instance, the observers were ranchers in Upper Owens Valley whose properties were partly overrun by the debris-flows of July 1952 from Cottonwood Canyon and Lone Tree Creek (Kesseli and Beaty, 1959; Beaty, 1963). Both accounts are consistent in indicating the following sequence of events:

1. About two hours after the heaviest rain in the mountains, that is, at roughly 1800 hours on the 26 July 1952, loud rumbling noises were heard apparently emanating from the lower canyons of the affected drainages. The observers were 3.5 to 4.5 km from the canyon mouths.
2. About 30 minutes later, masses of debris were seen advancing downslope on the upper parts of the fans. From a distance of 1.5 km the material, as viewed through binoculars, had the appearance of a rolling wall or rampart of boulders and mud 1 m high, without visible water.
3. The moving debris was accompanied by a low, 'boiling' cloud of dust, presumably thrown up from the dry fan surfaces as the viscous material advanced downslope.
4. The material flowed across the fans in a series of waves, or surges, each successive wave apparently overtaking the preceding one and progressing farther down the surface. This sporadic movement is assumed to have been controlled by irregular releases of fluid debris in the canyons as temporary dams gave way from time to time.
5. The advancing lobes tended to follow active stream channels until these shallowed on the lower parts of the fans. Here, the debris spread out and assumed the form of broad, thin, discontinuous sheets of mud (Figure 4.16).
6. The debris-flows on the fans were accompanied by noises likened by one of the observers to 'the sound of a thousand freight cars bumping together simultaneously.'
7. The material was moving over the more gentle parts of the fans at a speed estimated to have been 'about as fast as a man can dog-trot,' perhaps 130 to 200 m/minute^{-1}. On the lower fans the advancing debris was about 0.5 m thick and seemed to have greater fluidity than when first observed.
8. The flows lasted about 45 minutes to one hour.
9. After the debris-flows had come to rest, high water flow continued in the affected drainages for 24 to 48 hours.

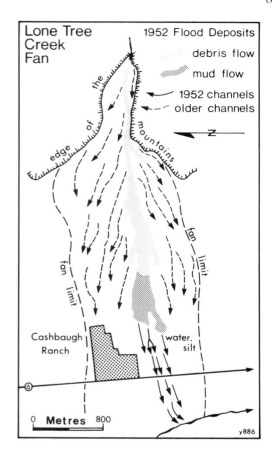

Figure 4.16. Sketch map of Lone Tree Creek fan showing deposits of the 1952 debris-flow. One of the eyewitnesses was at the Cashbaugh ranch. Traced from an aerial photograph taken in 1954. Modified by permission from Beaty, 1963

Allowing for variations in topographic and meteorologic circumstances, the appearance and behaviour of the White Mountains flows seem generally to be in agreement with those recorded from other parts of the world (Fisher, 1971; Rodine and Johnson, 1976; Pierson, 1981; Costa, 1984). In particular, the well-known Wrightwood mudflow of 1941 in southern California (Sharp and Nobles, 1953) exhibited many of the characteristics reported from the White Mountains. Earlier, F. J. Pack (1923) and E. Blackwelder (1928) had described similar debris-flow and deposition along the margins of a number of other ranges in semiarid western America, and their accounts correspond well with those of the present examples.

Two significant morphological effects were brought about on the fans by these flooding episodes: (1) considerable material was added to fan surfaces; and (2) active channels were deepened.

Debris deposition was in the form of long, relatively narrow strips extending radially from the apex to near the lower margin (Dutton's 'radial or sectorial slabs'). On the upper parts of the fans the debris was mainly confined to the active channels, which have depths of 3.0 to 5.0 m. The average width of the flows here was 30 to 50 m. In a few places, however, debris overtopped channel walls, and short lobes of material moved diagonally away from the main masses. The largest of these spilled from the active channel near the apex of the Lone Tree Creek fan (see Figure 4.16), and, maintaining a width of about 60 m, advanced some 240 m downslope.

In the middle third of the fans, where active channels shallow to less than 2 m, the debris-flows increased their width by overtopping channel walls and spreading or dividing laterally, in some instances changing from one to as many as three or four individual lobes (Figure 4.17).

The blocky flows ended about two-thirds of the way down the fan slopes. Beyond these points, sizable boulders become scarce, and the nature of the deposits suggests that the material had the consistency of viscous mud with a liberal supply of small rubble. In this form, the flows advanced in narrower tongues and fingers another 0.5 km down the slope. The change from material heavily

Figure 4.17. Low-altitude aerial view of the Milner Creek alluvial fan. Deposits of the 1952 debris-flow are conspicuous and illustrate the tendency of lower parts of flows to split into several narrow lobes or strands. Reproduced by permission of The American Journal of Science from Beaty, 1970

charged with large blocks to mudflow was gradual and much less abrupt than indicated diagrammatically in Figure 4.16.

Thickness of deposited debris decreases downslope from fan apexes. Just outside canyon mouths, thicknesses of 2 to 3 m are common; the largest boulders here are 1.5 to 3 m in length. In mid-fan locations, mean thickness approximates 1 m, and most boulders are about 0.5 to 1 m in diameter. The predominantly muddy material deposited at fan margins is thinner still, ranging from 15 to 30 cm.

In their courses down the fans, movement of the flows was strongly influenced by the underlying topography. The main masses were, with but few exceptions, confined to and guided by the preexistent active channels on the upper halves of the fans. Tongues of debris which formed when material topped channel walls tended to follow older stream cuts in their courses down the fans. This close adherence to rather shallow depressions suggests that the mobile debris was moving at such slow rates that its momentum was insufficient to overcome even minor surface irregularities.

After deposition of the main rubble masses, flows of muddy water came from the canyons and followed down approximately the centre of the flood deposits. The primary effect was dissection of the fresh debris and excavation of material from the floors of the older, pre-flood channels underlying the new deposits. Depth of incision into older fanglomerate varied from 2 m or more at fan apexes to 10 to 20 cm in mid-fan positions. Material thus removed was transported downslope, and the finer fraction was deposited as sand and silt in a braiding system of flood channels on the lower fan, and in broad, thin masses on the flatter land beyond.

Three aspects of debris-flow behaviour have long intrigued students of alluvial sedimentology. First, the ability of many subaerial flows to advance surprisingly long distances on slopes of very gentle inclination has seemingly lacked a sound theoretical and/or empirical explanation (Rodine and Johnson, 1976). Second, attempts to account for the *size* of individual boulders transported several kilometres on low slopes have led to a number of proposed models of support and movement (Hampton, 1979; Costa, 1984). Finally, the usual inability to investigate directly the nature of possible flow regimes in active debris-flows has given rise to considerable speculation, accompanied by imaginative interpretation of preserved fabrics in a variety of depositional forms (Bull, 1972; Enos, 1977; Postma, 1986).

Debris-flows of the White Mountains are characterized by (1) great flow lengths on gentle slopes, and (2) remarkably large boulders (>10 m) at impressive distances from the mountain front, especially on fans along the northwestern flank. Additionally, where clean exposures of the contacts between debris-flows and the underlying surface have been examined, little or no evidence of erosion of the overridden surface has been discovered, suggesting, but not *proving*, that laminar flow must have prevailed in at least the lowermost part of the mass (Johnson and Rodine, 1984). The deposits tend to be poorly sorted, usually contain large fragments set in a finer matrix, and often are essentially without internal structure within individual units, which are supposed to have been deposited by single surges. The structureless 'rigid plug', in short, appears to be present in most cases (Filipov, 1986). White Mountains flows possess all of the enigmatic features reported from other parts of the world.

The debris-flow, then, represents the final phase of the interconnecting 'conveyer-belt' transport system, with alluvial fans receiving periodic additions to their volumes by debris-flow deposition. Virtually all of the available evidence from the White Mountains supports the inference that the fans basically are accumulations of superimposed debris-flows, with deposition by other fluvial agents providing only very modest amounts of additional sediment.

RECURRENCE OF DEBRIS-FLOWS

Of interest at this point is a brief consideration of the possible recurrence interval of major debris-flows in the White Mountains. Nearly 20 years ago, relatively simple calculations, based upon a number of geomorphic and meteorologic suppositions, led to the conclusion that two or three events per millenium of the magnitude of the 1952 flow on the Milner Creek fan could have built the fan during its estimated lifetime of 700 000 years (Beaty, 1970). Repetition of debris-flows at the indicated rate would give rise to a mean recurrence interval of about 350 years.

More recently, utilization of Carbon-14 dating of plant materials buried by debris-flows on three of the White Mountains fans has produced an average recurrence interval of 320 years (Filipov, 1986). The assumption of a recurrence interval of 300–350 years in a given drainage system is thus bolstered by both qualitative and quantitative evidence. Accordingly, the not-unreasonable expectation of two or three 'average' flows per 1000 years in a single catchment basin would readily allow for the construction of an 'average' fan in about three-quarters of a million years.

Requisite Meteorological conditions

THE HISTORICAL PATTERN OF FLOODING

As indicated earlier, study of the historical record disclosed two pronounced characteristics of flooding in the general White Mountains region:
1. There has been an increase in the number of *reported* floods over the past 85 years.
2. There has been a strong clustering of major floods in the warmer part of the year.

There is no evidence in the available regional weather record to suggest that climatic factors could account for the increase in reported floods. Since the same pattern of flooding incidence has been detected in other parts of the western United States, it is here assumed that no particular meteorological significance should be attached to the recorded trend through time.

The *seasonal* pattern of flooding, on the other hand, is of considerable interest. Perusal of the record leads to recognition of three general flooding types in the region:
1. The *cloudburst flood*, occurring almost invariably in summer.
2. The *snowmelt flood*, taking place in late spring and early summer; sometimes augmented by thunderstorm rainfall.
3. The *wintertime flood*, resulting from warm rain on snow, rain on frozen ground, or heavy rain in the lowlands accompanied by snow at higher elevations.

Most of the recorded major debris-flows have been generated by intense summer thunderstorms, with almost two-thirds of those observed directly or reported in the press having occurred in July and August. The association of debris-flows with heavy summer rains in the White Mountains seems clear; a comparable seasonal distribution is found throughout the American Great Basin.

HOW MUCH RAIN IS 'ENOUGH'

Accurate assessments of flood-producing rain have been extremely rare in the White Mountains. The regional weather station network is sparse, and most of the long-term statistics come from valley-floor locations. Maximum official 24-hour values range from 30 to 95 mm (Kesseli and Beaty, 1959), but it has often been the case that weather stations closest to areas of heavy, flood-producing rains have received little or no precipitation on days on which flooding has occurred.

It is therefore necessary to rely upon 'accidental' measurements to arrive at even a crude estimate of how much rain might be 'enough' to trigger a debris-flow. Newspaper accounts of catches in such receptacles as watering troughs, wheelbarrows, wash tubs, and pails suggest that totals of as much as 100 to 140 mm in two or three hours from individual thunderstorms are not uncommon. Short-time downpours are difficult to evaluate, but it is possible that intensities of at least 80 to 100 mm in 30 minutes have been achieved. The heaviest precipitation *accurately* recorded anywhere within the general area occurred on 19 July, 1955, when 200 mm of rain fell in about two hours on the eastern side of the crest of the northern White Mountains (D. Powell, personal communication, 1957). The catch was made in a portable rain gauge, carried by the observer on the chance that a heavy shower might occur.

Obviously, rainfall amounts received at valley stations must be regarded as minima, and in most cases they should probably be at least doubled or tripled to correspond to precipitation in the highlands. For example, while hardly indicative of long-term trends, records kept during the summer of 1967 along several transects extending from the floor of Upper Owens Valley to stations on the crest show a systematic increase in precipitation with increasing elevation (Beaty, 1968); the valley-floor rain gauges caught only 10 to 20% of that received along the summit. Experience and observation in the White Mountains suggest that an intensity of at least 60 to 90 mm h^{-1} is necessary to produce movement of significant volumes of debris; lesser intensities may give rise to episodes of high-water flooding, with attendant erosion on alluvial fans, but debris-flows will not develop (Beaty, 1968). A rough rule, seemingly applicable to all of the drier American west, is that, as seen from an adjacent valley, precipitation within the mountains heavy enough to obscure topographic details for an hour or more is required to generate a debris-flow.

The synoptic situation leading to thunderstorm development in the region is one in which warm, moist, potentially unstable air enters the area, either from the Gulf of Mexico or the Pacific Ocean. Given a flow of such air across the mountains, intense summertime heating of the ground often leads to convective activity, and thunderstorms are the result. During the warmer months of the year, specifically July and August, the general atmospheric circulation pattern across western North America frequently favours import of the requisite warm, moist air masses; the concentration of debris-flow-producing floods in those two months is thus understandable.

A final aspect of debris-flow activity should be mentioned. The historical flows about which some minimal information is available *all* seem to have been the result of extremely concentrated convective rain. That is, the bulk of the precipitation from the thunderstorm cell responsible has fallen in only one or two catchment basins. The event of 18 July, 1984, on the eastern flank of the White Mountains which produced the accumulation of logs shown on the cover of the journal *Geology* (Beaty, 1985) provides an instructive example. That particular debris-flow came from the small drainage system of Busher Creek (Figure 4.1), the uppermost part of which is well below the crest of the range. Total area of the catchment basin is only some 17 or 18 km^2, yet the mass of material that surged across the highway on the floor of Fish Lake Valley, 1.5 km from the canyon mouth, was described by one witness as 'a two-metre wall of mud, water, and boulders'. The two closest weather stations, one on the valley floor, the other on the summit, recorded *no* precipitation on the day in question, and adjacent streams showed only minor flooding along the range front south of Busher Creek. Obviously, remarkably heavy rain must have fallen in an equally remarkably small area.

The point to be stressed is that only when all or most of the precipitation from a thunderstorm cell falls in a single catchment basin is a major debris-flow likely to develop. If the rain is scattered over several basins, the intensity of discharge in any one of them will probably be insufficient to initiate movement of significant amounts of unconsolidated material on its trunk canyon floor.

Summary

Large cloudburst floods and debris-flows are sporadic in both time and place in the White Mountains. The gradational work of such violent denudational and depositional agents is clearly favoured by canyons of steep gradient, but the necessary cloudbursts may dump their water anywhere within the range. The occurrence of a major debris-flow in a given drainage system thus appears to require a rather special set of circumstances: the affected canyon must have a steep profile and must have on its floor several metres of unconsolidated rubble, and all or most of the heavy rain from a single cell of intense precipitation must fall within its catchment basin. Given these conditions, morphologically significant flows will take place.

Once a debris-flow has occurred in a steep canyon, thereby removing most of the loose material in one great rush, a long period of comparative denudational inactivity will follow. Cloudburst floods of the high-water type there may be, and these can produce scouring in active channels on alluvial fans, but without the presence of unconsolidated alluvium and colluvium on the canyon floor, large debris-flows cannot develop.

It is difficult to determine how much time need elapse before debris accumulation on the floor of a canyon is sufficient to allow another major flow to take place. Evidence from the White Mountains suggests that 30 to 50 years is a *minimum* period (Beaty, 1960). The estimate is based in part on debris replacement in two canyons known to have flooded severely in 1918 and 1939; in these, most traces of removal of rubble by the reported torrents have been erased by infilling of alluvium and colluvium in four or five decades. Additionally, examination of the lower 0.5 km of Cottonwood Canyon in the summer of 1987 revealed accumulations of 1 to 3 m of alluvium and colluvium in parts of the canyon deposited since the major debris-flows of 1952. Whatever the required time lapse may be, once movement of debris from tributary canyons and processes of slope retreat have replenished that swept away by the most recent flood, development of another large flow would seem to be but a matter of time and chance.

The making of an alluvial fan on the flanks of the White Mountains is thus seen to be the result of a series of highly irregular processes. Spectacular episodes of debris-flow deposition alternate with periods of quiescence, during which unconsolidated material accumulates on canyon floors. From time to time cloudbursts have occurred, and imposing volumes of debris have been transferred from the mountains to the subjacent fans. During intervals between significant flows, while mass movements and other processes have been shifting debris from canyon walls to trunk canyon floors within the range, major morphologic changes on the fans have been minimal. In the American Great Basin, it seems, the spectacular appears to have been the normal (Beaty, 1974), for it is by spectacular processes that most of the gradational work has been accomplished.

Acknowledgements

The maps and diagrams were drawn by G. S. Young, Department of Geography, University of Lethbridge, and the manuscript was ably and cheerfully typed (and retyped) by L. Wehlage, departmental secretary. Logistical support from the University of California's White Mountain Research Station, Bishop, California, is acknowledged. All photographs are by the author.

References

Albers, J. P. and Stewart, J. H. 1972. Geology and mineral deposits of Esmeralda County, Nevada. *Nevada Bureau of Mines and Geology Bulletin*, **78**, 80 pp.

Bateman, P. C. 1965. Geology and tungsten mineralization of the Bishop district, California. *United States Geological Survey Professional Paper*, **470**, 208 pp.

Beaty, C. B. 1959. Slope retreat by gullying. *Geological Society of America Bulletin*, **70**, 1479–1482.

Beaty, C. B. 1960. *Gradational processes in the White Mountains of California and Nevada*. University of California, Berkeley, Ph.D. Thesis. 260 pp.

Beaty, C. B. 1963. Origin of alluvial fans, White Mountains, California and Nevada. *Association of American Geographers Annals*, **53**, 516–535.

Beaty, C. B. 1968. Sequential study of desert flooding in the White Mountains of California and Nevada. *United States Army Natick Laboratories, Technical Report*, **68–31–ES**, 92 pp.

Beaty, C. B. 1970. Age and estimated rate of accumulation of an alluvial fan, White Mountains, California, U.S.A. *American Journal of Science*, **268**, 50–77.

Beaty, C. B. 1974. Debris flows, alluvial fans, and a revitalized catastrophism. *Zeitschrift für Geomorphologie Supplementband*, **21**, 39–51.

Beaty, C. B. 1975. Sublimation or melting: observations from the White Mountains, California and Nevada, U.S.A. *Journal of Glaciology*, **14**, 275–286.

Beaty, C. B. 1985. Wood—mostly pinyon pine and junipers—brought to the apex of the alluvial fan by a debris flow in Busher Creek, east flank of the White Mountains, Fish Lake Valley, Nevada, on July 18, 1984. *Geology*, **13** (10). Cover photograph.

Blackwelder, E. 1928. Mudflow as a geologic agent in semiarid mountains. *Geological Society of America Bulletin*, **39**, 465–483.

Bryan, K. 1940. Gully gravure—a method of slope retreat. *Journal of Geomorphology*, **3**, 87–107.

Bull, W. B. 1968. Alluvial fans. *Journal of Geological Education*, **16**, 101–106.

Bull, W. B. 1972. Recognition of alluvial-fan deposits in the stratigraphic record. In Hamblin, W. K. and Rigby, J. K. (Eds), *Recognition of Ancient Sedimentary Environments*. Society of Economic Paleontologists and Mineralogists Special Publication, **16**, 63–83.

Buwalda, J. P. 1951. Transportation of coarse material on alluvial fans [abs.]. *Geological Society of America Bulletin*, **62**, 1497.

Cockerham, R. S., Corbett, E. J., de Polo, C. M., Ramelli, A. R., Lienkaemper, J. J., Pezzopane, S. K., Clark, M. M., Rymer, M. J., Gross, W. K.,

Savage, J. C., and Smith, K. D. 1987. The July 1986 Chalfant Valley, California, earthquake sequence: preliminary results. *Seismological Society of America Bulletin*, **77**, 280–313.

Costa, J. E. 1984. Physical geomorphology of debris flows. In Costa, J. E. and Fleisher, P. J. (Eds), *Developments and Applications of Geomorphology*. Springer-Verlag, Berlin. 268–317.

Dutton, C. E. 1880. *Report on the Geology of the High Plateaus of Utah*. United States Government Printing Office, Washington, D.C. 307 pp.

Eckis, R. 1928. Alluvial fans of the Cucamonga district, southern California. *Journal of Geology*, **36**, 224–247.

Enos, P. 1977. Flow regimes in debris flow. *Sedimentology*, **24**, 133–142.

Filipov, A. J. 1986. *Sedimentology of debris-flow deposits, west flank of the White Mountains, California*. University of Massachusetts, Amherst, M.S. Thesis. 207 pp.

Fisher, R. V. 1971. Features of coarse-grained, high-concentration fluids and their deposits. *Journal of Sedimentary Petrology*, **41**, 916–927.

Gilbert, G. K. 1882. *Contributions to the History of Lake Bonneville*. United States Geological Survey Second Annual Report. 167–200.

Hampton, M. A. 1979. Buoyancy in debris flows. *Journal of Sedimentary Petrology*, **49**, 753–758.

Hooke, R. LeB. 1967. Processes on arid-region alluvial fans. *Journal of Geology*, **75**, 438–460.

Johnson, A. M. and Rodine, J. R. 1984. Debris flow. In Brunsden, D. and Prior, D. B. (Eds), *Slope Instability*. John Wiley and Sons, New York. 270–317.

Kesseli, J. E. and Beaty, C. B. 1959. Desert flood conditions in the White Mountains of California and Nevada. *United States Army Quartermaster Research and Engineering Command, Technical Report*, **EP-108**, 107 pp.

Pack, F. J. 1923. Torrential potential of desert waters. *Pan-American Geologist*, **40**, 349–356.

Pakiser, L. C., Kane, M. F., and Jackson, W. H. 1964. Structural geology and volcanism of Owens Valley region, California—a geophysical study. *United States Geological Survey Professional Paper*, **438**, 68 pp.

Pierson, T. C. 1981. Dominant particle support mechanisms in debris flows at Mt Thomas, New Zealand, and implications for flow mobility. *Sedimentology*, **28**, 49–60.

Postma, G. 1986. Classification for sediment gravity-flow deposits based on flow conditions during sedimentation. *Geology*, **14**, 291–294.

Rinehart, C. D. and Ross, D. C. 1957. Geology of the Casa Diablo Mountain quadrangle, California. *United States Geological Survey Quadrangle Map*, **GQ99**.

Rodine, J. D. and Johnson, A. M. 1976. The ability of debris, heavily freighted with coarse clastic materials, to flow on gentle slopes. *Sedimentology*, **23**, 213–234.

Sharp, R. P. 1942. Mudflow levees. *Journal of Geomorphology*, **5**, 222–227.

Sharp, R. P. and Nobles, L. H. 1953. Mudflow of 1941 at Wrightwood, southern California. *Geological Society of America Bulletin*, **64**, 547–560.

Strand, R. G. 1974. *Mariposa sheet*. California Division of Mines and Geology, Geologic Map of California (2nd printing).

Wahrhaftig, C. and Birman, J. H. 1965. The Quaternary of the Pacific mountain system in California. In Wright, H. E. and Frey, D. G. (Eds), *The Quaternary of the United States*. Princeton University Press, Princeton, N.J. 299–340.

Woolley, R. R. 1946. Cloudburst floods in Utah; with a chapter on physiographic features. *United States Geological Survey Water Supply Paper*, **994**, 128 pp.

CHAPTER 5

Alluvial Fans in Japan and South Korea

Yugo Ono
Hokkaido University, Sapporo

Abstract

Alluvial fan studies, which in Japan began in the 1930s, are summarized, and the development of fans since the last interglacial time is described on the basis of field data obtained in Japan and South Korea. The alluvial fans in Japan are mainly located in the northern and central parts, along the margins of high mountains and Quaternary volcanos. Most of them are dissected fans which were formed during the last glacial age. Presently forming fans (Holocene fans) are relatively few in number, although frequent debris-flows make steep and small alluvial fans (debris-flow fans), especially in southwestern Japan.

Geomorphological studies on the alluvial fans development in north and central Japan and South Korea reveal that fan expansion occurred mainly in the early half of the last glacial age, between about 90 000 and 40 000 yr B.P. Fan expansion was associated with upstream valley filling. In the later half of the last glacial age, between 30 000 and 10 000 yr B.P., fan building was more limited. Rapid downcutting occurred in the upstream valleys around 10 000 yr B.P., resulting in fan expansion for a short time at the beginning of Holocene. After 9000 yr B.P., downcutting dominated even in the fan area. The alluvial fans in Japan and South Korea were, therefore, formed through glacial/interglacial climatic changes which controlled the balance of debris supply and discharge of the fan-building rivers.

Introduction

Japan is a country of alluvial fans (Senjochi; in Japanese). It is the only country where three textbooks have been published which deal exclusively with alluvial fans (Yazawa *et al.*, 1971; Ashida; 1985; Saito, 1988). Nearly 600 papers, studying alluvial fans directly or indirectly, have appeared between 1930 and 1970 in Japan in the fields of geomorphology, hydrology, and human geography. Especially since the 1970s, many additional contributions have been added from the field of civil engineering and hydraulics. Unfortunately, most of these studies are not known in the western countries because of the linguistic barrier, so that, for example no Japanese literature was cited in the monograph written by Rachocki (1981).

The purposes of this paper are, therefore, to introduce the previous and current studies on alluvial fans in Japan to English-speaking researchers, and to review the development of alluvial fans in Japan and South Korea since the last interglacial age.

Studies on Alluvial Fans in Japan

GEOMORPHOLOGICAL APPROACHES: TOPOGRAPHICAL ANALYSIS

Alluvial fan studies in Japan go back to the 1930s, when Teizo Murata described the faulted alluvial fans along the Median Tectonic Line in Southwestern Japan (Murata, 1931). He also was

Alluvial Fans: A Field Approach edited by A. H. Rachocki and M. Church
Copyright © 1990 John Wiley & Sons Ltd.

the first to recognize that the slope of the alluvial fan is steeper than that of the upstream river bed (Murata, 1932). His studies, which he summarized in Yazawa's book (Murata, 1971), present a purely topographical analysis of the shape of faulted alluvial fans. Vertical and horizontal deformations of alluvial fan surfaces by the movement of active faults which cut across them were analysed theoretically on maps, and the results were compared with the features in the field.

Map analysis of faulted alluvial fans is still important in Japan though it is, at present, combined with detailed geomorphological field work. In fact, many active faults in Japan (R.G.A.F., 1980) were identified by the surface deformation of alluvial fans which are cut by them.

GEOGRAPHICAL APPROACHES: DISTRIBUTION OF ALLUVIAL FANS

Most alluvial fans in Japan are dissected. According to Toya et al. (1971), the dissected alluvial fans total 314, while there are 95 presently active ones. The dissected alluvial fans were constructed mainly during the last glaciation. The ones on which the formative processes are still operating can be regarded as Holocene fans.

The difference in elevation between the dissected alluvial fan surface and the present river bed ranges from several metres to several tens of metres. When a part of the surface has been dissected by the river that constructed the fan, and two or more levels of fan surface have been created, the alluvial fan is terraced. Most of the dissected alluvial fans in Japan can be called 'fan-terraces' (Figures 5.1 and 5.2). The fan-terraces continue to the river terraces up and downstream from the fan area. In such cases, the highest alluvial fan surface generally corresponds to the filltop terrace surface (Howard, 1959).

Volcanic ashes often cover the fan-terrace surfaces and, sometimes, are even intercalated in fan-terrace gravel (Figure 5.2). Tephrochronology is always effective in determining the age of fan-terrace formation. The granulometry of fan-terrace gravel, which usually differs from that of the present fan gravel, indicates the change of river regime between the last glaciation and the present.

Figure 5.3 is the distribution map of dissected and contemporary alluvial fans in Japan compiled by Toya et al. (1971). The alluvial fans on this map have surface gradients steeper than 3%, and the sum of fan length and the maximum fan width more than 5 km. According to Saito's more exhaustive inventory (Saito 1982, 1984a,b, 1985), 586 alluvial fans in Japan are steeper than 2% and larger than 2 km^2 in area.

FACTORS CONTROLLING THE REGIONAL DISTRIBUTION OF ALLUVIAL FANS

The distribution of alluvial fans in Japan is strongly biased. Figure 5.3 shows clearly that they are located mainly in northeastern and central Japan, and they are rare in the southwest. The characteristics of alluvial fan distribution in each region of Japan are described in the following paragraphs.

1. Eastern Hokkaido. All alluvial fans are dissected. On the eastern foot of the Hidaka Range (H in Figure 5.3), their development is closely connected with alpine glaciation and the periglacial environment on this range during the last glacial age, as will be discussed in the next section. The alluvial fans on the foot of the Shiretoko Mountains (S in Figure 5.3) are related to volcanic activities.
2. Central Tohoku. There is a clear contrast between the eastern and western sides of the Kitakami Valley (K in Figure 5.3), where alluvial fans develop only on the latter. The absence of alluvial fans on the former is explained by the low relief of the Kitakami Mountains on this side of the valley. The alluvial fans in Southern Tohoku are mostly constructed on the foot of volcanos, where both dissected and active fans are located.
3. Kanto Plain. Many large, dissected alluvial fans occupy the western and northern parts of the plain, the widest one in Japan. The western part of Tokyo City stands on the Mushashino Upland, which corresponds to one of the

Figure 5.1. Typical fan-terrace in the Ina Valley: oblique air photograph looking westward (upstream) over the Mikoshiba (Mk) surface along Ozawa River

largest fans in the plain. Some fans are related to volcanos, especially in the north.

4. Chubu. Most of the contemporary, large alluvial fans have developed on the lower reach of rivers which flow from the Japanese Alps, central Japan. Steep gradient and high sediment load in these rivers explain the fan formation under the present climatic conditions. However, dissected fans cover a much wider area especially in the intermontane basins such as Ina Valley, which will be described in the following section.

5. Kinki. Though small, many dissected alluvial fans develop along the foot of fault block mountains in the region. The distribution of active fans is limited to the basins around Lake Biwa where debris-flows occur frequently.

6. Shikoku. Both dissected and active alluvial fans are located mainly in the fault valley along the Median Tectonic Line, where active faults run along the Shikoku Mountains.

7. Chugoku and Kyushu. Alluvial fans are rare and dissected fans develop mostly on the foot of volcanos.

The distribution of alluvial fans described above indicates that alluvial fans are located mainly at the foot of the high mountains, fault scarps, and volcanos; and that the dissected fans have developed mostly in north and central Japan, while the presently active alluvial fans are concentrated in central Japan.

Saito (1982, 1984a,b, 1985) analysed systematically factors that might determine the distribution of alluvial fans in Japan, including area, geology,

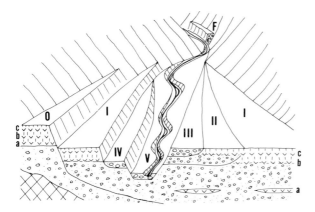

Figure 5.2. Schematic diagram of 'fan-terrace'. O—older alluvial fan surface (Oz surface in Ina Valley); I—alluvial fan surface corresponding to the widest fan expansion during the last glacial age (Mk surface in Ina Valley, Ko I surface in Tokachi Plain); II—late glacial alluvial fan surface (Ko II surface in Tokachi Plain); III—late glacial–Holocene fan surface (Md surface in Ina Valley, Ko III surface in Tokachi Plain); IV and V—Holocene river terrace surfaces; F—filltop river terrace in the upstream valley; a, b, and c—key tephra intercalated in volcanic ash or fan gravel; a—Pm I in Ina Valley, ASO-4 in Tokachi Plain; b—Pm-IV in Ina Valley, Spfa-1 in Tokachi Plain; c—AT in Ina Valley, En-a in Tokachi Plain

and relief ratio (Schumm, 1956) of the drainage basin, area of the intermontane basin, presence of a volcano, climate, distance from the valley mouth, topographic and geologic conditions of the area of fan construction, amount of uplift during the Quaternary, and existence of active faults. He concluded that:

1. The relief ratio, area of drainage basin, and topography of area of the fan construction are the most important factors;
2. Drainage basins areas of between 100 km^2 and 500 km^2 give the most favourable conditions for fan formation;
3. The geology of the drainage basin affects fan formation: for the drainage basin area less than 250 km^2, granite, metamorphic rocks, Tertiary tuff, and volcanoclastic rocks favour fan construction;
4. When the drainage basin area is less than 200 km^2, the existence of the active faults favours fan construction;
5. When the drainage basin area is more than 2 km^2, the topography of the area of fan construction and the climatic factors become important;
6. Alluvial fans are mostly formed on wide alluvial plains and in intermontane basins, while fan-deltas are relatively few on the Japanese coast;
7. The significance of climatic factors explains the abundance of dissected alluvial fans in northern and central Japan where the periglacial climate during the last glacial phase favoured fan construction.

The last conclusion will be examined in more detail in the last section of the paper.

CIVIL ENGINEERING APPROACHES

The smaller alluvial fans which are made by the accumulation of debris-flow deposits are called debris-flow fans. Their gradient is generally steeper than 1%, often exceeding 5%. As their area is less than 2 km^2, they are not included in the inventory of alluvial fans in Japan (Toya *et al.*, 1971; Saito, 1984b). They correspond to the transitional form between alluvial fans and cones. Civil engineers, including researchers in hydraulics and forestry, concentrate their attention on the study of the occurrence of debris-flows and the mechanism of formation of debris-flow fans.

Occurrence of Debris-flows

Table 5.1 shows the geology of drainage basins with the area of less than 10 km^2 in which debris-flows occurred between 1962 and 1977. It is clear that nearly half of the debris-flows occurred in drainage basins with granite and metamorphic rocks; 95% of them are located in southwestern Japan. Not only the geology, but also the intensive rainfalls which appear in company with typhoons and the activity of the polar frontal zone (Baiu, in Japanese) explain the frequency of debris-flows in southwestern Japan, where mean annual precipitation exceeds 2000 mm in wide areas, and daily maximum precipitation generally exceeds 200 mm and may attain 800 mm.

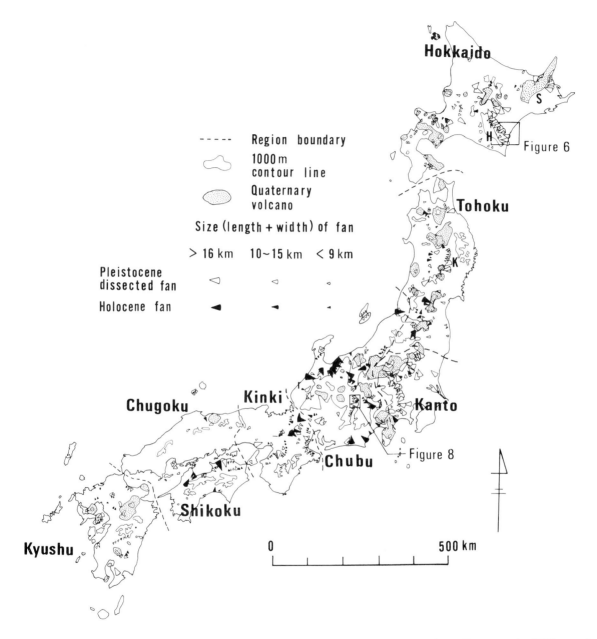

Figure 5.3. Distribution of alluvial fans in Japan (Reproduced by permission from Toya *et al.*, 1971; partly modified according to Saito, 1982)

Table 5.1. Geology of drainage basins which generate debris-flows between 1972 and 1977 in Japan. Drainage basins are less than 10 km² (after Ashida (Ed.), 1985)

Geology	Number of drainage basins	%
Granite	146	28.9
Other plutonics	9	1.8
Volcanics	68	13.4
Metamorphics	92	18.2
Palaeozoic and Mesozoic rocks	81	16.0
Tertiary rocks	68	13.4
Quaternary rocks	42	8.3
Total	506	100.0

In drainage basins with area less than 10 km², debris-flows occur mostly in those with area less than 0.5 km², and 92% of them occur in those less than 3 km². This means that the debris-flow fans are constructed at the mouth of small torrents of the first or second order. Therefore, the size of drainage area is one of the most important criteria which differentiate the occurrence of debris-flow fans from that of alluvial fans.

Studies of Debris-flow Fans

Since debris-flows are rapid mass movements that occur only occasionally, it is difficult for researchers to observe them. The study of the mechanism of sediment transport and deposition by debris-flows, therefore, requires an automatic field observation system which allows the recording of debris-flows at any time.

Such monitoring system for debris-flows was established by the Debris Flow Research Group of Kyoto University on the gullies and debris-flow fan at the foot of Mt. Yakedake Volcano (2455 m), in the Northern Japanese Alps (Figure 5.4). This group succeeded in taking motion picture film of many debris-flows which enabled the movement of debris-flows to be analysed. Periodic measurements of the topography of both the gully section and the debris-flow fan surface over 10 years gave much information on the mechanism of fan formation by debris-flows (Okuda et al., 1980; Suwa and Okuda, 1980, 1983, 1985).

The intensively-studied debris-flow fan (called Kamikamihori Fan), is located on the northern side of the Azusa River (Figures 5.4 and 5.5). The mean slope angle and the area of the fan are about 10% and 0.12 km², respectively. Stony and muddy debris-flows frequently reach the fan surface and accumulate deposits in a lobate form (Figure 5.5).

The microrelief of the fan surface corresponds to the depositional pattern of debris-flow lobes that are classified into two types: convex or 'swollen'-shaped and flat. The former has a steep frontal margin which exclusively consists of sand and gravel. The cross-section is convex and the longitudinal profile has several bulges. The inner sedimentary structure clearly shows an inverse grading where the larger debris is concentrated on the top and the snout of the lobe. This type of lobe develops mainly on the upper part of the fan. On the other hand, the flat-shaped type is a thinner lobe with flat surface which consists of sands with gravels scattered among them. Inverse grading appears at the frontal margin of these deposits also but the lower part of the lobe shows an alternation of gravelly horizons with high silt content (more than 10%) and sandy horizons with low silt content (less than 3%). The sandy horizons have a laminated structure which characterizes common flood deposits. Therefore, the flat type lobe, which mainly develops on the lower part of the fan, is formed by the alternation of debris-flows and floods. The flood deposits occupy 10–20% of the total sediment of the flat type lobe. They are not only transported directly by flood from the upper reach, but also reworked from the debris-flow deposits which have already been formed on the surface.

Periodic measurements of the fan surface have revealed topographical change which indicate the fan growth. Between 1978 and 1980, debris-flows were concentrated on the northern half of the Kamikamihori Fan, when only three debris-flows occurred on the southern half (Figure 5.5). This fact can be explained by the elevation of the fan surface, which is lower on the northern part than on the southern. The debris-flows are accumu-

Figure 5.4. Oblique air photograph of the Kamikamihori fan at the foot of Yakedake Volcano, Northern Japanese Alps (Reproduced by courtesy of Matsumoto Sabo Office, Ministry of Construction)

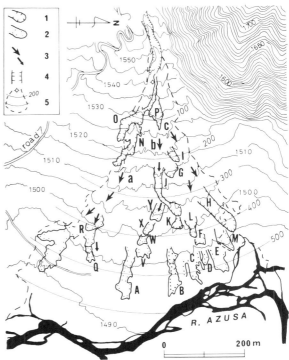

Figure 5.5. Distribution of debris-flow lobes and the direction of debris-flow on the Kamikamihori fan (Reproduced by permission from Suwa and Okuda 1985). 1. convex or 'swollen'-shaped debris-flow lobe; 2. flat debris-flow lobe; 3. direction of debris-flows: a—1970–71, 83; b—1972–79; c—1972–75, 1979; 4. gully of Kamikamihori-zawa; 5. extent of debris-flow fan (open circle indicates the fanhead and the italic number shows the distance from the fanhead in metres). Date of deposition of debris-flow lobes: V and W—17 Aug. 1978, X—4 Sept. 1978, Y—29 Sept. 1978, B–M—22 Aug. 1979, A—21 Sept. 1979, P—23 Aug. 1980, Q—5 and 7 Sept. 1983, R—22 Sept. 1983

lated mainly on the northern part of the fan surface. However, the sedimentation did not flatten the fan surface at all, and the cross-section of the debris-flow fan always shows an irregular surface. This fact proves that the fan surface is formed by the accumulation of debris-flow lobes. The upward growth of the fan between 1963 and 1983 was about 1.5 m.

Development of Alluvial Fans in Japan and Korea

Although debris-flow fans are growing actively under the present climate, most alluvial fans in Japan are dissected. These fans were deposited during the last glaciation. This section is devoted to describing their development since the last interglacial age (about 125 000 yr B.P.).

Two areas in Japan, Tokachi Plain and Ina Valley (Figure 5.3), were chosen for this purpose, because of the wealth of dating and geomorphological studies on Pleistocene alluvial fans there. One area in South Korea, Kuyre Basin, was added for comparison of the fan development between Japan and Korea.

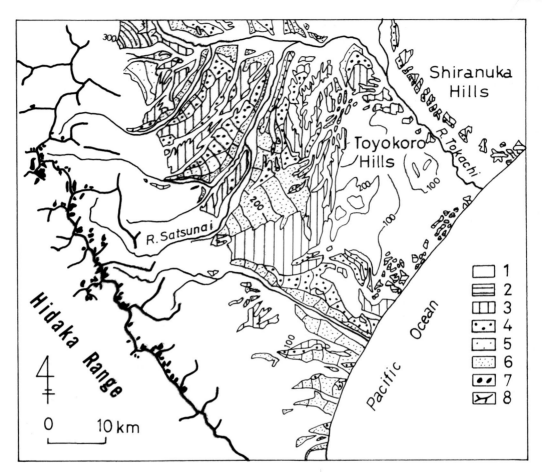

Figure 5.6. Distribution of alluvial fans in Tokachi Plain (simplified from Hirakawa and Ono, 1974; Hirakawa, 1977). 1. mountains, hills, and alluvial plain; 2. alluvial fan surfaces older than the Ko I; 3. Kamiobihiro I (Ko I) surface; 4. Kamiobihiro II (Ko II) surface; 5. Kamiobihiro III (Ko III) surface; 6. Kamiobihiro IV (Ko IV) surface; 7. distribution of glaciers during the last glacial age; 8. ridge of Hidaka Range higher than 1000 m

TOKACHI PLAIN

The Tokachi Plain is located in Hokkaido, the most northerly main island in Japan. It is a sedimentary basin which has been covered with marine and terrestrial strata since Plio–Pleistocene time. Tokachi River, flowing through the centre of the Plain, divides it into northern and southern parts. Although alluvial fans spread across the whole plain, the southern part (Figure 5.6) is more important for alluvial fan study because the fan-building rivers flow down directly from the Hidaka Range, which suffered alpine glaciation during the last glacial phase.

Pleistocene alluvial fans extensively occupy the Plain between the Hidaka Range to the west and the Toyokoro Hills to the east. Hirakawa and Ono (1974), and Hirakawa (1977) classified them into ten different fan surfaces. Among them, the older five are now regarded as surfaces older than the last glacial age (category 2 in Figure 5.6). They are much more dissected than the younger fans, and actually form fan-terraces. The fan surfaces named Kamiobihiro I, I' and II (Ko I, I', and II) correspond to those which were formed during the last glacial age.

The Ko I surface consists of a thick gravel which attains 20 m. It continues up- and downstream to the filltop terrace surface. The thickness of terrace gravel is as much as 50 m in the upper reaches of the rivers in the Hidaka Range. Though the original Ko I surface is not preserved widely, the period of its formation corresponds to that of the widest alluvial fan expansion in the Plain. It began at least before 70 000 yr B.P., most probably around 90 000–80 000 yr B.P., and ended about 40 000–35 000 yr B.P. (Figure 5.7). The age of the fan building, which is simultaneous with the valley filling-up and downstream, is known by key tephra: the Aso-4 ash (= HP I in Hirakawa and Ono, 1974) is intercalated in the fan gravel, and the Spfa-1 pumice overlies it. The thickness of volcanic ash between the top of fan gravel and the Spfa-1 is variable over the fan area, reflecting the microtopography of the fan surface. This fact suggests that the transportation and deposition of fan gravel did not end simultaneously on the whole surface. The time span needed for the formation of the Ko I surface likely involved several thousand years (Figure 5.7). The Ko I' surface is only locally developed fan-terrace surface, therefore, it is included in the Ko I in Figure 5.6.

The Ko II surface is an alluvial fan surface with thin gravel, and it continues up- and downstream to the fillstrath terrace (Howard, 1959) surface. The thin gravel of this surface overlies the thick gravel of the Ko I in the fan area and upper reach. A slight downcutting occurred after the formation of the Ko I. This created several channels (less than 5 m deep) on the Ko I surface, but the gravel of the Ko II not only filled them up, but also spread over the part of the Ko I which was not cut by them.

On the other hand, the Ko II surface grades to a lower level than the Ko I downstream. Therefore, the longitudinal profiles of the Ko I and the Ko II surfaces cross on the fan. Eustatic sea-level lowering between 18 000 and 15 000 yr B.P., clearly resulted in increase of the gradient of the Ko II surface in the lower reach of the Tokachi River, as far as about 40 km from the present river mouth. The Ko II surface was formed just before the fall of the En-a pumice, which is dated about 15 000 yr B.P.

Ko III is an alluvial fan or fan-terrace which was formed at about 9000 yr B.P. Along Satsunai River, one of the main tributaries of the Tokachi, this surface overlaps widely onto the Ko I and II surfaces at the fanhead (Figure 5.6). However, the gravel of the Ko III is also thin, and this surface continues to a fillstrath terrace up- and downstream as does the Ko II.

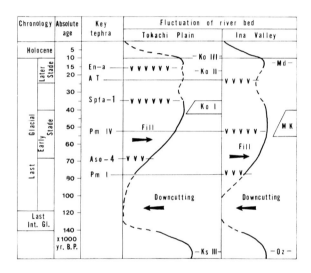

Figure 5.7. Fluctuation of river beds in the alluvial fan areas in Tokachi Plain and Ina Valley and chronology of fan development

All of these surfaces are cut by a Holocene surface Ko IV. The Ko IV is not a fan surface but a river terrace which develops narrowly along the present channel. In fact, it involves several terrace surfaces of different levels which were formed during downcutting from the level of the Ko III surface to the present river bed. The gravel is thin, and the longitudinal continuity of terrace surface at different levels is hardly distinguishable. As shown in Figure 5.7, the down cutting of the alluvial fans occurred during 9000 years in Holocene time.

INA VALLEY

In contrast to the Tokachi Plain, Ina Valley is a relatively narrow tectonic valley which is parallel to the Median Tectonic Line in the Chubu District, central Japan (Figures 5.3 and 5.8). The valley extends from north to south along the Tenryu River between the Kiso Range to the west and the Ina mountains to the east. The Kiso Range is tectonically active, and exceeds 2500 m in height, while the Ina Mountains have lower tectonic activity and relief. Hence, most of alluvial fan development is on the western side of the Tenryu, reflecting the abundant debris supply by the tributaries flowing down from the Kiso Range. On the eastern side, fan-building is limited to the junction of the Tenryu and the Mibu which flows down from the Akaishi Range which is higher than the Kiso Range and located further east. It is noteworthy that the main stream, the Tenryu itself, does not construct any alluvial fans in the valley. That is because the source area of the Tenryu, about 15 km north of the Ina valley, has only a low relief.

Ono and Masaki (1980a,b) classified the alluvial fans of the northern part of the Ina valley into seven surfaces. Figure 5.8 is a revised distribution map of the fan surfaces of this area. The Oizumi surface (Oz) is the highest alluvial fan surface in the northern Ina Valley. This fan surface is not only deeply dissected, but also dislocated by the active faults. A series of reverse faults (e and f in Figure 5.8) made the Oz surface incline upstream, so that it plunges upstream into the younger fan surface. On the other hand, the active faults which run along the foot of the Kiso Range (d and g in Figure 5.8) uplifted the Oz surface. The gravel of the Oz is thicker than 35 m and much weathered. The Oz surface represents the alluvial fan surface constructed during the penultimate glacial age (greater than 140 000 yr B.P.).

The most extensive fan surface in Ina Valley is called the Mikoshiba surface (Mk). It consists of a fan gravel exceeding 30 m in thickness (Figure 5.9). The fan surface is dislocated by active faults, but it slopes always downstream. The apparent horizontal dislocation by fault in Figure 5.8 is about 250 m. Ikeda and Yonekura (1986) estimated the horizontal slip rate of this fault as about 5.2 m/1000 yr under assumption that the dislocation was acheived by a horizontal shortening.

The Mk surface was partly covered with pumice IV of the Ontake Volcano (Pm-IV) which is dated about 57 000 yr B.P. (Takemoto et al., 1987), but the deposition of gravel continued on parts of the fan surface after this pumice fall. Therefore, the establishment of the Mk surface is estimated in Figure 5.7 to fall between 57 000 and 40 000 yr B.P. The accumulation of fan gravel under the Mk surface began about 90 000 yr B.P., before the fall of pumice I of the Ontake Volcano (Pm-I).

The Mk surface was dissected slightly before the construction of the younger fans marked by the Minamidono surface (Md). The fan gravel of the Md surface filled up the channel cut into the Mk, and overflowed onto the latter (cf. Figure 5.2). The thickness of the fan gravel of the Md surface is less than 7 m. Tephrochronological studies have revealed that the fan surface which corresponds to Ko II in Tokachi Plain is buried beneath the Md (Ono and Masaki, 1980a). Therefore, the Md surface is correlated to Ko III in Tokachi Plain (Figure 5.7).

The deep downcutting of the river bed, comparable to that which occurred after the formation of the Oz surface, began after the formation of the Md, namely since about 10 000 yr B.P. Holocene river terraces with thin gravel deposits have developed narrowly along the present channel. Alluvial fan construction is rare in Holocene time, although many steep and small alluvial cones fringe the foot of the Kiso Range. They generally overlie the Md surface.

KURYE BASIN

The Kurye Basin is a typical intermontane basin in South Korea, located in the southern part of the peninsula (Figure 5.10). Two rivers flow in the basin, Sumjin Kang flowing west to east at the southern margin of the basin, and Sisi Chon, flowing through the centre. Just like the Tenryu River, these rivers do not construct the alluvial fans. They are predominantly formed by the small

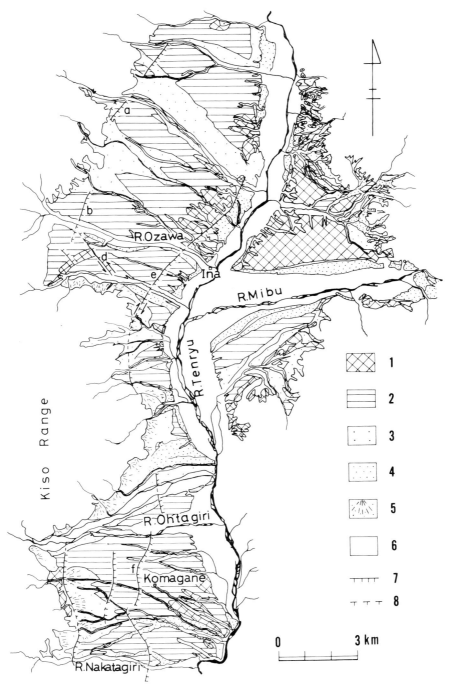

Figure 5.8. Distribution of alluvial fans in the northern part of Ina Valley. 1. Oizumi (Oz) surface; 2. Mikoshiba (Mk) surface; 3. Minamidono (Md) surface; 4. Holocene fan-terrace surfaces; 5. Holocene alluvial cones; 6. mountains, hills, and alluvial plain; 7. active fault; 8. inferred active fault

Figure 5.9. Fan gravel beds exposed on the left bank of Yotagiri River, central Ina Valley

tributaries of Sosi Chon coming down from the Chri Mountains, more than 1500 m high and located at the northeastern side of the basin. Chang Ho et al. (1986b) classified them into seven surfaces (Figure 5.10). The H1 and H2 surfaces are the remnants of the oldest fans in the basin. They are highly dissected and the fan gravel, which is deeply weathered, is not preserved well. Therefore, some authors interpreted them as pediment-like erosion surfaces (e.g. I Min-Hee and Chang Jae-Hoon, 1982; Chang Jae-Hoon 1983). However, the topographic similarity of these surfaces to those of the younger fan suggests that they are old alluvial fans.

M1 and M2 are the highest alluvial fan surfaces that are widely preserved. The M1 surface, which is slightly higher than the M2, consists of a moderately weathered fan gravel at least 11 m thick. Both surfaces are covered with a reddish soil, pedologically similar to that developing on higher terrace surfaces in southwestern Japan which are, at least, older than the last glacial. Chang Ho et al. (1986b), considered that M1 and M2 are the alluvial fan surfaces constructed during the penultimate glaciation.

The L1 and L2 represent the lower alluvial fan surfaces, and they correspond to the last glacial age. The L1 surface is associated with fan gravel as thick as 15 m at the fanhead near Pang Kwang Ri, where buried valleys exist beneath it (Figure 5.10). Although the thickness of the fan gravel generally does not exceed 5 m at fan margins, it is evident that the flatness of the fan surface is effected by the deposition of gravel, not by erosion as Chang Jae-Hoon (1983) thought.

The L2 surface, having a thinner gravel, partly overlaps the L1 near the fanhead, but cuts the

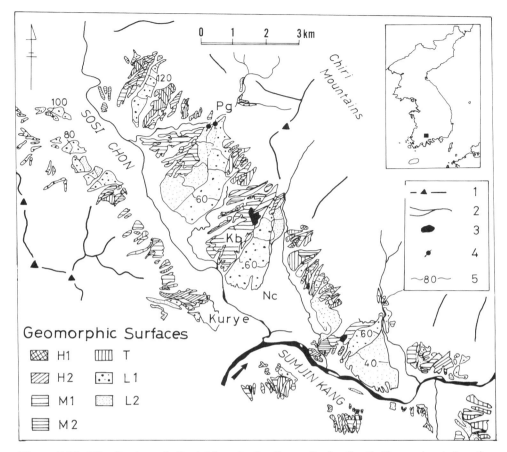

Figure 5.10. Distribution of alluvial fans in the Kurye Basin, South Korea: inset, location map. 1. main ridge and peak; 2. river; 3. reservoir; 4. buried valley beneath the fan surface; 5. contour line (partly drawn, contour interval 20 m). Pg: Pang kwang-ri, Kb: Kap san-ri, Nc: Naeng chon-ri. See text for explanation of the geomorphic (fan) surfaces

latter downstream. There is no dating for these fans, but a peat bed of about 0.5 m thickness which directly covers the L1 surface in the Sogcho area, on the eastern coast of the Korean Peninsula gave a ^{14}C date of 31 900 yr B.P. (Jo, 1980). A wood sample in a peat bed which also covers the L1 in the Kajo Basin, about 80 km northeast of the Kurye Basin, was recently dated older than 41 690 yr B.P. (Gak-12958). These dates suggest that the L1 surface was constructed during the early half of the last glacial age.

On the other hand, the L2 surface corresponds to the period of maximum sea-level lowering, since it plunges into the Holocene alluvial plain near the river mouth in the eastern coastal area (Chang Ho et al., 1986a). Downcutting of the river beds seems to have occurred during Holocene time except in the lowest reach of the river. Such a downcutting seems also to have occurred between the formation of the M2 and L1 surfaces in the intermontane basins. During this period, the steeply inclined T surface was made (Figure 5.10). This surface consists of debris-flow deposits which include huge boulders. It will be interesting to compare the palaeoenvironment of the formation period of the T surface with those of M and L surfaces, but further studies are required for this purpose.

Climatic Controls of Alluvial Fan Formation

The field data obtained in the Tokachi Plain, the Ina Valley, and the Kurye Basin reveal a conspicuous similarity of alluvial fan development since the Last Interglacial age. As indicated in Figure 5.7, the accumulation of gravel, which resulted in fan growth both vertically and horizontally, occurred nearly simultaneously in Hokkaido and in central Japan. The dissection of alluvial fans occurred exclusively in interglacial time, during the Last Interglacial and the Holocene Epoch. The downcutting of the fan surfaces during the glacial age was, on the other hand, so limited that the younger fan gravel which accumulated after about 30000 yr B.P. overlapped the older one which was deposited between 90000 and 40000 yr B.P. Alluvial fans in South Korea show a similar pattern of development.

These facts strongly suggest that the development of alluvial fans in these regions is controlled by climatic changes. However, what was changed by climate was a geomorphic process. In order to make clear process change during the Last Glacial age, analyses of gravels and studies of periglacial and glacial landforms were conducted in these regions.

The matrix of fan gravel (Ko I, Mk, and L1 surfaces) is more silty than that of the present-day river bed materials. The gravel size under these fan surfaces is also smaller than that in the present river bed (Ono and Hirakawa, 1975; Hirakawa, 1977; Ono and Masaki, 1980a).

In the Tokachi Plain and the Ina Valley, gravel transported for a long distance decreases remarkably in the last-glacial fans. This naturally lowers the roundness of the gravel. In both regions, the value of roundness of the pebble size gravel of granitic rocks, measured by Krumbein's chart (1941), is lower by 2 or 3 points than that of the present-day river bed. The effect of frost shattering on the debris production in the drainage basin is also taken into consideration. It is noteworthy that, in the Ina Valley, the decrease of roundness appears only for the fan-building rivers which have drainage basins with a substantial area above 2000 m. The present timber line in the Kiso Range is at about 2500 m elevation, and the depression during the last glacial age is estimated as about 1000 m on the basis of the distribution of glacial and periglacial landforms (Ono, 1984). The drainage basins of rivers flowing down from the higher part of the Range, thus included a wider periglacial zone in which frost and snow action supplied more debris to the river bed than at present. Alpine glaciation also contributed to the debris supply in source areas of the rivers.

The increase of debris supply in the mountains is evidenced by the existence of filltop terraces in the upper reaches of fan-building rivers. They consist of thick gravel beds which sometimes exceed 50 m. Certain outcrops expose angular gravel, supplied just from the adjacent valley walls, interfingering with more rounded gravel which was transported by the main stream. The lateral supply of debris from the valley walls undoubtedly decreased the roundness of the fan gravel.

Therefore, it was the increase of debris supply in the mountains and precipitation sufficient to maintain the river discharge to transport debris that expanded the alluvial fans during the early half of the last glacial age. However, the decrease of the gravel size in the fans and related filltop terraces suggests that the peak discharge must have been significantly diminished during the accumulation of the fan gravel.

But the climate of Japan and South Korea was not so dry as in the later half of the Last Glacial age. The richness of muddy periglacial slope deposits around the Hidaka Range, which show a random gravel orientation (Hirakawa, 1977) and the expansion of alpine glaciation in this period (Ono, 1984) indicate a cold and wet climate. Snow-melt water evidently played an important role in debris transportation in this period both in channels and on slopes.

A slight downcutting which occurred during the interstadial between about 40000 and 30000 yr B.P. suggests a decrease of debris supply and/or an increase of river discharge. But the small amount of downcuttings indicates that the interstadial change of fluvial regime was much less significant than that of the glacial/interglacial transition.

The later half of the last glacial age was characterized by an apparently stable fluvial condition in which neither deposition nor downcutting occurred remarkably on the alluvial fans. Only a thin gravel filled the shallow channels which were cut into the fan surface during the interstadial period, although it sometimes overlies the latter (cf. Figure 5.2). This fact suggests a remarkable decrease of both debris supply and river discharge under the cold and dry climate during the later half of the last glacial age.

The late-glacial aridity in Hokkaido, central Japan, and South Korea can be explained by the southward shift of the polar frontal zone, which actually attains to the northern part of Hokkaido (Suzuki, 1962). In the present climate, the peak discharges of rivers in Japan and South Korea appear in company with typhoons and with activity of the polar frontal zone. There is no doubt that the southward shift of the polar frontal zone with which is associated displacement of typhoon tracks certainly decreases the river discharge.

Figure 5.11 is a palaeogeographical reconstruction of Japan and South Korea at the maximum coldness during the last full glacial phase. Not only the southerly shift of the polar frontal zone, but also the geographical change of Japan and Yellow Seas—which resulted from the eustatic sea-level lowering—affected the climate of the area. A drastic decrease of snowfall was inferred from the study of glacial landforms (Ono, 1984). This made river discharge during the snow-melt season much less than in the early stage. The decrease of debris supply is also explained by the dryness, since the amount of debris production by frost shattering becomes smaller in dry conditions than in wet ones even if the number of freeze–thaw cycles increases (Ono and Matsuoka 1984).

New expansion of the alluvial fans occurred at about 10 000 yr B.P. This expansion, which made Ko III and Md surfaces in the Tokachi Plain and the Ina Valley respectively, was associated with a rapid and intensive downcutting in the upper reaches of the rivers. The increase of precipitation at the end of the last glacial age enabled the rivers to transport debris previously deposited in the upper drainage basin. The coarse texture of the gravel under the Ko III and Md surfaces

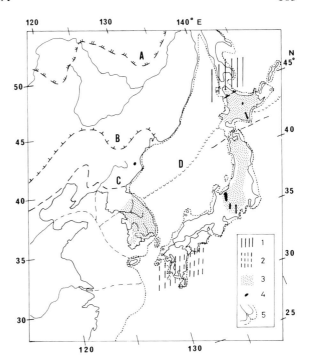

Figure 5.11. Palaeoenvironments of Japan and adjacent area at the maximum coldness during the last glacial phase. 1. present summer position of polar frontal zone; 2. last-glacial summer position of polar frontal zone; 3. area where last-glacial valley filling has been recognized; 4. last-glacial mountain glaciation; 5. present and last-glacial coast line and river course; A. present southern limit of continuous permafrost; B. last-glacial southern limit of continuous permafrost; C. last-glacial southern limit of discontinuous permafrost; D. last-glacial southern limit of sea ice

indicates frequent floods with high discharge. But the gravel accumulation in the fan area is quite limited. This is probably because the period of fan formation was short. After about 9000 yr B.P. downcutting began in the fan area and a deep incision has continued to develop right to the present day. Under Holocene climatic conditions, river discharge has generally been too large for debris supply to maintain fan formation in Japan and South Korea. Contemporary alluvial fans are located only in the areas which receive an excessive debris supply, such as active volcanos and mountain areas with many landslides; extensive spread of non/or poorly-vegetated areas are characteristic of their drainage basins.

In summary, the development of alluvial fans in Japan and South Korea since the last interglacial time indicates that:
1. Cold and dry climate in the later half of the glacial age, and the warm and wet climate of the interglacial age are both inadequate for alluvial fan formation;
2. Alluvial fans expanded most during the early half of the glacial age, which experienced a change from the interglacial warm and wet climate to a glacial cold and wet one;
3. The most recent fan expansion occurred during only a short period at the transition from the last glacial age to the Holocene Epoch when the climate changed from cold and dry to warm and wet.

Alluvial fans in the warm and wet monsoon climate of east Asia, therefore, represents a depositional landform which reflect directly the change of fluvial regime, both from interglacial to glacial, and from glacial to interglacial conditions.

References

Ashida, K. (Ed.) 1985. *Debris Flow Disaster on Alluvial Fans*. Kokon-shoin, Tokyo. 224 pp. (In Japanese)
Chang, J.-H. 1983. A geomorphological study on the gentle slope at the foot of mountains in Korea. *Kynghee University Geographical Studies*, **3**, 225–338. (In Korean)
Chang, H., Ono, Y., and Nagatsuka, S. 1986a. Geomorphic surfaces and soils in Kang nung area, east coast of Korea. *Association of Japanese Geographers, Abstracts*, **29**, 70–71. (In Japanese)
Chang, H., Ono, Y., and Nagatsuka, S. 1986b. Geomorphic evolution Kurye Basin, southern part of Korea. *Association of Japanese Geographers, Abstracts*, **29**, 72–73. (In Japanese)
Hirakawa, K. and Ono, Y. 1974. The landform evolution of the Tokachi Plain. *Geographical Review of Japan*, **47**, 607–632. (In Japanese)
Hirakawa, K. 1977. Chronology and evolution of landforms during the late Quaternary in the Tokachi Plain and adjacent areas. *Catena*, **4**, 255–288.
Howard, A. D. 1959. Numerical systems of terrace nomenclature: a critique. *Journal of Geology*, **67**, 239–243.
Ikeda, Y. and Yonekura, N. 1986. Determination of late Quaternary rates of net slip on two major fault zones in central Japan. *University of Tokyo, Department of Geography Bulletin*, **18**, 49–63.
I Min-Hee and Chang Jae-Hoon 1982. The alluvial fan and pediment on the development of the gentle slope in Korea. *The Kyunghee Geographical Review*, **119**, 11–17 (In Korean)
Jo, W. 1980. Landform evolution of the Sogcho area, eastern coast of Korea. In *Prof. K. Nishimura Retirement Memorial Volume*. Kokon-shoin, Tokyo. 71–75. (In Japanese)
Krumbein, W. C. 1941. Measurement and geologic significance of shape and roundness of sedimentary particles. *Journal of Sedimentary Petrology*, **11**, 64–72.
Murata, T. 1931. Theoretical consideration on the shape of alluvial fans. *Geographical Review of Japan*, **7**, 569–586. (In Japanese)
Murata, T. 1932. On the slope relation between the alluvial fan and the upstream valley. *Geographical Review of Japan*, **9**, 857–869. (In Japanese)
Murata, T. 1971. Purely geomorphological study of faulted alluvial fans. In Yazawa, D., Toya, H., and Kaizuka, S. (Eds), *Alluvial Fans—Regional Characteristics*. Kokon-shoin, Tokyo. 1–54. (In Japanese)
Okuda, S. *et al.* 1980. Observations on the motion of a debris flow and its geomorphological effects. *Zeitschrift für Geomorphologie Supplementband*, **35**, 142–163.
Ono, Y. 1984. Last Glacial paleoclimate reconstructed from glacial and periglacial landforms in Japan. *Geographical Review of Japan*, **57 B**, 87–100.
Ono, Y. and Hirakawa, K. 1975. Glacial and periglacial morphogenetic environments around the Hidaka Range in the Würm Glacial age. *Geographical Review of Japan*, **48**, 1–26. (In Japanese)
Ono, Y. and Masaki, T. 1980a. Last Glacial terrace formation on the western side of the Tenryu River, Northern Ina Valley, *Japan Geographical Association Abstracts*, **18**, 60–61. (In Japanese)
Ono, Y. and Masaki, T. 1980b. Abnormal sedimentation in the tephra layers in the Kami-Ina District, Central Japan. *University of Tsukuba, Institute of Geoscience, Annual Report*, **6**, 44–50.
Ono, Y. and Matsuoka, N. 1984. Mesure sur le terrain de la cryoclastie dans les Alpes japonaises sud et essai de gélifraction expérimentale. In *Mélanges Offerts à André Journaux*. Université de Caen. 199–214.
Rachocki, A. H. 1981. *Alluvial Fans*. Wiley, Chichester.
R.G.A.F. 1980. *Active Faults in Japan*. University of Tokyo Press.
Saito, K. 1982. Classification of alluvial fans in Japan by topographic and geological data of drainage basins. *Geographical Review of Japan*, **55**, 334–349. (In Japanese)
Saito, K. 1984a. Dominant factors influencing the exist-

ence and distribution of alluvial fans in Japan. *Tohoku Geographical Association Annals*, **36**, 1–12. (In Japanese)

Saito, K. 1984b. Dominant factors for the development of alluvial fans in Japan. *Hokkai-gakuen University Gakuenronshu*, **49**, 15–42. (In Japanese)

Saito, K. 1985. Comparison between dynamic equilibrium model and climatic linked model for alluvial fans in Japan. *Hokkai-gakuen University Gakuenronshu*, **52**, 35–81.

Saito, K. 1988. *Alluvial fans in Japan*. Kokon-shoin, Tokyo. (In Japanese)

Shumm, S. A. 1956. Evolution of drainage systems and slopes in badlands at Perth Amboy, New Jersey. *Bulletin Geographical Society of America*, **67**, 597–646.

Suwa, H. and Okuda, S. 1980. Dissection of valleys by debris flows. *Zeitschrift für Geomorphologie Supplementband*, **35**, 164–182.

Suwa, H. and Okuda, S. 1983. Deposition of debris flows on a fan surface, Mt. Yakedake, Japan. *Zeitschrift für Geomorphologie Supplementband*, **46**, 79–101.

Suwa, H. and Okuda, S. 1985. Measurement of debris flows in Japan. *IVth International Conference and Field Workshop on Landslides, Tokyo Proceedings*, 391–400.

Suzuki, T. 1962. Southern limit of periglacial landform at low level and the climatic classification of the latest ice age in Japan. *Geographical Review of Japan*, **35**, 67–76. (In Japanese with English abstract)

Takemoto, H. *et al.* 1987. Stratigraphy and correlation of the younger Ontake Tephra Group. *The Quaternary Research*, **25**, 337–352. (In Japanese)

Toya, H. *et al.* 1971. Distribution of alluvial fans in Japan. In Yazawa, D., Toya, H., and Kaizuka, S. (Eds), *Alluvial Fans—Regional Characteristics*. Kokon-shoin, Tokyo. 97–120. (In Japanese)

Yazawa, D. *et al.* (Eds) 1971. *Alluvial Fans—Regional Characteristcs*. Kokon-shoin, Tokyo. (In Japanese)

CHAPTER 6

Humid Fans of the Appalachian Mountains

R. Craig Kochel
Southern Illinois University, Carbondale

Abstract

Alluvial fans are abundant throughout the Appalachian Mountains of the eastern United States. Most of these humid fans are small, irregularly-shaped landforms formed dominantly by debris-flow processes which occur in first or second order basins. Larger, more regular fans formed dominantly by fluvial processes also occur in the Appalachians and are best exemplified by fans along the perimeter of the Shenandoah Valley in west–central Virginia. These two fan types commonly coexist in the same region.

Fan morphology, sedimentology, and dominant depositional processes are controlled by: (1) location in the drainage network; (2) recovery rates on hillslopes; (3) source basin lithology; (4) depositional frequency; and (5) post-depositional modifications. Activity on most of these fans occurs during the incursion of tropical moisture into the region, producing locally intense rainfall. Although depositional events at a site are infrequent (typically on the order of a few thousand years) fans are activated every few years in various parts of the Appalachians.

Introduction

Alluvial fans display a variety of geomorphic styles depending largely upon the dominant depositional processes which, in turn, are controlled by climate and lithology. Kochel and Johnson (1984) provided a generalized comparison of fans formed in a variety of climates, including arid, humid glacial, humid tropical, and humid temperate, and showed that the physical characteristics of fans may vary greatly due largely to the influence of climate. Important factors that affect fan morphology include: (1) the nature of dominant depositional processes, generally whether fan sediments are delivered by fluvial processes (traction dominated transport) and hyperconcentrated flood flows, or by debris-flow processes; (2) the frequency of depositional events; (3) the rate of recovery or revegetation of hillslopes and fan surfaces following depositional events; (4) source basin lithology; (5) the degree of topographic restriction of the depositional site where fans are constructed; and (6) post-depositional modifications of fan sediments by geomorphic processes operating in the vicinity. These factors will be considered, among others, in this survey of alluvial fans formed in the humid temperate region of the Appalachian Mountains in the United States.

Alluvial fans are widespread, significant landforms in piedmont regions of the Appalachian Mountains from New England to North Carolina (Figure 6.1). Appalachian fans occur in two distinct geomorphic styles. Relatively small fans occur predominantly in high relief, low order

---

Alluvial Fans: A Field Approach edited by A. H. Rachocki and M. Church
Copyright © 1990 John Wiley & Sons Ltd.

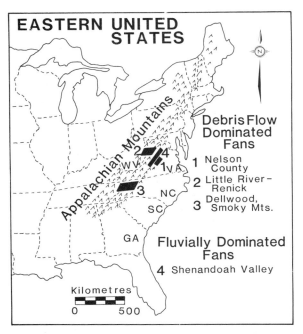

Figure 6.1. Map of the eastern United States showing the location of the major Appalachian Mountain study areas discussed in the text

drainage basins widespread throughout the Appalachian Mountains. These small fans are dominantly constructed by infrequent episodes of debris-flows and/or debris avalanches (Kochel and Johnson, 1984). They are active today under Holocene climatic conditions and will be termed 'debris-flow dominated' fans or debris fans in this paper.

The other geomorphic style of Appalachian fan appears to be largely relict and is presently undergoing a phase dominated by dissection. Holocene activity appears to be confined to fan channels which account for only a minor fraction of the fan's surface area. These will be termed 'fluvially-dominated' fans here because they are composed mostly of braided stream deposits and hyperconcentrated flood sediments of the type described by Costa (1988) and Smith (1986). Fluvially-dominated fans are larger than debris fans and may record periods of extensive fan progradation under climatic conditions somewhat different than the present. Alternatively, some of the differences from debris-flow dominated fans may be related to varying sediment yields from their respective basins due to differences in lithology or hydrology in the basin upstream from the fan.

This paper will present a brief regional view of the characteristics of humid fans in the Appalachians. The bulk of the paper, however, will describe models for the two geomorphic fan styles and discuss the climatic, geomorphic, and sedimentological variables that may explain variations in these fans. The models for the fluvially-dominated fans and debris-flow dominated fans will be developed by discussing selected sites that exemplify traits of each fan style where sufficient field work has been done to elaborate on their geomorphic and sedimentological attributes (Figure 6.1, boxed sites). The model for the debris fans will be based largely on numerous Holocene fans studied in west–central Virginia (Kochel and Johnson, 1984), east–central West Virginia (Hack and Goodlett, 1960), and in western North Carolina (Mills, 1982a,b, 1983, 1987). The fluvial fan model will be based on the extensive pre-Holocene fans that occur in an 80-km belt along the western flank of the Blue Ridge Mountains in the Shenandoah River valley of west–central Virginia.

Debris-flow Dominated Fans

The most widespread style of alluvial fan in the Appalachian Mountains is constructed dominantly by debris-flow/debris-avalanche processes. Debris fans occur in all piedmont settings along the Appalachians. Most debris fans exhibit geomorphological and sedimentological evidence of repeated depositional activity by debris-flow processes throughout the Holocene, including historical times. Appalachian debris avalanches can be triggered by initial failure at many sites on the slope (Williams and Guy, 1973) and continue rapidly downslope until the valley floor is intersected (Figure 6.2). At the valley floor, debris-avalanche sediment is either deposited as a hummocky, poorly-sorted mass at the slope base, or mixed with sufficient runoff from tributary channels to

HUMID FANS OF THE APPALACHIANS 111

Figure 6.2. Appalachian debris avalanches: (A) Avalanche chute and uppermost portion of debris fan formed in August, 1984, on Anakeesta Ridge in the Smoky Mountains, Tennessee. Note the abrupt margins separating the area of complete denudation from the undisturbed forest. The entrained colluvial sediment cover varied in depth from less than 1 m along the chute edges to 2 m in the central hollow (photo taken in 1985); (B) Three debris avalanche scars coalescing at the apex of an irregular debris fan in Nelson County, Virginia, caused by Hurricane Camille rains in 1969 (photo taken in 1981). The coarsest boulders were deposited on the fan apex and make up the hummocky terrain at the base of the bedrock-floored chutes. Debris from the 1969 flows thins rapidly away from the apex. The fan surface slopes at about 13°

continue across debris fans as debris-flow or hyperconcentrated flood-flow.

GEOMORPHOLOGY AND DEPOSITIONAL PROCESSES

Debris-flows and Climate

Systematic records indicate that Appalachian debris-flows have been numerous over the last 50 years (Table 6.1). Episodes of debris avalanching have invariably been caused by intense rainfall events. Synoptic weather patterns associated with these storms indicate that most debris-flow-producing storms can be directly linked to the incursion of warm, moist tropical air masses over the mountains between May and November. Most of the large debris-flow events have been caused by the remnants of hurricanes or tropical storms (such as Camille in 1969 and Juan in 1985). Other high intensity rains have been produced by convective storms along the interface between extratropical systems and tropical air, intensified by interactions with the mountains.

The heavily-forested slopes of the Appalachian Mountains are generally quite stable under most rainfall and snowmelt conditions. Figure 6.3 provides an idea of the rainfall magnitude necessary to produce widespread debris-flows on these slopes: the lower line probably represents a threshold of minimum rainfall intensity necessary to cause debris-flows in the Appalachians for various storm durations. However, caution should be used in interpreting relationships like these because the data are subject to considerable error for several reasons. First, the relationship of declining intensity with duration is partially a function of physical limits inherent in meteorological conditions (Caine, 1980). Second, there may be significant error in the rainfall data: many of these points are from non-recording rain gauges, hence a storm may be given a 24-hour duration because the gauge was checked only once a day, while the rainfall may have accumulated over a much shorter duration. Additional problems result because

Table 6.1. Major historical appalachian debris-flows

Date	Location	Rainfall Data Amount (cm)	Duration (hrs)	Source*	Reference
8/4–5/38	Webb Mt., TN	28–38	4	T	Moneymaker, 1939
8/17–18/40	Watauga, TN	–	–	T–H	U.S.G.S., 1949
8/17–18/40	Grandfather Mt., NC	–	–	T–H	U.S.G.S., 1949
8/17–18/40	Radford, VA	–	–	T–H	U.S.G.S., 1949
6/17–18/42	Northcentral, PA	90	12	?	Eisenlohr, 1952
6/17–18/49	Little River, VA	24	3–4	T	Hack and Goodlett, 1960
6/17–18/49	Petersburg, WV	40	24	T	Stringfield and Smith, 1956
9/1/51	Smoky Mts., NC–TN	10–20	1	T	Bogucki, 1970; 1976
6/30/56	Cove Creek, NC	30	1	T	Bogucki, 1970
8/19–20/69	Nelson Co., VA	80	8	T–H	Williams and Guy, 1973
8/19–20/69	Spring Creek, WV	10–40	8	T–H	Schneider, 1973
8/10/76	Dorset Mt., VT	10	6	T–H	Ratte and Rhodes, 1977
6/19–20/77	Johnstown, PA	25–30	–	ET	Pomeroy, 1980
8/10–11/84	Smoky Mts., TN	10	24	?	this study
Various dates	Adirondack Mts., NY	–	–	–	Bogucki, 1977
Various dates	White Mts., NH	–	–	–	Flaccus, 1958

*T = demonstrable tropical air mass
T–H = hurricane
ET = extratropical cyclone
? = difficult to determine

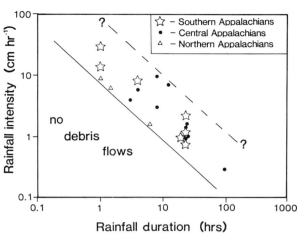

Figure 6.3. Intensity-duration relation for major debris-flow producing storms in the Appalachians. Debris-flows occur at lower rainfall intensities as the duration of the storm increases. See discussion in text. Data were compiled from the following sources: Bogucki, 1970, 1976, 1977; Costa, unpublished; Eisenlohr, 1952; Eschner and Patric, 1982; Hack and Goodlett, 1960; Moneymaker, 1939; Neary and Swift, 1984; Patric, 1981; Pomeroy, 1980; Ratte and Rhodes, 1977; Schneider, 1973; Stringfield and Smith, 1956; Williams and Guy, 1973

rain gauges often were not at the sites of the debris-flows and great spatial variation is common in mountainous terrain during intense rainfalls. In particular, gauges are seldom at high elevations where the debris-flows originate, hence they usually fail to record the orographically elevated rainfall amounts. Third, considerable spread in the data is expected due to differences in antecedent moisture conditions and due to lithological factors.

The threshold of rainfall required to initiate debris-flows in the Appalachians varies with lithology, vegetation, and topography. However, catastrophic rainfall is required to produce significant slope failures in these areas where slopes and source basins are typically covered by mature forest. Only catastrophic storms are capable of sufficiently elevating pore pressures at the soil–bedrock interface, or within the soil, on these steep slopes with thin colluvial cover, to produce a debris avalanche. An example of the kind of storm capable of initiating extensive debris-flow activity on Appalachian fans is Hurricane Camille of 1969. Camille dumped over 70 cm of rain in less than 8 hours on parts of Nelson County, Virginia (Williams and Guy, 1973; Kochel and Johnson, 1984). This intense storm resulted from the combined effect of the hurricane remnants moving over the Blue Ridge from the southwest, collision with an extratropical frontal system moving into the area from the northwest, orographic effects, and the addition of moisture from strong southeasterly winds originating from another tropical storm off the southeast coast of the United States (Schwarz, 1970).

Other severe rainfall events capable of producing debris-flows and catastrophic floods in the Appalachians have been the result of hurricane landfall, but many have also been simply associated with tropical moisture. The most severe storms of this class are referred to as 'terrain locked' storms (Lee and Goodge, 1984) or Appalachian convective clusters (Michaels, 1985). The key ingredient in these storms is the uninterrupted supply of warm, moist tropical air such as occurs with the light southeasterly winds common in the Appalachian piedmont. When these moist winds are orographically lifted over the Appalachians, and, if they encounter weak upper-air steering currents, then convective storms of great proportions and intensity may result (Michaels, 1985). Currently, weather records indicate that such intense storms are relatively rare in the Appalachians. However, when one considers the relatively low density of recording weather stations and the localized nature of these intense events, it seems likely that many go undetected. They may in fact be much more common than officially recorded. An increased network of rainfall gauges is needed in mountainous terrain to better understand these processes.

Hurricane Camille was perhaps one of the best examples of the combined influence of terrain locking and tropical moisture, resulting in the production of hundreds of debris-flows in the central Appalachians in 1969. Extensive debris avalanching in this area of concentrated rainfall resulted in numerous debris-flows onto debris fans that many workers had considered to be

Figure 6.4. Examples of debris fans activated in 1969 in Nelson County, Virginia. (A) Coarse boulders and broken trees spread in a fan shape where debris-flows could spread out in an unconfined valley along Muddy Creek; (B) More typical confined valley where an elongated debris fan was covered by poorly-sorted, bouldery debris. This fan surface in Davis Creek was fully forested before 1969. (Photos reproduced by courtesy of Virginia Division of Mineral Resources)

Fan Morphology

Appalachian debris fans are small, typically less than 1.0 km² in area, and usually occur in first or second order basins near drainage divides. Because of their basin head locations, debris fans are laterally constrained by valley walls and tend to be irregular or elongate in plan view, rather than displaying typical fan shape (Figure 6.5). Variations in debris fan shape appear to be controlled by: (1) bedrock resistance; (2) bedrock structure and its effect on stream order; and (3) grain-size of the debris deposited on the fan. Debris fans developed in areas of uniformly resistant rocks tend to be the smallest and have the most irregular shape. These fans generally form in highly confined, narrow valleys and are exemplified by fans in Nelson County, Virginia, a region underlain by granite–gneiss of the Lovingston Formation (Kochel and Johnson, 1984). Debris fans in Little River basin in western Virginia (Hack and Goodlett, 1960) have been built on to the floodplains of stream valleys of significantly higher stream order. Low order tributaries enter high order axial valleys due to the trellis drainage pattern developed in response to differential erosion of the folded sandstones and shales underlying the area. Little River fans appear to be affected little by floods in the main axial valley because of armouring by their exceedingly coarse debris.

Debris fans in the Dellwood, North Carolina region tend to be elongate, but less constrained and larger than their Virginia counterparts (Hadley and Goldsmith, 1963; Mills, 1982a,b, 1983). Mills (1982a, 1987) noted that the Dellwood fans are segmented. His studies of the weathering characteristics of these fans (Mills, 1982b) indicated that the segments represented a continuum of fan building episodes attributed to piracy processes similar to those described by Rich (1935). Piracy and preservation of older fan segments concomitant with renewed fan sedimentation at lower levels attests to the reduced lateral confinement of the Dellwood fans compared to those in Nelson County. Due to the restricted nature of fan depositional areas in Nelson County, subsequent episodes of fan building usually result in destruction of most remnants of earlier fans.

The dependency of debris-fan size upon topographic confinement is evident in Figure 6.6A. Bull (1964) described a well-defined relationship between fan area and basin area for fans where debris-flow processes are important in a semiarid region of the southwestern United States. Exponents of this relationship averaged between 0.8 and 1.0 (Bull, 1964). In contrast, the exponent for the relationship between fan area and basin area of Nelson County fans is about 0.2 with a low standard deviation. This indicates the severe topographic constraint and shows that fan area is relatively constant regardless of basin area. Mills (1987) reported exponents of 0.6, but with a high standard deviation of 0.53, for Dellwood fans showing that, although constrained, these fans vary considerably.

Most Appalachian debris fans are relatively thin: typically these deposits are less than 30 m thick. Fan profiles are generally very steep, averaging 10° to 17° compared to slopes of 4° to 5° for the fluvially dominated fans common along the western slopes of the Blue Ridge in the Shenandoah Valley of Virginia. The thin deposits and steep profiles of these debris fans is attributable largely to the dominance of debris-flows as their major depositional process. Transportation of coarse clasts by Bingham fluids and debris-flow processes (Johnson, 1970) yields considerably steeper depositional slopes than are observed on fans where deposition is dominated by fluvial processes. Similar ranges of depositional slopes of debris-flow dominated fans have been reported from southwestern Canada by VanDine (1985).

Depositional Frequency

Inspection of Table 6.1 suggests that debris-flows are relatively frequent in the Appalachian Mountains: there have been tens of events documented during the last 50 years. Thus, the average return interval for a debris-flow event

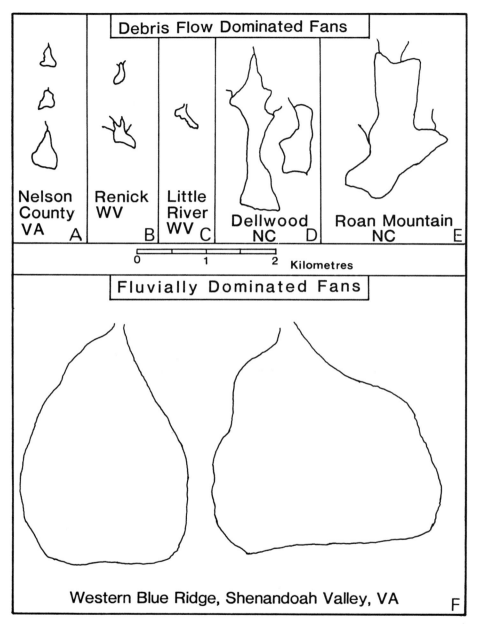

Figure 6.5. Comparative maps of debris-flow dominated fans and fluvially dominated fans. Debris fans are extremely irregular and variable in size, but considerably smaller than the fluvial fans in the eastern Shenandoah Valley. (A) Fans in Davis Creek and Ginseng Hollow, Nelson County, Virginia (from Kochel and Johnson, 1984, and this study); (B) Fans in the vicinity of Renick, West Virginia (from Wilson, 1987); (C) Fan 770 in Little River, Virginia (from Hack and Goodlett, 1960); (D) Examples of fans near Dellwood, North Carolina (from Mills, 1982b); (E) Typical fan at Roan Mountain, North Carolina (from Mills, 1983); (F) Two examples of unconfined, fluvially dominated fans near Waynesboro, Virginia, in the Shenandoah Valley

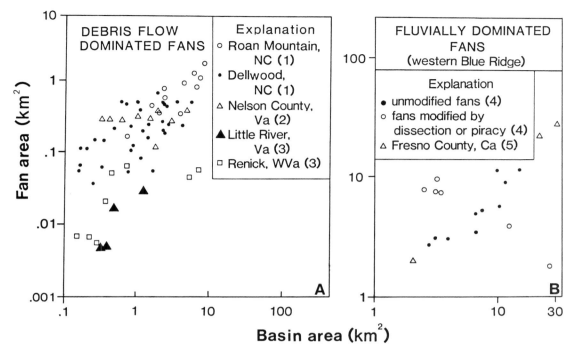

Figure 6.6. Relation of fan area to drainage basin area upstream from the fan apex. (A) Appalachian debris-flow dominated fans discussed in text. Note the ill-defined relation for the confined fans in Virginia and West Virginia and the somewhat better relation for less confined fans in North Carolina. Data sources: (1) Mills, 1987; (2) Kochel and Johnson, 1984; (3) Wilson, 1987; Simmons, 1987; (B) Appalachian fluvially-dominated fans in the Shenandoah Valley. A well-defined relation exists for most sites except those marked by open circles. These fans have been altered after their formation by erosional processes and/or piracy. Note how near the trend reported for fans in California the Virginia fans plot. Data sources: (4) Kochel and Johnson, 1984; Simmons, 1988; (5) Bull, 1964

somewhere in the Appalachians is in the order of every five to seven years. In addition, many debris-flows probably have gone unrecorded due to their association with terrain locked storms that have occurred in particularly remote areas. However, when the recurrence intervals of debris-flows on specific fans are considered, their frequency appears to be very low. Kochel and Johnson (1984) were successful in obtaining samples of palaeosols developed on fans in Davis Creek, located in Nelson County, Virginia, for radiocarbon dating. They found that at two separate sites, only three or four major debris-flow events had occurred within the last 11 000 years, from which it might be inferred that the recurrence interval at any particular location of a storm like Camille is roughly 3000 to 6000 years. These long recurrence intervals may reflect, in part, the real frequency of recurrence of the special meteorological conditions required to produce such catastrophic rainfalls at a specific site.

There must also be a recovery period between events for revegetation to occur and for residual colluvial regolith to accumulate once again before a subsequent rainfall can trigger another debris-flow. Subsequent rains that occur before such recovery is complete would result in sheetwash and flash flooding on the fan, but the likelihood for slope failures and debris-flows would be minimal. Lenses of moderately well-sorted sand and fine gravel occur in some Nelson County and Dellwood fans and may record floods that occurred between debris-flow events. Testimony to the importance of the temporal ordering of events and their associated geomorphic effects is available by comparing the results of successive storms that occurred in 1969, 1972, and 1985 in the same area of Virginia. Extensive and widespread rainfall occurred in conjunction with Hurricane Agnes in 1972 and with Hurricane Juan in 1985, resulting in larger floods on the James River (to which these areas are tributary) than was produced by the Camille storm in 1969. However, no significant debris-flows resulted from these later storms in the Nelson County area devastated in 1969. Part of the reason for the paucity of debris-flows in the latter storms was due to the lower intensity and longer duration of rains in 1972 and 1985 than in 1969. However, the Juan flood did trigger numerous debris-flows in 1985 in areas not affected previously by Camille. This suggests that most of the potentially unstable material had been released by Camille and hillsides had not recovered sufficiently by 1985 to result in slope failures in these areas. Similar observations were made by Newson (1980) in a comparison of the geomorphic effects of two large storms in Great Britain.

SEDIMENTOLOGY

Appalachian debris-fans are composed of very poorly-sorted sediments that range in size from boulders several metres in diameter to clay. These deposits occur as both matrix supported and clast supported units, depending upon the amount of water that was mixed with the debris as the flow travelled across the fan surface. Most of the fans observed in the Nelson County area of central Virginia are matrix supported (Kochel and Johnson, 1984), while those described by Hack and Goodlett (1960) in western Virginia are generally clast supported. Deposits in the southern Appalachians (Dellwood and Roan Mountain, North Carolina) are generally dominated by clast supported bedding, although matrix supported units are common. Sediment supply to the fans may differ between areas due to significant differences in source basin lithology, weathering processes, and basin hydrology. The coarse, matrix-free nature of fans in western Virginia may be due to their occurrence in folded sedimentary rocks dominated by resistant sandstones which yield only thin soils composed of sandy loam. Nelson County debris fans have a more readily available supply of matrix because of the thicker, clay-rich soils formed on the granitic rocks of central Virginia.

The hydrology of floods in these two areas may also play an important role in the nature of sedimentation on these debris fans. The volume of runoff associated with debris-flows in the first order basins in Nelson County may be considerably less than the amount available from similar unit rainfalls in the higher order basins of Little River in western Virginia. Debris-flows in higher order basins generally contain larger amounts of water; thus, hyperconcentrated sediment flows may dominate over debris-flows in these areas. A systematic survey of the hydrologic, climatic, and lithologic controls on central Appalachian fans is currently in progress and will address these questions.

Stratification in Appalachian debris fan deposits is lacking, but contacts between units are generally sharp. Individual units are thick, ranging from a few tens of centimetres to 2 m. Palaeosols are partly preserved at some of the contacts between debris-flow units, indicating that considerable time had elapsed between successive debris-flows during which significant soil formation occurred. Because each stratum in a debris fan is produced by a discrete flood event, their textural characteristics tend to differ greatly. As a result, abrupt changes in texture are common in vertical sections through these deposits. Abrupt changes in weathering characteristics and mineralogy of the sediments is also common due to weathering processes and the great differences in age that normally exist between successive strata (Kochel and Johnson, 1984).

Sedimentary structures are virtually absent in Appalachian debris-flow deposits, aside from occasional occurrences of weakly imbricated clasts and reverse grading. Surface features such as debris levees have been observed at a number of sites in photographs taken immediately after the 1969 flows, but are generally lacking. Hack and Goodlett (1960) described numerous levees, boulder lobes, and channels in the Little River area associated with the flood of 1949 in western Virginia. These forms persist today as well as similar features produced by older debris-flows at those sites. Floodplains downstream from the Little River debris fans exhibit extremely irregular surface topography dominated by boulder levees and abandoned channels produced by earlier flows (Figure 6.7). Lower magnitude floods are incapable of redistributing this coarse material in the fluvial system. Hence, material generated by debris-flow remains as a lag deposit on the downstream floodplain until an event of similar magnitude occurs that is competent to entrain the coarse sediment again. An example of this kind of event was the Hurricane Juan flood of November, 1985. The Juan flood caused considerable erosion and deposition on valley floors in eastern West Virginia and western Virginia and was also of sufficient intensity to cause significant numbers of debris-flows (Kochel and others, 1987).

Debris fans studied in Virginia and North Carolina show very little change in their sediments from proximal to distal reaches. These deposits are generally thickest and extremely coarse at the fan apices where the coarsest debris avalanche material is deposited as a hummocky mass (see Figure 6.2B). Most debris-flow units change little in thickness and grain size downfan from this region. The lack of change can be explained because material is carried by debris-flows which retain their competence throughout their travel over the small debris fans. There is little opportunity for sorting to occur before the material is deposited in valley bottom streams marginal to the distal parts of the fans (Costa, 1984).

AGE RELATIONSHIPS AND GEOMORPHOLOGY

Fans in topographically confined areas such as those in Nelson County, Virginia, appear to have been largely constructed during the Holocene Epoch. Radiocarbon dates of the basal debris-flow units resting upon bedrock at each of three localities in Virginia studied in detail indicate that the oldest fan deposits were deposited about 11 000 yr B.P. (Kochel and Johnson, 1984). This date corresponds approximately with the period when palynological data (Watts, 1979; Delcourt and Delcourt, 1984) indicate that the summer position of the polar front had retreated sufficiently far north after the late Wisconsinan glacial maximum to allow tropical moisture to reenter the central Virginia region (Kochel, 1987). It is unlikely that debris-flow activity occurred here as a major process during the glacial intervals of the Pleistocene Epoch because the atmospheric circulation would then have been unfavourable for the movement of significant tropical air masses into the region (Bryson, 1966; Bryson et al., 1970).

Episodes of debris fan deposition probably occurred throughout the Quaternary Period in Virginia during interglacial intervals. Evidence of the earlier fans is limited because subsequent fan building in these confined valleys probably destroyed older deposits. Evidence of pre-Holocene fans exists in several wider valleys like Ginseng Hollow (Figure 6.8A) where there was enough room for the Holocene fan to form adjacent to the older fan. Here, the Holocene fan, including deposits from the Camille event, is entrenched into an older fan. The older fan remnant (Figure 6.8B) appears to be significantly older than the Holocene fan on the basis of weathering indices.

Debris fans described by Mills (1982a; 1983) in the Roan Mountain and Dellwood areas of North Carolina exhibit multiple ages based upon recognition of fan segments preserved at various topographic levels and relative differences in weathering. Fluvial incision of the fans generally occurs along fan margins between debris-flow events. Subsequent debris-flows follow the incision, and so extend fans laterally and at a lower elevation in a manner similar to that described for terrace formation and piracy in piedmont regions (Rich, 1935; Ritter, 1967). The most recent episode of fan construction by debris avalanching is represented by the lowest topographic segment on these fans.

During 1985 we conducted a survey of fans in North Carolina and sampled organic materials for radiocarbon dating. The basal debris-flow units of the most recent segments of several fans in the Dellwood area were sampled to determine when the current phase of debris avalanching began.

Figure 6.7. Examples of irregular floodplain topography in areas affected by debris-flows upstream. These are from Little River, Virginia, and have been modified by the flood of 1949 (Hack and Goodlett, 1960). (A) Stepped floodplain surface showing boulder berm at left; (B) Segmented boulder levees and abandoned channels on floodplain surface

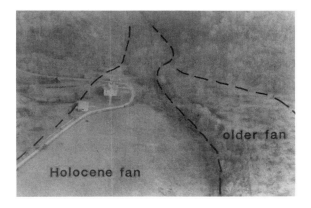

Figure 6.8. Ginseng Hollow in Nelson County, Virginia, is a site where relatively little valley confinement has resulted in the preservation of a pre-Holocene fan remnant, now dissected with an incised Holocene fan. (A) Oblique aerial photo shows the older fan, on which an orchard is located. The lower fan was totally covered by debris-flow sediment in 1969; (B) Thick, oxidized, red soil developed on clast-supported sediment in the higher, older fan. Exposure is just opposite the house near the fan apex. These clasts are extremely rotted from a long period of intense weathering, indicating a Pleistocene age

Basal units generally dated between 16 000 yr B.P. and 18 000 yr B.P. Older dates in the range of 22 000 yr B.P. to 25 000 yr B.P. were also obtained for several debris-flow layers in fans of the Dellwood area. This indicates that, although debris-flow activity may have been retarded during glacial maxima, it may have still been possible occasionally for tropical moisture to invade the southernmost regions of the Appalachians in North Carolina and Tennessee. Historical debris-flows have not been observed in the Dellwood or Roan Mountain areas of North Carolina, but have been frequent in the neighbouring Smoky Mountains of Tennessee and North Carolina. In the Smoky Mountains, radiocarbon dating indicates that major debris-flows occur at a site on average about once every 400 to 1600 years, which is considerably more frequent than farther north, in the Nelson County area in Virginia.

Fluvially Dominated Fans

At certain sites within the Appalachian Mountains, extensive alluvial fans occur that appear to be largely composed of sediments deposited dominantly by tractive bedload transport mechanisms in braided streams. The most widespread and well exposed of these fluvially dominated fans occur along the western flank of the Blue Ridge Mountains in the southeastern Shenandoah Valley of Virginia (Figure 6.1). These deposits were identified by Hack (1965) in his reconnaissance geomorphic mapping of the Shenandoah Valley. He labelled them undifferentiated terrace, pediment, and fan sediments but did not conduct detailed studies of these features.

The gravels along the western Blue Ridge are deposited in distinctly fan-shaped geometries (Figure 6.5F), they exceed 200 m in thickness in some areas, and they can be linked to upstream source basins. They will be called fluvial fans in this paper. Their distal reaches are graded to the various terrace levels of streams following the axis of the Shenandoah Valley, such as South River and South Fork Shenandoah River. These fans formed at the topographic break that marks the abrupt change in lithologic resistance encountered by streams issuing from the Blue Ridge. Basin headwaters are underlain by resistant lithologies that include Precambrian crystalline rocks and basalts and lower Cambrian quartzites, while the southeastern Shenandoah Valley is underlain by a thick section of weak, steeply-dipping Cambro-Ordovician carbonate rocks. Fan progradation occurred as coarse bedload was rapidly deposited by competent streams in flood as they left the confines of high gradient, narrow mountain channels and spread out onto the rolling, carbonate piedmont areas.

GEOMORPHOLOGY AND DEPOSITIONAL PROCESSES

Fan Morphology

Appalachian fluvial fans are restricted in their occurrence to areas where there are distinct highland source areas bordered by extensive lowland basins, like the situation in the southern Shenandoah Valley (Figure 6.9A). This type of geomorphic setting is not unlike the settings of intermontane fans located in the northern Rocky Mountain region of western Montana.

The fans along the Blue Ridge are considerably larger than Appalachian debris fans, ranging between $2\,km^2$ and $18\,km^2$. Although they have typical fan shapes in plan view, fans have often merged to create an extensive alluvial apron or bajada along the mountain front. Individual fan boundaries can generally be determined from studies of topographic maps and field checking because the fans are bounded by streams draining along their margins. Recent dissection has made the exact location of all fan boundaries uncertain in some cases. Figure 6.6B illustrates that the relationships between basin area and fan area is much better developed than for most debris fans. The area relationship is very similar to that described for arid fans by Bull (1964), as shown by the exponent of approximately 0.7. This reflects the absence of topographic constraints upon areal fan growth, unlike the situation of debris fans formed in basin head areas.

Most of the fluvial fans are fed by streams of the third or higher order, whereas most of the

Figure 6.9. Morphology of Shenandoah Valley alluvial fans. (A) View on a fluvially dominated fan a few kilometres north of Vesuvius, Virginia. Note the extensive scree slopes higher on the Blue Ridge Mountains and the sharp break in slope between colluvial slopes and the fan surface. The fan surface slopes toward the lower left at about 4°. Photo was taken at upper midfan; (B) View of the wall of a manganese quarry near the apex of a fan about 15 km southwest of Waynesboro, Virginia. Lines drawn on the photo indicate the approximate boundaries of the upper, unweathered gravel, the lower, weathered gravel, and the carbonate residuum (see Figure 6.11A for details of the stratigraphy). The tree in front of the outcrop is about 6 m tall

debris fans in Nelson County occur at the base of first order channels. The higher order basins along the Blue Ridge can therefore collect greater volumes of runoff which are more conducive to normal fluvial or hyperconcentrated flood-flows rather than to debris-flows. Blue Ridge basins are underlain dominantly by the resistant quartzite of the Antietam Formation, limiting the supply of fine-grained sediment necessary for the production of debris-flows. The combination of increased catchment area and decreased matrix availability relative to the nearby debris fans in

Nelson County, Virginia, may explain the absence of debris-flow processes on these fans.

Longitudinal profiles of the fluvial fans along the Blue Ridge show much gentler gradients than those that occur on debris fans in Virginia and elsewhere in the Appalachians, averaging 4° to 5°. The lower gradient reflects the dominance of fluvial tractive transport rather than debris-flow modes of sediment transport onto these fans. Fan thickness is highly variable, but is generally much greater than that observed in debris-fans. Drillers' logs indicate that it is not uncommon for these fluvial fans to contain up to 250 m of gravel in the mid-fan and distal regions. Manganese quarry wall exposures show that proximal gravel thickness generally averages between 10 and 25 m Figure 6.9B).

Depositional Processes and Climate

Mature forests with well-developed soils occur on the fluvial fan surfaces today, suggesting that they have not received any appreciable depositional activity on their surfaces during late Holocene time. Mature spodosols occur everywhere on the fan except in the active channel areas, which account for less than 2% of the fan surface. Therefore, it appears that active sedimentation on these fans is either relict from past climates (i.e. Pleistocene) or occurs with such rarity that extensive soil formation has occurred since the last depositional events. Support for the former hypothesis was provided by observations made of the impact of Hurricane Juan rainfall of November 1985. Five-day rainfall totals from Juan exceeded 30 cm in many parts of the Blue Ridge, resulting in extensive flooding in basins tributary to the fans and in downstream areas along South River and Shenandoah River. Geomorphic effects of these floods were significant, but generally limited to channel widening, channel scour, and bar reorganization within the confines of the narrow floodplains of the streams along the margins of the large fans (Kochel, 1988). The fan surfaces themselves were unaffected by this flood. The recurrence interval of the rainfall for the Juan storm was estimated at between 50 and 100 years (Virginia State Climatology Office, personal communication, 1986) and the resulting flood on the major rivers in the area was about a 100 year flood. Therefore, it is unlikely that rainfalls expected in the present climatic regime can contribute substantially to fan development in these areas.

The Blue Ridge fans appear to be undergoing a phase of active dissection today. Modern floods seem to be transporting sediment in fan channels toward and beyond the distal portions of the fans. Current geomorphic processes appear to be actively degrading these fans along the Blue Ridge Mountains. The coarse sediment delivered from the resistant upland basins on to the fans has armoured the fan surfaces and streams subsequently have shifted to the fan margins (Hack, 1965). The shift of fan drainage to the lateral margins could have resulted from a number of processes, including: (1) armouring of the channels with resistant lithologies from the mountains; (2) lateral forcing due to exceedingly rapid aggradation near the fan axis; and (3) entrenchment long after the main period of active fan aggradation had ended. Present streams appear to be eroding these fans very slowly along their margins, carrying sediment from the fans to the axial streams along the Shenandoah Valley. Recent studies by Simmons (1988) showed that the Juan flood was competent to move all sizes of sediment available in the active channels. In addition, the distal portions of some of the fans have been truncated by lateral migration by valley axis streams such as South River, indicating that these fans were formed earlier than others graded to lower levels and not truncated (Figure 6.10).

SEDIMENTOLOGY

Unlike the discrete event-dominated stratigraphy of Appalachian debris fans, there does not appear to have been lengthy periods of inactivity during the deposition of sediment in these fluvially-dominated fans. Two major episodes of fan building can be discerned from the observation of extensive exposures of fan gravels in quarry walls (Figure 6.9B). These episodes are separated by a well-developed and laterally extensive palaeosol (Figure 6.11A) which represents a significant

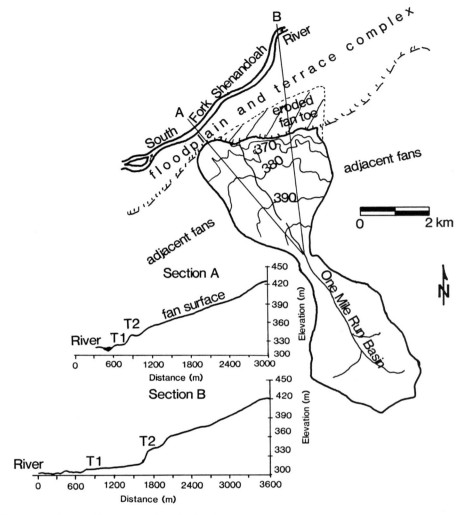

Figure 6.10. Topographic sketch map and longitudinal cross-sections of a fan in the Shenandoah Valley between Waynesboro and Elkton. Evidence of the old age of this fan includes its being graded to a high terrace level and post-depositional dissection. Erosion by the South Fork of the Shenandoah River has removed a significant portion of the northern toe of this fan. This can be seen clearly by comparing the two fan profiles. Fans appear to be of varying ages along the eastern Shenandoah Valley because others are graded to lower terrace levels

hiatus in fan building. No radiocarbon dates have been obtained from these fans, but observations of clast weathering indicate that the lower gravels are very old. Gravels above and below the palaeosol are dominantly clasts of resistant Antietam Quartzite. The upper gravels are extremely rigid and show little disintegration. Weathering rinds on the upper gravels are virtually absent. The lower quartzite gravels, however, are sometimes ghosts that can be penetrated with a dull knife or broken by hand and disaggregated into sand. The extreme disintegration of quartzite clasts below the palaeosol indicate a Pleistocene or perhaps even Tertiary age for the lower gravels. The older period of fan progradation appears to have been more extensive because the younger gravels have extended only to approximately the lower mid-fan region.

Figure 6.11. Stratigraphy of Shenandoah Valley alluvial fans. (A) View of a quarry wall exposure near the apex of a fan 15 km southwest of Waynesboro (see Figure 6.9B for stratigraphic context). Material above the person's head is unweathered upper gravels composed of imbricated clasts of Antietam Formation quartzite. A well-developed spodosol occurs at the fan surface, covered with mature forest. The lower gravel is intensely weathered, although not all the clasts have been decomposed to sand. The upper 50 to 70 cm of the lower gravel contain a well-developed palaeosol that can be recognized on many fans. Person's hand rests on the palaeosol; (B) Example of the gently dipping, horizontal stratification common in the thick upper gravels of the mid-fan region. This site is in a quarry near Vesuvius. Note the alternating layers of poorly-sorted gravel and moderately well-sorted sand. Gravels display strong imbrication perpendicular to palaeoflow direction

The sedimentology of the fluvial fans along the western slopes of the Blue Ridge is considerably different than that observed in the debris fans found throughout the Appalachian Mountains. The fluvially-dominated fans exhibit well-defined stratification dominated by thick, horizontal beds. These gravels are generally better sorted than debris fan sediments, but are still poorly sorted. Individual beds vary considerably in their relative sorting (Figure 6.11B), from poorly-sorted gravels to well-sorted sand. Large-scale channel cut-and-fill structures are also common in these deposits. Cobbles generally show well-developed, unidirectional imbrication of the type common in fluvial deposits formed by tractive processes, with long axes oriented normal to palaeoflow direction. Sediment size is variable ranging mostly from sand to small boulders rarely over 0.5 m in diameter. These sediments are considerably finer-grained than debris fan sediments. They can be entrained on gentler slopes than required for the transportation of coarser grained sediment observed in the debris fans.

The fluvially dominated fans show considerable variation in grain size and thickness from proximal to distal facies (Figure 6.12). Fan thickness appears to increase to a maximum in the mid-fan region and then slowly thin distally until the fans merge with high terraces of the axial streams of the Shenandoah Valley. Fan sediments also fine distally. Near fan apices, it is not uncommon to find an abundance of large boulders several metres in length where streams were still confined in their narrow upland canyons. Extensive deposits of quartzite scree (Hack, 1965; Hupp, 1983) occur on steep slopes above many of these fans along the Blue Ridge. Sediments in distal regions are dominated by pebble gravels, fine cobbles, and sand. Interfan areas, which probably contained small lakes, show increased percentages of sands and muds.

All of the above characteristics indicate a fluvial braided stream origin for the gravels, not unlike the modern channels seen on the fans but from streams having greater competence that allowed them to affect large areas of the fans during depositional events. Clast supported gravels, commonly found in these deposits,

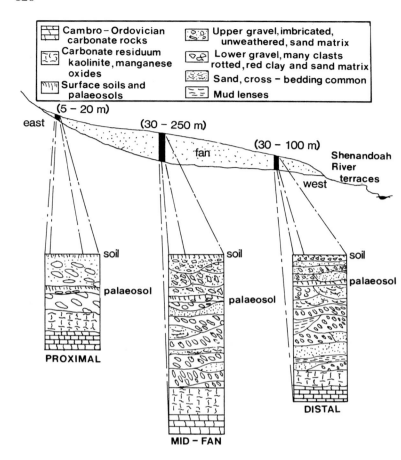

Figure 6.12. Schematic diagram of downfan variation in facies in the fluvially dominated Shenandoah Valley fans. Thicknesses in the columns are not to scale, but the range of observed or inferred (from drillers' logs) thicknesses are given in parentheses. Proximal fan facies are dominated by poorly-sorted, coarse-grained, angular to subangular bouldery material. Mid-fan facies contain interbedded sand and subrounded cobble gravel. Distal fan facies are dominated by cobble to granule gravels and well-stratified sheet sands

appear to have been deposited by hyperconcentrated flood-flows of the type described by Costa (1988) and Smith (1986). It is not unreasonable to envision hyperconcentrated flows being common in these piedmont settings, where catastrophic floods are likely to entrain large quantities of sediment. Debris-flows may be uncommon on these fans because of the lithological and hydrological controls in these basins. The lower concentration of matrix in the fan gravels may be due to lower production of muddy sediment in the source areas underlain by quartzite. In addition, much of the runoff from these basins may be lost rapidly into the subsurface in proximal fan areas. Evidence for the importance of groundwater flow through these gravels is provided by the numerous major industrial wells tapping fan gravel aquifers in distal fan areas along the eastern edge of the Shenandoah Valley.

Summary and Conclusions

Alluvial fans are important piedmont landforms along the slopes of the humid temperate Appalachian Mountains. Considerable variation exists in the morphology, sedimentology, age, and dominant depositional processes common to these fans. In general, there are two major styles of Appalachian fan based on their sedimentology and depositional processes. Debris-dominated fans are ubiquitous throughout the Appalachians and appear to have dominated Holocene fan construction in most areas. Fluvially-dominated fans appear to be restricted to special physiographic situations dictated by geologic control, best expressed along the Blue Ridge in the eastern Shenandoah Valley, and appear to be experiencing a phase of degradation today.

Debris fans occur in a range of sizes, but generally are less than 1 km^2 in area and exhibit irregular shapes due to their topographic confinement in low order tributary basins. Appalachian debris fans are composed of very poorly-sorted gravels with both matrix and clast support. The gravels occur as thickly bedded, irregular depositional units associated with individual debris-flow events. These units are typically distinct because of interevent weathering and soil formation that occur between episodes of debris-flow which tend to be separated by several thousands of years. Most debris fans contain deposits associated with historical debris-flows, but all contain evidence of repeated deposition during the Holocene. All historical debris-flows were triggered by catastrophic rainfall events apparently associated with interactions between the mountains and tropical air masses in the form of terrain locked storms and/or tropical cyclones.

Debris fans in the Dellwood and Roan Mountain areas of North Carolina appear to be less confined than their central Virginia counterparts and tend to be areally more extensive. Because of their less confined nature, multiple fan segments of varying age are generally found. Weathering characteristics and radiocarbon dates of these fans indicate that deposition has occurred to some extent at least as far back as the Late Pleistocene, but that numerous flows have occurred on their surfaces since the major climatic amelioration that began there about 16 000 yr B.P. (Delcourt and Delcourt, 1984).

Central Appalachian debris fans appear to be represented by two types, differentiated primarily on their sedimentology. Fans in central Virginia (Nelson County) are small, steep, and generally composed of matrix supported gravels. These fans are highly irregular due to constraints imposed by their location in first or second order basins. If older fan sediments did exist, they have largely been removed by the reworking by Holocene flows which commenced about 11 000 yr B.P. This date may mark the retreat of the summer polar front to the central Virginia latitude, which allowed significant tropical moisture flow, capable of producing debris-flows into the area. Fans here have formed largely as the product of debris-flows from mobilization of mud-rich colluvial soils formed on Precambrian crystalline rocks.

Western Virginia and eastern West Virginia fans occur in narrow valleys in the folded sedimentary rocks of the Appalachian Ridge and Valley province, and generally contain a greater abundance of coarse clasts derived from the erosion of source areas characterized by thin, colluvial soils developed on resistant Palaeozoic sandstones. The coarse texture of the deposits on these fans effectively armours their surfaces so that subsequent debris-flows have constructed fan segments immediately adjacent to older deposits. The western Virginia fans have prograded into higher order basins, hence are not as confined as those to the east in Nelson County.

The Shenandoah Valley of west–central Virginia exhibits the best examples of Appalachian fans that appear to have been constructed largely by fluvial processes. These fans are unconfined topographically and have prograded out across rocks of low resistance from their source basins in the Blue Ridge. The fluvial fans are significantly larger than Appalachian debris fans, are much thicker, and generally display regular fan shapes in plan view. Their sediments tend to be better sorted, well stratified, and show distinct downfan changes in facies typical of deposits dominated by fluvial tractive transport processes. Occasional less well-sorted beds occur which may be ascribed to hyperconcentrated floodflows. Debris-flow deposits were minor in the numerous exposures in these fans, perhaps due to the lack of available matrix in their resistant, quartzite-dominated source basins.

Observations of soils, clast weathering, and fan morphology suggest that the fluvial fans in the Shenandoah Valley are inactive and may have been inactive throughout much or all of the Holocene Epoch. They currently appear to be undergoing a phase of dissection. Evidence to support the dissection model was provided by the recent Hurricane Juan flood that caused extensive sediment mobilization that was confined to the channels, which occupy only a minute fraction of the fan surfaces.

Controls on alluvial fan morphology and sedimentology in the Appalachian Mountains appear to be complex. However, the major factors that account for their variation seem to be: (1) source basin lithology and weathering products, which ultimately affect depositional processes; (2) topographic confinement of the fan depositional areas; (3) source basin hydrology and morphometry; (4) regional variations in patterns of intense rainfalls; and (5) the relative frequency of depositional events of debris-flows or water flood on these fans.

Acknowledgments

Partial support for this research was provided by the Office of Research and Development Administration, Southern Illinois University. I thank numerous students for their assistance in the field and their comments; in particular I am grateful for the generous help provided by David Simmons and Gregory Wilson. Betty Atwood kindly did the word processing. I thank Frank Ungaro for help with some of the figures. John E. Costa kindly provided an unpublished list of Appalachian rainfalls. Patrick Michaels provided considerable data and help from the Virginia State Climatology Office. Finally, Sam Valastro performed the radiocarbon analyses at the University of Texas at Austin.

References

Bogucki, D. J. 1970. *Debris slides and related flood damage associated with the September 1, 1951, cloudburst in the Mt. LeConte–Sugarland Mountain area, Great Smoky Mountains National Park.* Ph.D. Dissertation, University of Tennessee, Knoxville. 165 pp.

Bogucki, D. J. 1976. Debris slides in the Mt. LeConte area, Great Smoky Mountains. *Geografiska Annaler*, **58A**, 179–192.

Bogucki, D. J. 1977. Debris slide hazards in the Adirondack province of New York State. *Environmental Geology*, **1**, 317–328.

Bryson, R. A. 1966. Air masses, streamlines, and the boreal forest. *Geographical Bulletin*, **8**, 228–269.

Bryson, R. A., Barreis, D. A., and Wendland, W. M. 1970. The character of late-glacial and post-glacial climatic changes. In Dort, W. and Jones, J. K. (Eds), *Pleistocene and Recent Environments of the Central Great Plains*. Lawrence, University of Kansas Press. 53–74.

Bull, W. B. 1964. Relation of alluvial-fan size and slope to drainage-basin size and lithology in western Fresno County, California. *United States Geological Survey Professional Paper*, **450-B**, 51–53.

Caine, N. 1980. The rainfall-duration control of shallow landslides and debris flows. *Geografiska Annaler*, **62A**, 23–27.

Costa, J. E. 1973. *Large rainfalls and runoff in the Appalachians*. Unpublished data.

Costa, J. E. 1984. Physical geomorphology of debris flows. In Costa, J. E. and Fleisher, J. P. (Eds), *Developments and Applications of Geomorphology*. New York, Springer-Verlag. 268–317.

Costa, J. E. 1988. Rheologic, geomorphic, and sedimentologic differentiation of water floods, hyperconcentrated flows, and debris flows. In Baker, V. R., Kochel, R. C., and Patton, P. C. (Eds), *Flood Geomorphology*. New York, Wiley. pp. 113–122.

Delcourt, P. A. and Delcourt, H. R. 1984. Late Quaternary paleoclimates and biotic responses in eastern North America and the western North Atlantic Ocean. *Paleogeography, Paleoclimatology, Paleoecology*, **48**, 263–284.

Eisenlohr, W. S. 1952. Floods of July 18, 1942, in north–central Pennsylvania. *United States Geological Survey Water Supply Paper*, **1134-B**, 59–158.

Eschner, A. R. and Patric, J. H. 1982. Debris avalanches in eastern upland forests. *Journal of Forestry*, **80**, 343–347.

Flaccus, E. 1958. White Mountain landslides. *Appalachia*, **24**, 175–191.

Hadley, J. B. and Goldsmith, R. 1963. Geology of the eastern Great Smoky Mountains, North Carolina and Tennessee. *United States Geological Survey Professional Paper*, **349-B**, 118 pp.

Hack, J. T. 1965. Geomorphology of the Shenandoah Valley, Virginia and West Virginia, and the origin of the residual ore deposits. *United States Geological Survey Professional Paper*, **484**, 84 pp.

Hack, J. T. and Goodlett, J. C. 1960. Geomorphology and forest ecology of a mountain region in the central Appalachians. *United States Geological Survey Professional Paper*, **347**, 66 pp.

Hupp, C. R. 1983. Geo-botanical evidence of Late Quaternary mass wasting in block field areas of Virginia. *Earth Surface Processes and Landforms*, **8**, 439–450.

Johnson, A. M. 1970. *Physical Processes in Geology*. San Francisco, Freeman and Cooper. 577 pp.

Kochel, R. C. 1987. Holocene debris flows in central

Virginia. In Costa, J. E. and Wieczoreck, G. F. (Eds), *Debris Flows/Avalanches: Process, Recognition and Mitigation. Geological Society of America, Reviews in Engineering Geology*, **7**, 139–155.

Kochel, R. C. 1988. Geomorphic impact of large floods: review and new perspectives on magnitude and frequency. In Baker, V. R., Kochel, R. C., and Patton, P. C. (Eds), *Flood Geomorphology.* New York, Wiley. pp. 169–187.

Kochel, R. C. and Johnson, R. A. 1984. Geomorphology and sedimentology of humid-temperate alluvial fans, central Virginia. In Koster, E. and Steel, R. (Eds), *Gravels and conglomerates. Canadian Society of Petroleum Geologists Memoir*, **10**, 109–122.

Kochel, R. C., Ritter, D. F., and Miller, J. 1987. Role of tree dams in the construction of pseudo-terraces and variable geomorphic response to floods in Little River Valley, Virginia. *Geology*, **15**, 718–721.

Lee, L. G. and Goodge, G. W. 1984. Meteorological analysis of an intense 'east-slope' rainstorm in the southern Appalachians. *American Meteorological Society, 10th Conference on Weather Forecasting and Analysis. Proceedings.* 30–37.

Michaels, P. J. 1985. Virginia climate advisory. *University of Virginia, Charlottesville*, **9**(2), 30 pp.

Mills, H. H. 1982a. Long-term episodic deposition on mountain foot-slopes in the Blue Ridge province of North Carolina: evidence from relative age dating. *Southeastern Geology*, **23**, 123–128.

Mills, H. H. 1982b. Piedmont-cove deposits of the Dellwood quadrangle, Great Smoky Mountains, North Carolina, U.S.A.: morphometry. *Zeitschrift für Geomorphologie*, **26**, 163–178.

Mills, H. H. 1983. Piedmont evolution at Roan Mountain, North Carolina. *Geografiska Annaler*, **65A**, 111–126.

Mills, H. H. 1987. Debris slides and foot-slope deposits in the Blue Ridge province. In Graf, W. L. (Ed.), *Geomorphic Systems of North America. Geological Society of America, Centennial Special Volume*, **2**, 29–37.

Moneymaker, B. C. 1939. Erosional effects of the Webb Mountain, Tennessee, cloudburst of August 5, 1938. *Tennessee Academy of Science Journal*, **14**, 190–196.

Neary, D. G. and Swift, L. W. 1984. Rainfall thresholds for triggering a debris avalanching event in the southern Appalachians. *Geological Society of America, Abstracts with Program*, **16**(6), 609.

Newson, M. 1980. The geomorphological effectiveness of floods—a contribution stimulated by two recent events in mid-Wales. *Earth Surface Processes*, **5**, 1–16.

Patric, J. H. 1981. Soil–water relations of shallow forested soils during flash floods in West Virginia. *United States Department of Agriculture, Forest Service Research Paper*, **NE-469**, 20 pp.

Pomeroy, J. S. 1980. Storm induced debris-avalanching and related phenomena in the Johnstown area, Pennsylvania, with reference to other studies in the Appalachians. *United States Geological Survey Professional Paper*, **1191**, 24 pp.

Rich, J. L. 1935. Origin and evolution of rock fans and pediments. *Geological Society of America Bulletin*, **46**, 999–1024.

Ritter, D. F. 1967. Terrace development along the front of the Beartooth Mountains, southern Montana. *Geological Society of America Bulletin*, **78**, 467–484.

Ratte, C. A. and Rhodes, D. D. 1977. Hurricane-induced landslides on Dorset Mountain, Vermont. *Geological Society of America Abstracts with Program*, **9**(3), 311.

Schneider, R. H. 1973. *Debris slides and related flood damage resulting from hurricane Camille, 19–20 August, and subsequent storm, 5–6 September, 1969, in the Spring Creek drainage basin, Greenbriar County, West Virginia.* Ph.D. Dissertation, University of Tennessee, Knoxville. 131 pp.

Schwarz, F. K. 1970. Unprecedented rains in Virginia associated with the remnants of Hurricane Camille. *Monthly Weather Review*, **98**, 851–859.

Simmons, D. W. 1988. *Sedimentology and geomorphology of humid temperate alluvial fans along the west flank of the Blue Ridge Mountains, Virginia.* M.S. Dissertation, Southern Illinois University, Carbondale. 107 pp.

Smith, G. A. 1986. Coarse-grained, nonmarine volcaniclastic sediment: terminology and depositional processes. *Geological Society of America Bulletin*, **97**, 1–10.

Stringfield, V. T. and Smith, R. C. 1956. Relation of geology to drainage, floods, and landslides in the Petersburg area, West Virginia. *West Virginia Geological and Economic Survey, Report of Investigations*, **12**, 19 pp.

United States Geological Survey 1949. Floods of August, 1940, in the southeastern states. *Water Supply Paper*, **1066**, 544 pp.

VanDine, D. F. 1985. Debris flows and debris torrents in the southern Canadian Cordillera. *Canadian Geotechnical Journal*, **22**, 44–68.

Watts, W. A. 1979. Late Quaternary vegetation of the central Appalachians and the New Jersey coastal plain. *Ecological Monographs*, **49**, 427–469.

Williams, G. A. and Guy, H. P. 1973. Erosional and depositional aspects of Hurricane Camille in Virginia, 1969. *United States Geological Survey Professional Paper*, **804**, 80 pp.

Wilson, G. 1988 *Reconnaissance survey of debris fans in the Appalachians and paleoclimatic implications of debris flows.* M.S. Dissertation, Southern Illinois University, Carbondale. 165 pp.

CHAPTER 7

The Chandigarh Dun Alluvial Fans: An Analysis of the Process–Form Relationship

A. B. Mukerji
Panjab University, Chandigarh

Abstract

Following the formation of Chandigarh Dun in early to mid-Pleistocene time by movements on the Himalayan Main Boundary Fault and its offshoots, alluvial fans have developed there in upper Pleistocene and Holocene time. Faults control fan position and evolution. The fans generally conform with the usual conical form, but individuals have been constrained in their morphological development by the neighbouring features and the presence of interfan streams. Consequent streams have developed on the fans, some draining the south slope of the Himalaya front, and some rising on the fan.

The fans are composed of boulder gravels that have been deposited by both mudflows and streamfloods, the former being dominant near the fanheads. In this, they are similar to the underlying Boulder Conglomerate (Upper Siwalik Formation), which represents an early Pleistocene fan-building period. However, there are significant differences of detail between the sedimentary units.

The recent fans have formed during alternating periods of semiarid (glacial) and subhumid monsoon (interglacial) climate during the Upper Pleistocene and Early Holocene Epochs. Summer thunderstorm rainfall has characterized both climates, producing debris-flows; however, major streamfloods in the interglacials have caused entrenchment at the fanheads. Continued tectonic uplift on the Barsar Fault, at the fanheads, has created much greater total fan relief than is accounted for by bedding thickness. A reduction in tectonic activity in the later Holocene Epoch has allowed the axial streams to entrench themselves to the point that the fans are no longer active.

Introduction

For quite some time now there has been a growing awareness among geomorphologists of the widespread occurrence of alluvial fans in diverse climatic, tectonic, and topographic settings. This realization has been documented extensively in recent literature (*cf.* Rachocki, 1981). Although alluvial fans occur widely in India they have received very little attention and those occurring either within or at the foot of the Himalaya have been almost completely ignored (but see Nossin, 1971; Joshi, 1977; Dutta, 1984). In the Chandigarh Dun (which will be referred to as the Dun), which is typical of other duns located in the Himalayan piedmont, the alluvial fans are the most striking geomorphic features. The Dun (Figure 7.1) is a longitudinal, structural, gravel-filled valley bounded by the lower Himalaya to the north and the Siwalik Hills to the south, a kind of

Alluvial Fans: A Field Approach edited by A. H. Rachocki and M. Church
Copyright © 1990 John Wiley & Sons Ltd.

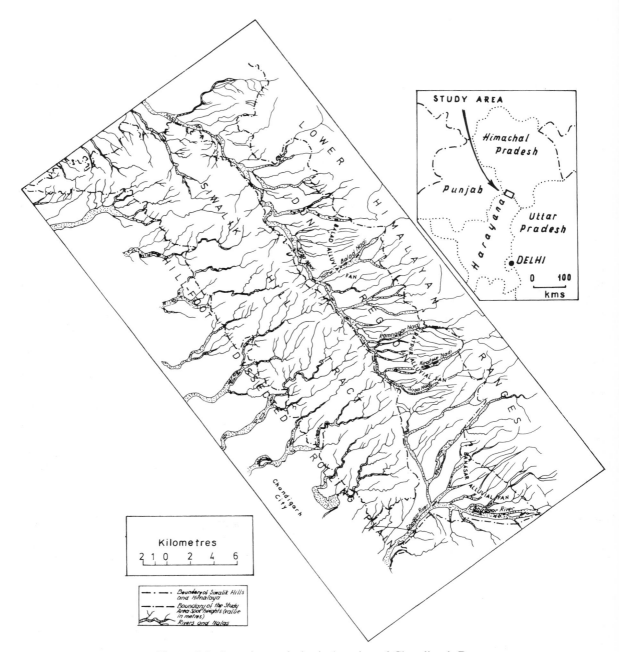

Figure 7.1. Location and physical setting of Chandigarh Dun

cul-de-sac providing a restricted setting for the development of alluvial fans.

There are several reasons why the Dun fans draw our attention. They have responded to glacially induced climatic changes in an area which has never been covered by glaciers nor directly influenced by periglacial or proglacial conditions. They are located in a tectonically unstable region affected by Pleistocene and Holocene neotectonic movements. Since the formation of the fans, the region has been experiencing a modified subtropical monsoon climate (Cwg of Koppen) which is in essence subhumid. The annual rainfall is 1000 to 1200 mm, of which 700 to 900 mm is concentrated in the summer monsoon between June and September. Rain occurs in high-intensity downpours or in protracted showers spread over several days. Finally, the fans are now relict 'landscape fossils'. The streams which formed them are deeply entrenched and cannot further modify their surface configuration.

Geological and geomorphological investigations of the Chandigarh Dun and its analogues in other parts of the Himalayan piedmont have been very few and of very limited scope. Although there have been publications on the Dun terraces (Sen, 1955: 176–185; Khan, 1970: 51; Nossin, 1971; Nakata, 1972: 139–149; Karir, 1985: 27–28), there have been only perfunctory references or statements on the Dun fans (Nossin, 1971; Karir, 1985: 26–27). Gravel surfaces have been noted (Nakata, 1972: 144–145), but not the alluvial fans.

The fans exhibit a number of interesting phenomena, including radiating, incised distributary channels; channels—other than the trunk stream—emerging from the mountains to cross the fans; segmented long profiles; total sediment thickness less than total relief; recent fanhead trenching; and evidence that tectonic movements have played a role in fan development.

It is surprising that the alluvial fans, which constitute the dominant geomorphic feature of the Chandigarh Dun, have been almost completely ignored. They are obviously of great interest in view of Pleistocene and Holocene tectonic activity and climatic fluctuations experienced in the Lower Himalaya and the Dun. These events are reflected in the morphology and genesis of the fans.

Geological Antecedents

The geological context of the alluvial fan landscape of the Chandigarh Dun has two components: rock formations and faults. The formations reflect the lithology and stratigraphy of the Dun and are important in explaining some of the properties of the alluvial fans. The faults have played an effective role in the initiation of fan formation, in the fanhead entrenchment, and in the segmentation of the fans.

From the Himalaya in the north to the Siwalik Hills in the south one encounters three formations (Figure 7.2). To the north of the Main Boundary Fault the Subathu Formation represents the Pre-Siwalik rocks. To the south of the fault and at a lower elevation lies the belt of the Nahans, forming the uppermost Formation of the Lower Siwaliks. The Middle Siwaliks are missing near and along the Barsar Fault (Chaudhri, 1976: 232) but to the south of the Fault, one observes an uninterrupted expanse of the Boulder Conglomerates which constitute the uppermost Formation of the Upper Siwaliks. At places, the Pinjaur Formation, a member of the Upper Siwaliks, underlies the Conglomerate.

The Pinjaurs are Upper Pliocene in age. The beds are composed of red clays, grey sandstones, and minor beds of yellow sandstones. The Boulder Conglomerates range in age from Lower to Middle Pleistocene (Gansser, 1964: 48; Khan, 1970: 50). They are composed of subangular to subrounded granules and pebbles. The rock types include quartzites, shales, limestones, cherts, slates, schists, and a high proportion of sandstones, derived from rocks to the north of the Main Boundary Fault (Gill, 1985: 45). In most places they lie almost horizontally or with gentle dips. The Boulder Conglomerates are considered to be earlier fan deposits of the upper Siwalik basin, associated with the steep and sudden rise of the Lower Himalaya (Gansser, 1964: 48; Gaur and Chopra, 1984: 354), that preceded the development of the recent fans.

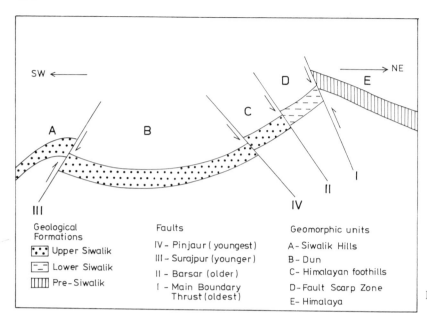

Figure 7.2. Generalized geological schema across Chandigarh Dun

Both the Pinjaurs and Boulder Conglomerate are only partly consolidated, loose, and friable. They are partly stratified and reveal pseudobedding. Both are eroded intensively by streams although the latter better retains a sculptured outline. At most places the recent fans have developed unconformably on the Boulder Conglomerate. In a few localities the thin Conglomerate beds have been eroded and the recent fans sit on the Pinjaurs.

As many as four faults have been recognized in the Dun (Figure 7.2). For most of their lengths they are steep, strike faults (Gansser, 1964: 48). They are thrust faults and, except for the Surajpur Fault, which dips toward the southwest, they dip northeast toward the Himalaya (Khan, 1959: 65). They are aligned northwest–southeast, parallel to the grain of the Himalaya and the Siwalik Hills. Thus, they control the alignment of the fans.

The Main Boundary Fault is located between the Nahans (Lower Siwaliks) and the Subathus (Pre-Siwaliks) (Sen, 1971: 257). It demarcates the northern limit of the Upper Siwaliks and separates the main Himalaya from the Siwaliks, both in geologic and topographic aspects (Valdiya, 1964: 275; Nossin, 1971: 26). It was formed during the Early Pliocene, but the major movement began in the Lower Pleistocene. This suggests that the other three faults belong to the Middle and Upper Pleistocene and even the Holocene Epochs.

Although all four faults have been influential in the formation of the Dun graben and the Dun geomorphic landscape, the Barsar and Pinjaur Faults have been particularly effective in the initiation and subsequent development of the fans. They are among the many faults which diverge from the south side of the Main Boundary Fault with generally east–west strike (Gansser, 1964: 247; Khan, 1970: 48; Valdiya, 1973: 84). The Barsar Fault hades toward the northwest (Khan, 1970: 54). It is aligned with a large number of springs and wells, and the apices of the fans. The topographic displacement expressed in the fault-line scarp amounts to 15 to 20 m. The Pinjaur Fault is of post-Middle Pleistocene age. It has produced a topographic displacement of 10 to 15 m. It finds striking geomorphic expression in the segmentation of the alluvial fans and terraces in the Dun and in the terrace scarps. The Surajpur Fault appears to be a hinge and thrust fault (Khan, 1959: 66). It is not a continuous lineament but forms, more or less, the southern edge of the fans and limits their expansion in that direction.

Age of the Dun Fans

Although it is difficult to estimate the absolute age of the Dun fans, relative age can be inferred from the following facts and arguments:
1. The termination of Siwalik sedimentation, on the evidence of palaeontology, occurred within the Middle Pleistocene Epoch or at the end of it: the Upper Siwaliks extend essentially up to the Mindel glacial phase.
2. The Surajpur Fault, younger than the uplift of the Siwalik Hills, bounds the Dun graben to the south while the reactivated branches of the Main Boundary Fault form its northern boundary. Movement along the Main Boundary Fault took place during late Pliocene to Middle Pleistocene time and within this period the Siwalik Hills were folded (Valdiya, 1964: 277). Movement along the branch termed the Barsar Fault was a little earlier than on the Surajpur Fault. This fault is revealed not only in the steeper northward slope of the Siwalik Hills but also in the placement of the Boulder Conglomerate beds of the Upper Siwaliks at much lower topographic levels in the Dun. In effect, therefore, the Dun graben, the locale of the fans, has been formed between the Barsar and Surajpur Faults in Early to Middle Pleistocene time.
3. The gravel filling of the Dun and coeval alluvial fan formation began immediately after the formation of the graben and would therefore have occurred within the Upper Pleistocene Epoch.

These appearances are summarized in Figure 7.3. It is concluded that the recent Dun fans were formed after the Dun was created by the fault movements. Hence, they are of Upper Pleistocene and Holocene age. Formation began during the Mindel–Riss Interglaciation about 0.73 My ago (Gaur and Chopra, 1984: 353; Joshi, 1977: 200–201).

Morphological Attributes of the Dun Fans

GENERAL MORPHOLOGY

The fans have formed on the Himalayan front on the northeast side of Chandigarh Dun (Figure 7.1). Three features present strikingly classical morphology: Balad, Kiratpur, and Banasar fans (Figure 7.4). They have areas of 20.8, 26.5, and 22.0 km^2 respectively and are, on the whole, much larger than the fans of the Himalayan piedmont in the northern part of the Brahmaputra Valley (*cf.* Dutta, 1984: 19). The difference is associated with the larger basins of the Dun fan axial streams. These fans also have gentler gradient, lower fan height, and smaller topographic break-of-slope angles. Further, the drainage basin relief above the fan and stream gradient at the head of the fan are positively correlated with the size of the fan. Some of these relations have earlier been confirmed for other fans (Mukerji, 1976b: 197–198).

Alluvial fans, amongst the most regular of landforms, resemble a sector of a cone. Hence, in planimetric shape, they tend to approximate a sector of a circle. A measure of similarity is the 'fan conicality index' (FCA) introduced by Mukerji (1976b) as

$$FCI = \frac{\text{Area of the actual fan}}{\text{Area of the equivalent ideal fan}}$$

in which the ideal fan is defined by the enclosing circle given by the maximum radius of the actual fan. Hence, the area of the ideal fan is $\pi r^e_{max} a/360$, where a is the angle of divergence between the two fan margins as measured at the fan apex. Hence, $FCI < 1$. For the principal Dun fans the values are Balad, 0.63; Kiratpur, 0.57; Banasar,

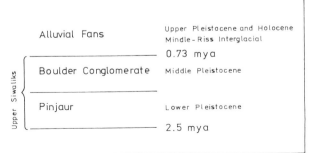

Figure 7.3. Stratigraphic units associated with the Chandigarh Dun alluvial fans

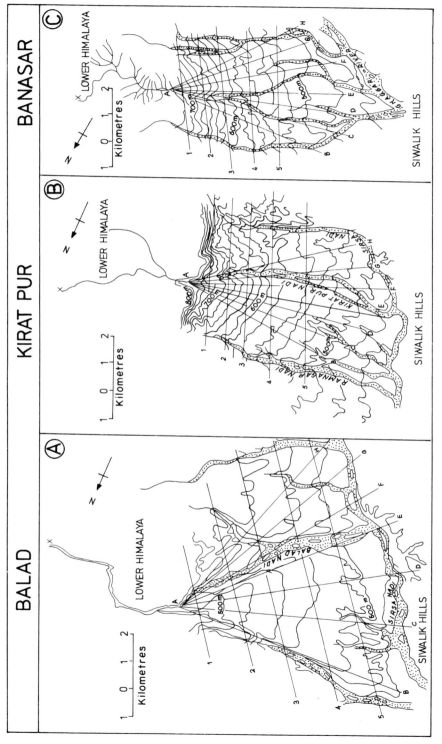

Figure 7.4. Topographic maps of the principal fans in Chandigarh Dun. The labelled lines indicate radial profiles illustrated in Figure 7.6 and transverse profiles illustrated in Figure 7.7

0.56. In each case, it can be seen (Figure 7.1) that the fans are longer and narrower than would be expected for a strictly circular segment. This is the consequence of their forming part of a continuous hem along the Himalayan Front, so that adjacent fans and interfan streams constrain full lateral development.

The most remarkable feature of the fans is the deep entrenchment of the axial (main) stream right across the fan. Entrenchment is deepest in the proximal portion of the fan (Figure 7.5), and so these are classical fanhead trenches. The deeper entrenchment there is related to the larger amount of tectonic uplift along the Barsar Fault near the apex of the fan, the steeper gradient, and the massive wedge of easily eroded, loosely consolidated sand, gravel, pebbles, and cobbles of the fan deposits and Boulder Conglomerate.

PROFILES AND SLOPE

In order to examine further the fans' configuration, radial longitudinal profiles have been drawn for the three major fans. Profiles have been constructed (Figure 7.6) for the section lines shown on Figure 7.4. The average fan surface gradients are 0°48′, 2°16′, and 2°42′ for the Balad, Kiratpur, and Banasar fans respectively.

The longitudinal profiles reveal that the Kiratpur and Banasar fan surfaces are far more regular than that of the Balad; the first two have subdued, concave slopes, whereas the maximum irregularity is displayed in the Balad profile in the form of long humps and shallow depressions. In different profiles, the slope forms vary in some measure: in all, the overall profiles are not smooth but gently concave curves with planar and concave segments. In most profiles, the upper reaches are slightly steeper and more concave than the middle and distal segments. The degree of concavity varies among the fans.

The Kiratpur and Banasar fans can be classified as steep fans (Blissenbach, 1954: 175–190), yet they are relatively large. The generalization that the length of the fan radii and the fan surface gradient are inversely correlated with each other (Williams, 1973; Reineck and Singh, 1973: 434) is not valid for the Dun fans. This is partly explained by the large differential uplift along the mountain front fault with consequent steepening of the fanhead channel and depositional gradients (*cf.* Bull, 1977: 254).

In the literature on alluvial fans there is very little discussion of the transverse profiles. Theoretically, an alluvial fan, considered as part of a cone, should exhibit a convex transverse slope. The gradient would be gentler near and along the axis than at increasing distances from it near and along the lateral margins. This probability is practically increased by the presence, on many alluvial fans, of streams located in the interfan depressions and flowing along their lateral edges, effecting lateral erosion.

The Dun fans (Figure 7.7) exhibit subdued, convex transverse slopes along the entire length of the axis. The convexity is accentuated in the proximal zone and becomes subdued in the middle and distal areas of the fan. The declining convexity is associated with the large reduction in deposition from the apex toward the distal segments, and corresponding reduction in lateral differences of deposit thickness. The locus of maximum deposition in any segment lies along the axial stream. When the fan is active, the distributary channels abstract a part of the sheet flow of the fan surface and transfer sediments from the central to the lateral, and from the proximal to the distal parts, the former being less in amount than the latter.

In the proximal parts the distributaries, existing and reconstructed, flow more toward the toe-edge than diverging laterally, and hence are not able to distribute the sediments from the locus of maximum deposition in the apex region toward the lateral edges (Figure 7.8). Consequently, the apex experiences the cumulative effect of surplus deposition along the axial stream belt and the thickness of the deposit there is several times larger than along the lateral edges. This results in the steep convexity of the transverse profile in the proximal segment. In the distal segment the diverging distributaries move sediments from the axial belt toward the lateral margins to a greater extent over a much wider fan surface, thus reducing the difference in elevation and thickness between the

axial and lateral edge zones. This gives rise to the subdued convexity.

Figure 7.8 presents a schematic summary of Dun fan geometry emphasizing the transverse profiles.

LONGITUDINAL PROFILES OF THE AXIAL STREAMS

Two segments of the axial streams can be identified: (1) the segment that extends from the origin of the stream to the apex of the fan; and (2) the segment that stretches from the apex of the fan to the toe-edge (Figure 7.9). The steep slope of the catchment segments is related to the uplift along the Barsar and Main Boundary Faults, and the consolidated and steeply dipping Lower and Pre-Siwalik formations over which the streams flow. The irregularities in the profiles are clearly related to structural and lithological variations.

In contrast, the fan segments are controlled by lithological and sedimentological factors. The structural surface upon which the fans are built in the Dun is formed of gently dipping, open folds. The stream profiles indicate general concavity, and are rather smooth in Kiratpur and Banasar fans and humped in Balad fan, following the fan surface. They do not show any discernible break except on Banasar fan. The concavity and general smoothness imply that the streams are in a state of equilibrium. The microirregularities and humps are related to the random variations in the channel flow, channel shifting, and channel deposition. In particular, the prominent hump on Balad fan is associated with the tendency for the axial stream to become braided/anastomosed in mid-fan (Figure 7.10).

Stream Characteristics

If the alluvial fans are the dominant geomorphic feature of the Dun landscape, the streams are the preponderant feature of the fans. The fineness of the topographic texture of the fans is really the result of the large number of streams flowing on them (Figure 7.10).

On the basis of source area there are two stream classes: those originating from the Himalaya and those originating on the fan. The streams belonging to the former group originate from the south-facing Himalayan slopes and, apart from the axial stream, are wholly consequent in genesis. These are younger than the age of the alluvial fan. Although they dissect the fan they do not contribute to the alluvial building of any part of the fan except, perhaps, in a limited measure at the slope base. These Himalayan streams are mostly first-order members of the system (Figure 7.11) and are characterized by their striking length. In most cases they merge either with the

THE CHANDIGARH DUN ALLUVIAL FANS

Figure 7.5. A composite view of the Kiratpur fan showing the proximal, middle, and distal segments. Note the striking, angular topographic contact between the Himalaya and the fan surface

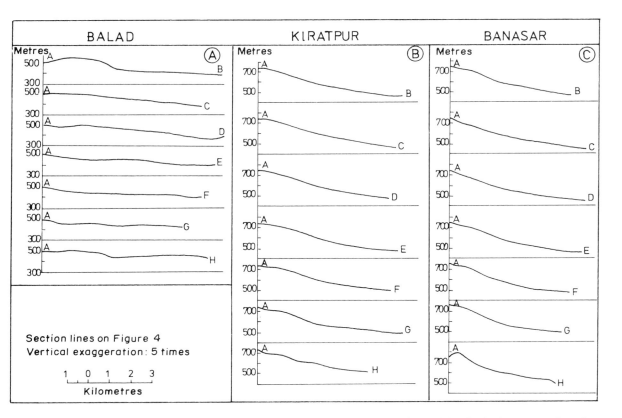

Figure 7.6. Longitudinal profiles of the Dun fans. Section lines given on Figure 7.4. Vertical exaggeration: 5×

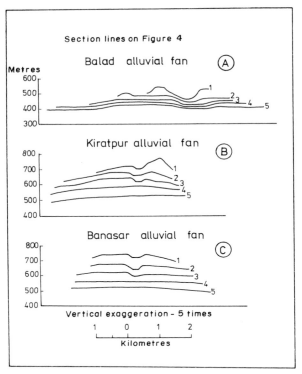

Figure 7.7. Transverse profiles of the Dun fans. Section lines given on Figure 7.4. Vertical exaggeration: 5×

streams forming the lateral boundaries of the fans or with the Sirsa River, flowing along the fan toe.

In the field, on topographical maps, and on large-scale aerial photographs the Himalayan streams reveal only a moderate degree of integration. There are several first-order streams which are longer than the second-order streams with which they join. The fan channels are typically straight, a character that is related mainly to the steep gradient of the fan surface. Some, however, display anastomosing tendencies (Figure 7.10).

The fan streams originate in proximal, mid-fan, and distal parts. These streams, also consequent, are younger than their Himalayan counterparts and reveal a more regular, concave longitudinal profile. They have further added to the fineness of the topographic texture of the fans. A large number of these streams merge with the larger fan-boundary streams and a few join among themselves.

Both types of stream have seasonal, intermittent flow of water in their channels. Most are rain-fed. No data are available for discharge. Another characteristic of all streams but the axial one is a decline in discharge in the distal parts: this is related to the heavy infiltration of water into the gravel channel beds in the proximal and, to some extent, mid-portions of the fans.

It is widely observed that the streams on the fans have a characteristic radial pattern (Figures 7.10 and 7.11). Most of the fans have their streams directed away from the axial channel. Many of the streams, if extended upstream, would converge on a common point near the fan apex. There could be two explanations for this pattern. During fan construction, the distributaries radiated out from the axial stream at the apex of the fan (Figure 7.12A) and hence maintained a live contact for the distribution of the discharge. During this stage the distributaries, through transportation and deposition of their sediment load, effected enlargement of the fan. In the post-entrenchment stage the axial stream has severed the active linkage of the distributaries and abstracted the entire flow. The distributaries now receive only rainwater from the fan and are in various stages of decay and filling.

During the latter development, the axial stream may either maintain its central position on the fan (Figure 7.12B), or experience not only entrenchment but, more importantly, a marked relocation from the apex or near it (Figure 7.12B). This results in the axial stream breaching its bank and bypassing the radiating distributaries. The contemporary channel pattern on the Balad fan appears to reflect this second model of development, whereas the other two fans reflect the first model. In both models there are two kinds of reconstructed fan surface streams: those which originated, according to the postulation, from the axial stream near the apex, and those which originated on the fan surface.

In the field one is repeatedly struck by the typically V-shaped gorge form above the apex of the fan and the deep, box-like cross-profile immediately below it (Figures 7.13A,B), where the axial stream is deeply entrenched into the fan. In this reach, one observes stretches of braiding

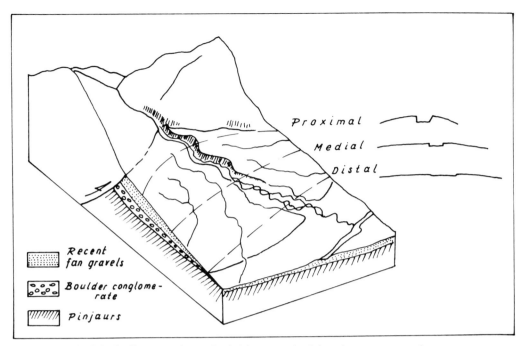

Figure 7.8. Block diagram of an alluvial fan, emphasizing the transverse slope segments

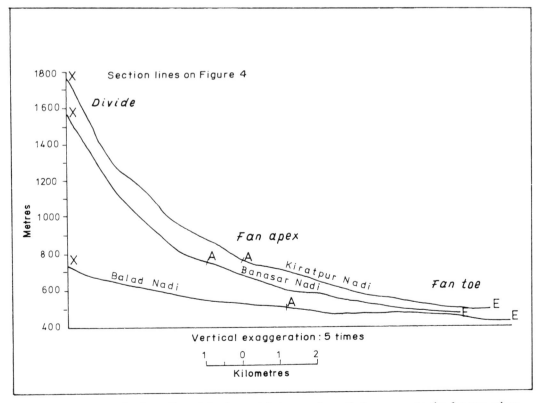

Figure 7.9. Longitudinal profiles of the fan axial streams from their sources to the fan toe-edges. Vertical exaggeration: 5×

channels (Figure 7.10) related to decline in discharge and to decline in gradient. The valley walls lose their height, initially 15 to 20 m, as one proceeds from the apex to the toe-edge while the flattish, wide bed persists (Figure 7.13B). All the streams, regardless of their origin, have a low hydraulic radius and are inefficient.

Figure 7.10. Aerial photograph of Balad Fan. Note the radiating distributary channels, deeply entrenched axial stream, and anastomosed channels along the axial stream. In the lower right are streams originating on the fan

Regardless of their origin, in the Himalaya or in any part of the fan, the streams are boldly etched and most of them are deeply entrenched into the fan. The depth of entrenchment is greatest—8 to 20 m—in the proximal fan portion (Figures 7.5, 7.10, 7.13B) near the apex, and smallest—1 to 3 m—in the distal portion. In years of unusually high discharge, the water level has risen up to about 2 m. Therefore nowhere, in any part of the year, do the streams experience overbank stage and flood the adjacent fan surface.

Fan Deposits

STRATIGRAPHY

Fan deposits overlie the Boulder Conglomerate (Figure 7.14A) or the Pinjaur Beds (Figure 7.14B) at different places. However, complete stratigraphic exposures are infrequent and the thicknesses of the fan deposits and of the exposed underlying beds vary greatly. In general the Boulder Conglomerate underlies the proximal and higher middle parts while the Pinjaurs are exposed in the distal fans. The fan deposits are in general grey and earthy coloured, unlike the yellowish, reddish sediments of the weathered Boulder Conglomerate and the Pinjaurs.

The recent fan deposits are as much as 20 to 25 m thick near the apex, as prominently revealed in the entrenched segment (Figure 7.15). Thickness declines to 3 to 6 m in the mid-fan (Figure 7.14A) and to 1 m or less in the distal segment (Figure 7.14B). There is, thus, a steep decline in the thickness from the proximal to distal segments. The thickness of the fan deposits is much less than the relief of the fans, which is from 100 (Balad) to over 200 m. Hence, the dip on the erosional surface of the underlying Boulder Conglomerates is a significant factor in the present relief.

SEDIMENTOLOGY

The fan deposits, regardless of their location, are characterized by very angular to subangular fluvial megaclasts (cobbles and boulders), many of them strikingly large, and clasts of very low sphericity, many of them having rod shapes (Figure 7.16A). Recent fan gravels have boulders larger than those of the Boulder Conglomerate (Gansser, 1964: 245); one frequently encounters boulders twice the average size of those of the older beds. Sand and silt, forming the matrix, are common everywhere. There is a clear break in size-classes, lithologies, and shape–class series between the fan sediments and those of the Boulder Conglomerate.

The fan materials have largely been derived from the Pre-Siwalik formations of the southern,

THE CHANDIGARH DUN ALLUVIAL FANS 143

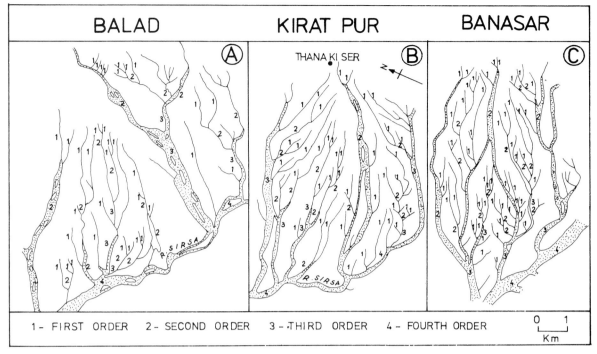

Figure 7.11. Stream orders on the Dun Fans. Note that the axial stream is arbitrarily given order 1 at the fan apex for comparison of the confluence pattern across the fan

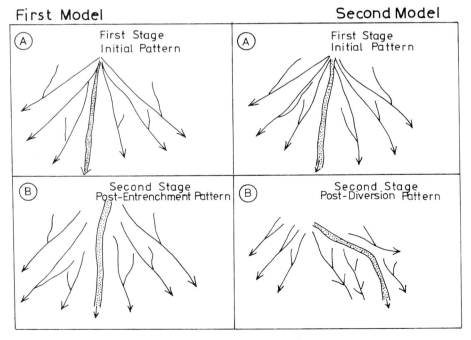

Figure 7.12. Models of evolution of Dun fan stream patterns

Figure 7.13. Channel morphology near the fan apex. (A) Straight, gorge section north of the Barsar Fault on Kiratpur Fan axial stream. Note the V-form of the gorge, consolidated Lower Siwalik rocks, and the gravel filled channel bed; (B) Wide valley just downstream of the gorge. The narrow, meandering channel is choked with gravels. On the left can be seen the sloping fan surface (*cf.* Figure 7.5) extending toward the Siwalik Hills, a continuous feature in the background

Figure 7.14. Stratigraphy of the alluvial fans. (A) Distal fan deposits overlying the clay facies of the Boulder Conglomerate. Note the rilled face to the right in the middle ground and the bed dipping toward the Himalaya; (B) Silty and clayey Pinjaur bed underlying thin fan gravel deposits in a distal segment. Note the vertical walls and their recession by mass-wasting processes

lower ranges of the Outer Himalaya (excluding the Siwalik Hills) to the north of the Main Boundary Fault. Clay and sand were derived from the shales of the same zone. Only a little amount was picked up during the process of valley filling from the underlying Upper Siwalik Boulder Conglomerate and Pinjaur Beds. The megaclasts and the smaller sediments have been transported by the high-energy traction and rolling processes associated with episodic, high velocity flood discharges that characterize the monsoon climate, and by debris-flows. What is happening now gives ample evidence of the fan-forming processes of the past (Figure 7.16A).

The Boulder Conglomerate, in contrast, is composed mainly of pebbles and cobbles, with some boulders, of sandstones, limestones, quartzites, and schists (Kharakwal, 1969: 211). Most of the megaclasts are subspherical and subrounded (Chaudhri and Gill, 1979: 85–88). These properties of the sediments suggest that they have travelled over longer distances, originating in the zone located to the north of the provenance of the recent fan deposits. The Boulder Conglomerate deposits are weathered to a much greater extent,

THE CHANDIGARH DUN ALLUVIAL FANS

Figure 7.15. The downstream end of the gorge section. The Boulder Conglomerate bed can be seen on the exposed walls along the gravel filled channel

Figure 7.16. Channel sediments in the axial stream. (A) Heterogeneous debris of megaclasts dumped in the bed in the middle segment; the large boulders measure from 200 to 550 mm in a-axis length; (B) Very wide, sand and gravel filled channel bed of the axial stream in the distal segment

as evidenced by red and dark yellow colouration, than the fan deposits.

There are marked differences in the fan deposits in the proximal, middle, and distal segments. In the proximal segment of the Kiratpur fan, Nakata (1972: 143) described sandstone boulders measuring up to 1 m in diameter and estimated the thickness of the deposit at up to 50 m in places. Broken clasts and megaclasts and sand dominate the deposits. (Nakata failed to recognize the deposit as a fan and instead termed it the Jhajra surface.) Turbulent, high-energy waters and occasional debris-flows may be suggested as the principal modes of transportation and deposition.

In distal segments pebbles and cobbles are the preponderant sediments along with sand, silt, and clay as secondary constituents providing the matrix (Figure 7.16B). The pebbles and cobbles display a well-organized fabric indicating sliding and rolling in moving waters. Lower energy fluvial transportation and deposition are suggested. A good proportion of discharge was lost through percolation into the channel bed gravel deposits.

The fan stratigraphy reveals, then, debris-flow deposits and water-laid deposits emplaced at different times and in different fan segments. One finds here corroboration of Lawson's (1913) findings: the boulder and gravel dominated apex deposits relate to debris-flows, and the smaller sediments (cobbles, pebbles, and coarse sand) characterize the distal parts. Dun fans answer very well to the description of coarse-grained deposits formulated by Fisher (1971: 919).

Formation of Dun Fans

An attempt is made in the following discussion to analyse the genesis of the Dun fans and to evaluate the role of tectonic and climatic conditions in this process. Three factors, steep slopes in the catchment basins in the Lower Himalaya, sparse vegetation dominated by semi-xerophytic species, and a climate characterized by summer thunderstorms and high intensity rainfall, have favoured the development of the Dun fans; while the first expresses the tectonic conditions, the second and third reveal the climatic context. The tectonic–climatic conditions prevailing during the period of formation of the Dun fans were somewhat similar to those existing during the formation of their geologically older analogue, the Boulder Conglomerates.

During the period of Dun fan formation there was uplift of the Himalaya north of the Main Boundary Fault, and of the foothills north of the Barsar Fault. Although the role of crustal movements in the development of the Dun fans and terraces has been earlier recognized (Nakata, 1975: 114), it has not been investigated in detail, particularly with respect to the effects of faulting. The uplift was accompanied by severe denudation of the south-facing slopes of the Himalaya and the southern flanks of the foothills by seasonal floods which have characterized the monsoon pluvial phases (Kulshrestha, 1976: 44).

Tectonic processes, which have initiated and maintained fan formation to times well within the Holocene Epoch, consisted principally of faulting. From the location of the fan apices, the deep fanhead trenches, and the gorges in the Barsar Fault zone, it is evident that this fault was responsible for the initiation of the Dun fan-forming processes. Once the topographic setting was formed the Lower Himalayan catchment basin of the fan axial stream became a zone of erosion and the Dun basin immediately below the break of slope became a zone of deposition. The two, then, together can be considered a tectonically initiated alluvial fan system (Figure 7.17).

The axial stream extending across the fault zone has provided the route for the flow of debris. These flows undergo a sudden decline of energy

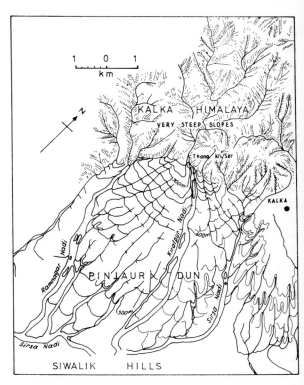

Figure 7.17. Kiratpur Fan, illustrating the basin alluvial system

at the break of slope in the topography and in the channel thalweg at the point of debouchement onto the adjacent plain. The average gradient of the stream within the mountains is approximately 1 in 6, representing a fall of 950 m over a distance of about 5 km: on the fans, the gradients are 1 in 20 or less (Figure 7.9). At the break of slope, the channel abruptly widens out. On the Kiratpur fan the channel widens out from less than 3 m in the gorge just above the apex (Figure 7.13A) to more than 30 m just below it (Figure 7.13B). Both the break of slope and sudden widening of the channel lead to reduction in velocity, capacity, and competence and finally to deposition of sediment. The break of slope and the channel widening are geological, the former related to tectonic and the latter to lithological conditions.

The Dun and the adjacent Lower Himalayan ranges have been subject to several episodic climatic phases since the Mindel–Riss Interglaciation (Allchin and Goudie, 1974: 360; Allchin,

1976: 471 and 485). Rainfall of the summer monsoon type was already established during the Mindel Glacial phase and continued through the later stages of the Pleistocene Epoch (Vishnu-Mittre, 1979: 34). The arid (glacial) phases brought rainfall of 125 to 250 mm, whereas almost 900 to over 1000 mm occurred in the more humid (interglacial) phases. These generated episodes of debris-flow deposits and water-laid deposits.

Fan deposition occurred, then, in semiarid to subhumid climatic conditions in which thundershowers, intense rainfall, and flash floods were characteristic. The climate was still relatively cold, indicating the lingering glacial phases in the adjacent high altitude zones (Mukerji, 1976a: 12–13). However, the Mindel–Riss Interglaciation, crucial for the initiation of the fan-forming processes, was more humid and longer than the succeeding glacial and post-glacial phases (Mukerji, 1976a: 10). That warm and humid conditions prevailed during the Mindel–Riss Interglaciation has been established on palaeobotanical and prehistorical evidence as well (Joshi, 1977: 4; Vishnu-Mittre, 1979: 29, 32 and 34; Mohapatra, 1979: 110–111).

A series of debris-flows characterized the period of initiation of the fans, and recurred in the subsequent Upper Pleistocene and Holocene episodes. With every debris-flow deposit near the apex, long tongues of debris radiated out while the axial stream channel filled up and was raised. The break of slope was maintained by spasmodic neotectonic movements, while the debris-flows continued to deposit and raise the stream channel.

The intense thunderstorm rainfall of the monsoon climate resulted in the formation within the mountains of floods heavily charged with sand, pebbles, and boulders which followed the debris-flow deposition. The floods caused deepening and widening of the channel at the mouth of the gorge and in the apex zone of the fan, changes in the course of the axial stream, and creation of anastomosed channels. During the late Holocene, with a longer interval between tectonic movements, the entrenchment of the axial stream has become too deep for the floods or debris-flows to top the earlier fan surface.

It is recognized that the debris-flow deposits which have contributed in major degree to the formation of the fans are related to heavy thunderstorm rain and flood surges out of the catchment basins (cf. Beaty, 1974: 48). This kind of rainfall is characteristic of both the semiarid and monsoon climate regimes which have alternated in the study area.

Conclusions

Located in the restricted topographical setting of the Dun graben valley, the Dun alluvial fans are surprisingly large and steep, and have impressive total relief which considerably exceeds the thickness of recent sediments. While the longitudinal profile of the fans is typically concave, the transverse profile is convex. The convexity becomes increasingly subdued with distance from the apex. On the fan surfaces are displayed systems of drainage lines which are radial but do not today originate from the apex. Two alternative models have been evolved to explain how the distributaries have developed this pattern.

The Dun fans are mainly composed of debris-flow deposits sitting with an angular unconformity on either the Boulder Conglomerate or the Pinjaur formations. The morphology, lithology, and fan formation are related to tectonic–climatic episodes which occurred during the period extending from the Middle Pleistocene Epoch and Mindel Glaciation up to the late Holocene Epoch.

Dun fan formation was initiated soon after the creation of the Dun graben by neotectonic movements along the Main Boundary and Barsar Faults early in the Mindel–Riss Interglaciation. By this time the typical summer monsoonal regime of rainfall had been established. Throughout the formation and development of the fans climate changed in episodic patterns, arid, semiarid, and humid, oscillating in succession during the Upper Pleistocene and Holocene Epochs. The uplift created the break-of-slope and climatic oscillations generated periods of high-energy, torrential streams and debris-flows. These conditions continued to exist during the Upper Pleistocene

(Riss and Würm Glaciations) and within the Holocene Epoch.

Only when the axial streams became deeply entrenched after a reduction in uplift rate on the Himalayan Front did the fan-forming processes cease to operate. Stream entrenchment occurred when debris-flow deposition near the fan apices failed to keep pace with fluvial downcutting. Fan formation completely ceased sometime during the late Holocene when the axial streams became so deeply entrenched that even the usually high floods could not reach the fan surface. There was no more deposition and the fans gradually changed into fossil geomorphic features.

The episodic oscillations of climate and neotectonic movements which have created the Dun fans, beginning from the Mindel–Riss Interglaciation and continuing into the late Holocene Epoch, conform with the model proposed by Beaty (1974: 49). They have been repeated in the formation of the fans in all the duns located between the Siwalik Hills and the Himalaya.

Acknowledgements

The author is grateful to Dr. Baldev Singh Karir and to Dr. Man Mohan Nath Kaul for helping him in conducting the fieldwork, to Professor Dr. R. S. Chaudhri for consultations, and to Mr. Mehar Singh and Mr. O. P. Sarna for cartographic work.

Editorial Note: This paper has been edited from a longer manuscript. The abridgement has been reviewed by Professor Mukerji.

References

Allchin, B. and Goudie, A. 1974. Pushkar: prehistory and climate change in western India. *World Archaeology*, **3**, 358–368.

Allchin, B. 1976. Palaeolithic sites in the plains of Sind and their geographical implications. *Geographical Journal*, **142**, 471–489.

Beaty, C. B. 1974. Debris flows, alluvial fans, and a revitalized catastrophism. *Zeitschrift für Geomorphologie*, **21**, 39–51.

Blissenbach, E. 1954. Geology of alluvial fans in semi-arid regions. *Geological Society of America Bulletin*, **65**, 175–190.

Bull, W. B. 1977. The alluvial fan environment. *Progress in Physical Geography*, **1**, 222–270.

Chaudhri, R. S. 1976. The problematic stratigraphical aspects of the Cenozoic sediments of north-western Himalaya—a critique. *Himalayan Geology*, **6**, 221–239.

Chaudhri, R. S. and Gill, G. S. 1979. Fabric and sedimentology of the Upper Siwalik Boulder Conglomerate exposed in the neighbourhood of Pinjaur (Kumaun Hills). *Geological Society of India Journal*, **20**, 83–89.

Dutta, L. H. 1984. Alluvial fans in the northern Brahmaputra Valley: a geomorphological interpretation. *Institute of Indian Geographers Transactions*, **6**(2), 1–15.

Fisher, R. V. 1971. Features of coarse-grained, high-concentration fluids and their deposits. *Journal of Sedimentary Petrology*, **41**, 916–927.

Gansser, A. 1964. *Geology of the Himalayas*. New York, Interscience.

Gaur, R. and Chopra, S. R. K. 1984. Taphonomy, fauna, environment and ecology of Upper Siwaliks (Plio-Pleistocene) near Chandigarh, India. *Nature*, **308**, 353–355.

Gill, G. T. S. 1985. Petrology of the Siwalik sequence exposed between the River Ghaggar and Markanda, northwestern Himalaya. *Centre of Advanced Study in Geology, Publications (N.S.)*, **1**, 43–58.

Joshi, R. V. 1977. Development of fluvio-glacial fans and cones in the Sub-Himalayan region during Quaternary Period and their archaeological significance. *Studia Geologica Polonica*, **52**, 195–205.

Karir, B. S. 1985. *Geomorphology and Stone Age Culture of North Western India*. Delhi, Sundeep Prakashan.

Khan, E. 1959. Stratigraphy, structure and correlation of the Upper Shivaliks east of Chandigarh. *Palaeontological Society of India, Journal*, **4**, 61–74.

Khan, E. 1970. The geology of the area between Chandigarh and Subathu. *Palaeontological Society of India, Journal*, **14**, 47–65.

Kharakwal, A. D. 1969. Petrological study of the Upper Siwaliks near Chandigarh. *The Indian Mineralogist*, **10**, 210–221.

Kulshrestha, B. D. 1976. Pleistocene climate of India. *The Geographer*, **23**, 17–48.

Lawson, A. C. 1913. The petrographic designation of alluvial fan formations. *University of California Publications in Geological Science, Bulletin*, **7**, 325–334.

Mohapatra, G. C. 1979. Pattern of cultural growth in prehistoric North-west India. In Chopra, S. R. K. (Ed.), *Early Man in Northwest India*. New Delhi, Allied Publishers. 107–135.

Mukerji, A. B. 1976a. Choe Terraces of the Chandigarh Siwalik Hills, India: a morphogenetic analysis. *Revue de Geomorphologie Dynamique*, **25**, 1–19.

Mukerji, A. B. 1976b. Terminal fans of inland streams in Sutlej–Yamuna Plain, India. *Zeitschrift für Geomorphologie*, **20**, 190–204.

Nakata, T. 1972. *Geomorphic History and Crustal Movements of the Foot-hills of the Himalayas*. Sendai, Tohoku University.

Nakata, T. 1975. On Quaternary tectonics around the Himalayas. *Science Reports, Tohoku University, 7th Series*, **25**(1), 111–118.

Nossin, J. J. 1971. Outline of the geomorphology of the Doon Valley, northern U.P., India. *Zeitschrift für Geomorphologie Supplementband*, **12**, 18–50.

Rachocki, A. H. 1981. *Alluvial Fans*. New York, Wiley. 161 pp.

Reineck, H. E. and Singh, I. B. 1973. *Depositional Sedimentary Environments*. Berlin, Springer-Verlag. 439 pp.

Sen, D. 1955. Nalagarh Palaeolithic culture. *Man in India*, **35**(3), 176–185.

Sen, D. P. 1971. Geology of Nahans around Kalka. *Himalayan Geology*, **1**, 251–258.

Valdiya, K. S. 1964. A note on the tectonic history and the evolution of the Himalaya. *New Delhi, 20th International Geological Congress, Proceedings*, 270–285.

Valdiya, K. S. 1973. Tectonic framework of India: a review and interpretation of recent structural and tectonic studies. *Geophysical Research Bulletin*, **2**, 80–114.

Vishnu-Mittre 1979. Environment of early man in north-west India: palaeobotanical evidence. In Chopra, S. R. K. (Ed.), *Early Man in Northwest India*. New Delhi, Allied Publishers. 20–66.

Williams, G. E. 1973. Late Quaternary piedmont sedimentation, soil formation and palaeoclimates in arid south Australia. *Zeitschrift für Geomorphologie*, **17**, 102–125.

CHAPTER 8

Morphology of the Kosi Megafan

K. Gohain
Oil India Ltd., Duliajan

and

B. Parkash
University of Roorkee, Roorkee

Abstract

Morphological features and processes of the Kosi river bed and megafan, with a radius of about 60 km in the Indogangetic plains, have been studied using Landsat images, air-photos, large scale topographic maps, and field surveys. Kosi River, within embankments on both sides in the plains, has developed four topographic levels corresponding to distinct discharges. The river shows downstream changes in channel pattern *viz.* gravelly–sandy braided (zone 1), sandy braided (zone 2), straight (zone 3) to meandering (zone 4), each zone being characterized by distinctive bed features and channel processes. High flow, receding flow, and wind activity are main processes which shape the river bed.

Photomorphological investigation of the Kosi megafan permits its division into Old Alluvial Plain and Young Alluvial Plain, and the latter can be divided further into Abandoned Channel Areas, Interchannel Areas, Low-lying Plain, Distal Plain, and Proximal Triangular Plain on the basis of surficial sediments and processes. The modern groundwater-fed streams and wind activity rework the older sediments of the megafan extensively.

Introduction

The Indogangetic plains are covered over large areas by alluvial megafan (Geddes, 1960). Some preliminary studies of the Kosi (Gole and Chitale, 1966) and Ganga (Hilwig, 1972) river megafans are available. However, no detailed descriptions of morphology or morphological processes acting on these fans are available. This paper describes some morphological features, including topographic levels, channel patterns, bed features, and topological analysis of the active Kosi river bed and different subunits of the Kosi megafan, which is the best developed megafan in North Bihar, India and Nepal (Figure 8.1). An attempt has also been made to describe the processes shaping the river bed and fan.

The Kosi is one of the largest braided streams of the world. The river originates in Tibet at a height of about 5500 m and, after flowing through the Nepal Himalaya, it emerges into the Gangetic plains. The river is also known as 'Sapt Kosi' (meaning literally seven Kosi) with seven tributaries, of which the major ones are Sun Kosi, Arun, and Tamur. These three tributaries meet at Tribeni (Figure 8.2). From Tribeni the river flows

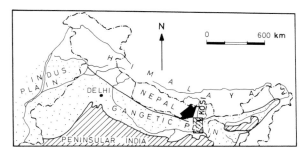

Figure 8.1. Regional setting of the Kosi megafan. Arrow indicates the area shown in Figure 8.2

through a narrow gorge for about 10 km until it debouches onto the plains at Chatra. Beyond Chatra the river runs through the alluvial piedmont region of Nepal for a distance of 40 km to Bhimnagar, after which it flows along the boundary between Nepal and North Bihar (India) for a distance of 40 km and then enters India. The river follows a curved path on the western extremity of its fan from Bhimnagar to the Mansi–Koparia railway line. Further downstream the river takes an easterly turn and flows parallel to the Ganga River for a distance of about 160 km before joining it a few kilometres downstream of Kursela (Figure 8.2).

The modern alluvial plain of the Kosi between Chatra and Bhimnagar is a steeply sloped triangular area with its apex at Chatra and the eastern and western boundaries being marked by a set of faults. Kosi River has formed a subcircular megafan with a radius of about 60 km downstream of Bhimnagar. The Kosi fan covers an area of about 16 000 km² in North Bihar (India) and parts of Nepal and lies between altitudes of 152 m a.s.l. in the north and 34 m a.s.l. in the south.

The fan is flat country like any other floodplain, the monotony of the landscape being broken by local features like the old courses now occupied by much smaller streams (locally known as *dhars*), old channel course lakes (locally known as *chaurs*), oxbow lakes, and, in places, dunes-like mounds along the abandoned Kosi courses.

The Kosi has been shifting its course drastically causing much destruction to life and property and has been, at times, described as 'sorrow' of Bihar. In fact, it has shifted 120 km westwards from a

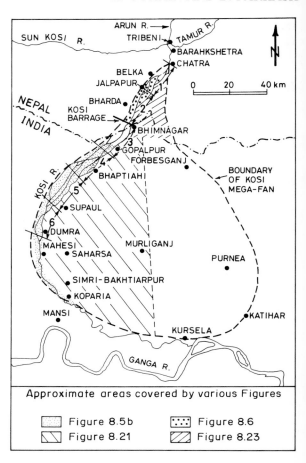

Figure 8.2. Map showing important locations on the Kosi megafan. Boundaries of reaches (1, 2, etc.) used in topological studies of the river also marked

course near Purnea in 1731 to its present position in about 220 years and, in the process, it has swept across almost the whole surface of the fan from one end to the other. The rapidly shifting behaviour of the river was first reported by Mookerjea (1961) in the form of a map showing positions of the river at different times in the past and then brought to prominence by publication of this map in Holmes' (1965) popular book. The name Kosi was derived from the ancient name *Kaushiki* and is related to its extremely shifting nature. *Kaushiki* was the legendary ascetic low caste woman who after being left by her *brahmin* (a person of the highest or priestly caste among the Hindus) lover, became frivolous and went to various places in quest of pleasure.

Regional Setting

GEOTECTONIC SETTING OF THE DRAINAGE AREA AND THE FAN

The drainage basin of the Kosi falls in the Nepal Himalaya which has been a subject of study by Bordet (1961), Gansser (1964), Hagen (1969), and Akiba *et al.* (1973). Rocks in the area are of Precambrian to Quaternary age and the major part of the watershed is occupied by metasediments of various grades. The major north dipping thrusts—the Main Central Thrust and Main Boundary Thrust—are present in the area and are active even at present (Nakata, 1982).

The development of the Kosi drainage is interesting. The tributary Arun originates in the Tibetan plateau, cuts through the Higher Himalaya and then flows through the Lesser Himalaya indicating that it was in existence prior to the rise of the Higher Himalaya during the Tertiary orogeny. Thus it has an antecedent drainage pattern.

The Kosi alluvial fan occurs in the eastern part of the Gangetic plain (Figure 8.1). On the north side, the Gangetic plain in the area of study is bounded by a set of almost E–W trending faults and some cross faults along which the Siwalik sediments (mid-Miocene to lower Pleistocene) are in contact with Gangetic plain sediments (Figure 8.3). Of these faults, F–F' is a prominent sinistral fault causing an offset of the Siwaliks by about 20 km. These faults are part of the prominent foothill fault (Himalayan Front Fault of Nakata, 1982) running along the southern foothills of the Himalaya and by analogy E–W faults are considered to be thrusts. South of the Gangetic plain in this region, Archaean rocks are exposed.

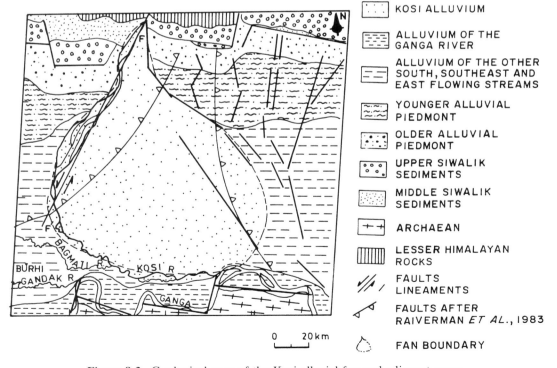

Figure 8.3. Geological map of the Kosi alluvial fan and adjacent areas

The contact between the two is marked by E–W faults and some cross faults and probably these are normal faults.

Fault F–F' striking northeast along the eastern margin of the Kosi fan and a set of faults trending northwest along the western margin of the fan have been confirmed by field observations and suggest that the Kosi fan area occupies a graben. However, geophysical studies of the Ganga basin (Sastri *et al.*, 1971; Karunakaran and Rao, 1976; Raiverman *et al.*, 1983) indicate that the major portion of the Kosi megafan lies on the Monghyr–Saharsa Ridge. Also, faults bounding the Ridge approximately parallel the Recent bounding faults of the Kosi fan. Thus, it seems that, though the Monghyr–Saharsa Ridge is generally a horst, recent movements due to reactivation of the basement faults are such that the area is acting like a graben.

HYDROLOGY, DISCHARGE, AND SEDIMENT LOAD CHARACTERISTICS

The catchment and fan areas receive 80 to 85% of the total rainfall from the southerly monsoons which break towards the end of May and continue until October. The average annual rainfall in the drainage basin increases from 1780 mm on the foothills to 3560 mm on the southern slopes of the Himalaya. It decreases to 250 mm at the northern end of the catchment. The fan receives an average rainfall of about 1450 mm. Mookerjea and Aich (1963) estimated that 74% of the discharge of the Kosi can be accounted for by precipitation in the form of rainfall. Because of non-availability of snow-survey data, contributions from snow-melt to the total runoff cannot be evaluated accurately.

The Kosi drains a catchment area of about 59 000 km^2 up to Chatra, 10% of which lies within the perpetual snow zone of the Himalaya. The catchment area of the main tributaries Sun Kosi, Arun, and Tamur constitute 32, 58, and 10% of the total catchment area respectively. Of the total annual runoff of the Kosi (52×10^9 m^3), 44, 37, and 19% of the runoff are contributed by the Sun Kosi, Arun, and Tamur respectively. The Kosi exhibits a highly variable discharge. The peak flows normally recorded from July to October following heavy rainfalls in the catchment areas and may be ten times as large as the mean discharge in a single year. During 67% of the time, flow is less than the mean annual discharge of 1600 m^3 s^{-1} (Figure 8.4).

The Kosi River carries a very high concentration of suspended load ranging between 2000 and 5000 mg l^{-1} and it may reach a maximum of 10 000 mg l^{-1} during monsoon months, which compares with some of the highest loads observed in the world. The mean annual sediment load of the Kosi is about 130 million cubic metres. Of the total sediment load carried by the Sapt Kosi, 42% comes from the Sun Kosi, 36% from the Arun, and 22% from the Tamur. Suspended sediment of the Sapt Kosi is commonly a function of discharge. It increases with the increase in flow, but the rate of increase falls at higher volumes of flow.

In 1953, a river training programme was initiated on the Kosi. Earthen embankments have been completed on both banks from Chatra to Koparia. Also, a barrage was constructed at

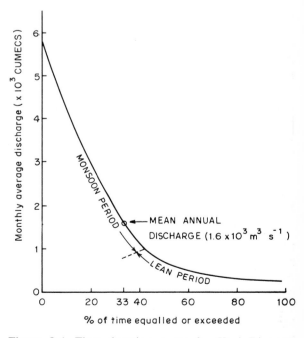

Figure 8.4. Flow duration curve for Kosi River at Barahkshetra

Bhimnagar in 1963 for feeding a system of canals irrigating almost the whole of the fan and these have a capacity of 425 m^3 s^{-1}.

Morphology of the River Bed

The term 'river bed' is defined as the area through which different channels of the Kosi have been migrating and depositing sediments in the recent past. The river has been confined between embankments since 1959. These embankments were constructed with due regard for the known activity of the river (except in the case of the Low-lying Plain which should have been kept within embankments, as discussed later). Thus the river bed width more or less coincides with the region between the embankments.

The river bed morphology has been studied by field investigations and interpretation of black–white air-photos with a scale of 1:25 000 taken on 13 to 15 November, 1973. The black–white air-photo coverage was available for the river reach between Gopalpur and Koparia (Figure 8.5A), a distance of 100 km. Modular multiband scanner (M^{2}S) air-photo coverage taken on 10 January, 1978 (scale 1:35 000) included the reach 10 km downstream of Chatra to Koparia. These air-photo studies were supplemented by work on a Landsat false colour composite photograph with a scale 1:250 000 (30 March 1975) for the area. Also topographic sections and maps of the river bed for the years 1956 to 1977 for the reach between Chatra and Koparia have been examined.

TOPOGRAPHIC LEVELS AND CHANNEL HIERARCHY

Four different topographic levels for the river reach between Gopalpur and Koparia have been recognized on the air-photos (Figure 8.5B). The results of photo-interpretation have been qualitatively confirmed by ground checks keeping in view the highly unstable nature of the river bed. M^{2}S air-photos have been used for the study of the topographic levels between Chatra and the Bhimnagar barrage (Figure 8.6).

Figure 8.5B shows the topographic levels and larger bed features of the river reach between Gopalpur and Koparia. These levels are best developed for the braided reach of the river. Characteristic aspects of the four levels are as follows:

Level 1—the active channel course with low bars;

Level 2—0.5–0.9 m higher than water surface on level 1 during the lean (low flow) period; no vegetation, submerges with a small increase in flow;

Level 3—about one metre higher than level 2, sparsely vegetated, submerged during high flows of floods;

Level 4—between 0.5 and 0.8 m higher than level 3, it comprises the surface of the islands and banks.

Levels 2 and 3 are relatively unstable and are dissected intensively every year during floods. Level 4 is covered by grasses and shrubs along the entire reach and parts are covered by dense vegetation including tall trees. Settlements and extensive agricultural activities are seen on this level. It is flooded during peak floods when the discharge at Barahkshetra (Figure 8.2) exceeds 8400 cumecs (0.3 million cusecs). At times, shallow (0.5 m) sheets of water covering large areas of this level are left behind by receding floods.

Topographic differentiation into levels has been reported from valley braided systems (Williams and Rust, 1969; Cant and Walker, 1978) and glacial outwash rivers (Boothroyd and Ashley, 1975; Church and Gilbert, 1975). The different levels in these rivers have been considered to represent various stages in progressive downcutting (Miall, 1977). The Kosi River with the artificial embankments can be considered to be similar to confined braided systems. However, the presence of various levels in the Kosi river bed is due to its response to dominant discharges. A flow-duration curve (Figure 8.4) using the average monthly discharge shows a distinct change in slope at about 1000 cumecs corresponding to flow of the 'lean' months of November to February when water submerges only levels 1 and 2. Levels 3 and 4 correspond to the monsoon discharges and extreme flood discharges respectively.

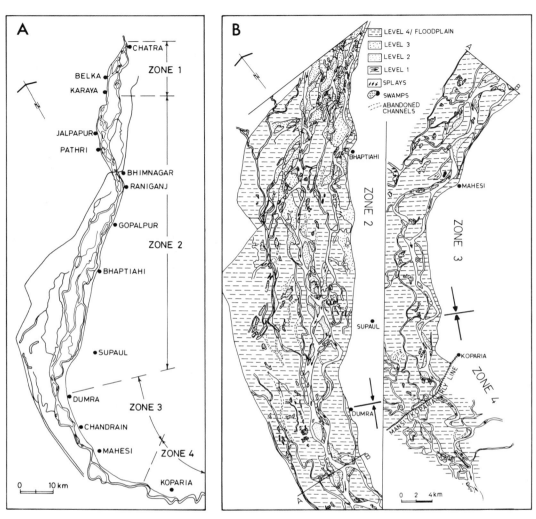

Figure 8.5. (A) Zones along the course of the Kosi River, and the confining embankments; (B) Morphological map of a part of the Kosi River bed, showing the four topographical levels

Channels are divided into primary and secondary channels. The primary channels are deep (from level-4 surface to low water surface depth is about 2 m) and carry water even at low stages (Figure 8.5B and 8.6). All along the course from Chatra to Dumra one or two primary, braided channels are present and further downstream there is a single, primary, straight, or meandering stream. The maximum distance (measured across the direction of flow) between the easternmost and westernmost subchannels of the primary channels is 2.2 km and usually it is much less. A subchannel is defined as a part of the river bed which has only bars with level 2 and bars/islands with levels 3 and 4 are not present. At a typical monsoon discharge of 5000 cumecs, levels 2 and 3 mid-channel bars are submerged and the primary channel becomes a single, wide channel. Subchannels of a primary channel may show a sinuosity up to 1.5. Three types of secondary channels are recognized. The first type are small meandering channels usually with widths less than 50 m. The second type are straight channels with alternate bars, having widths up to 100 m and average

Figure 8.6. Topographical levels interpreted from M^2S air-photos of Kosi River upstream of the Bhimnagar Barrage

depths of 1.5 m, and sinuosity of less than 1.2. These streams carry considerable discharge during monsoon months and during the rest of the year, a small stream (depth 0.50 m) meanders along the thalweg of the channel. The first and second types of secondary channel are confined to level 4 and carry water at flood stages only. The third type of channel develops over levels 2 and 3 during the falling stage of floods and these drain the wet sands. These streams are shallow and short, and modify the bars of these levels extensively.

CHANNEL PATTERN TRANSITIONS

On the basis of field and air-photo examination, the course of Kosi River can be divided into four zones (Figure 8.5A).

Zone 1 extends for about 20 km downstream of Chatra. The mean slope for the zone is 0.00045. One primary braided channel is present.

Zone 2 extends downstream of zone 1 for about 95 km to Dumra. It has a slope of 0.00048. This is the main braided zone of the river, having two primary channels separated by fairly stable islands with lengths and widths up to 28 km and 4.5 km respectively. Two or more secondary, straight channels and a few meandering channels on level 4 are present. Abandoned channels are also observed in this zone. Due to the effect of the barrage in this zone, the Kosi forms a single primary channel from 10 km upstream of the barrage to Bhaptiahi. Different types of bars separate the subchannels and no islands are present.

Zone 3 is a 40 km reach from Dumra to a few kilometres upstream of Koparia. The mean slope

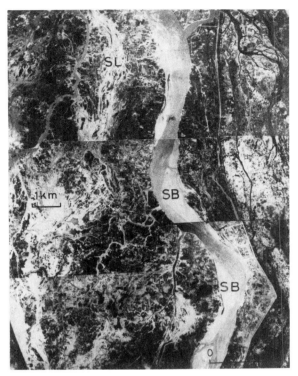

Figure 8.7. Air-photo mosaic showing the straight nature of Kosi River in zone 3. Splay channels (SL) seen in the floodplain. SB indicates a side bar

is 0.0001. One straight (sinuosity of about 1.01–1.16) primary channel is present for most of this zone (Figure 8.7). In the upper part of the zone the primary channel is divided into two unequal channels by an island of 7 km in length and 4 km in width. Along the channel for the whole reach, large triangular-shaped side bars are present. Islands up to 1.25 km in length and 0.5 km in width are present in the channel and small mid-channel bars are also exposed at low stage flow. In contrast to the downstream zone 4, the floodplain in this zone does not show any abandoned meandering channels. The area west of the channel is characterized by numerous small channels (some with meandering character) draining into the main channel. These small channels are incised into the floodplain, which is covered with muddy sediments. During rising floods, these channels carry water from the Kosi into the floodplain

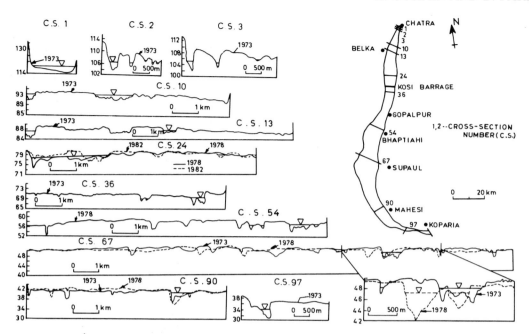

Figure 8.8. Bed profiles of Kosi River. Figures at the left side of the sections represent bed levels in metres above mean sea level

areas. During receding floods and low-stage flows, these channels drain the floodplain into the main Kosi channel. Thus these channels resemble channels of tidal flats and are considered to constitute a drainage pattern similar to a reticular pattern (Howard, 1967). Mud ponds and swamps are present in the floodplain area.

Zone 4 extends for a distance of 160 km downstream of zone 3 to Kursela (Figure 8.2). The mean slope for this zone is 0.00005. The stream meanders all along its course. Point bars and side bars are common and mid-channel bars emerge at low flow stage. Abandoned channels with chute cutoffs and neck cutoffs are common in the floodplain. In this zone, the Kosi runs parallel to Ganga River without breaking through its levees and thus forms the typical 'Yazoo' type drainage pattern (Howard, 1967).

The profiles of the Kosi river bed at different localities show that in most cases the channel cross-sections are asymmetric (Figure 8.8). Entrenchment of the channel near Chatra is obvious. Comparison of profiles for different years indicates accretion of the bed by a fairly uniform thickness of sediment at section 24, whereas lateral erosion and sedimentation are indicated in sections 67 and 90. Cross-section 67 shows erosion of a new channel by removal of sediment over a distance of 500 m up to a depth of 6 m from 1973 to 1978.

The width and width/depth ratio are small for the main subchannel of the primary channel of the Kosi near Chatra in the gravelly reach of zone 1 (Figure 8.9). These values increase in the downstream direction but drop in the distal part of zone 3. Almost an exactly opposite trend is shown by the depth. These inferences are based on the assumption that the average profile at a section remains the same with time, as the data used in Figure 8.9 are derived from various cross-sections (including some given in Figure 8.8) measured in different years (1973, 1978, and 1982).

The subchannels of the braided Kosi River in zone 1 carry gravels and level 4 is covered with sand to silt sized sediments. In zone 2 subchannels carry sandy sediments and level 4 is plastered with

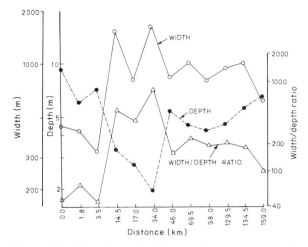

Figure 8.9. Variation of width, depth and width/depth ratio of the main channel of Kosi River with distance from Chatra in the downstream direction (note that the abscissa is not metrically scaled)

silt and a minor amount of mud. In these two zones, levels 2 and 3 are usually comprised of sediments similar to those in the channels. Thus zones 1 and 2 may be referred to as gravelly–sandy braided and sandy braided zones respectively. In zones 3 and 4, channel transport is mainly fine and true development of a floodplain covered with mainly muddy sediments takes place. In fact, level 4 of zones 1 and 2 passes into a true floodplain in zone 3 with an increase in the amount of clay in the surficial sediments. Type 1 secondary channels are most abundant in the floodplains.

Bed Features of the Channel Environment

The Kosi river bed can be broadly divided into two environments: channel and overbank. Different features recognized from channel environment are banks, islands, different types of bars (mid-channel bars, side bars, alternate bars, point bars, and linguoid bars), and bed form like megaripples and ripples. Brief description of these features are given below.

BANKS

Kosi River, as it flows over a coarse (gravelly or sandy) tract, does not have any permanent bank in its course from Chatra to Dumra (for all locations along the river refer to Figures 8.2, 8.5A, and 8.6). The heights of banks at low stage ranges up to 6.5 m near Chatra. Here, the right bank of the river is comprised of either the Siwalik rocks or a gravelly terrace, producing steep banks. The height of banks in the braided reach (zones 1 and 2) varies typically between 1.2 and 2.2 m. Further downstream, the banks consist of modern sands or old Kosi sands overlain by 1.3 m to 1.8 m of silty or muddy sediments in zones 3 and 4. At Koparia, the low-stage bank is 4.5 m high.

Low stage-flow causes removal of sand from the lower part of sandy banks. The banks first develop a shear fracture and then slump into the water (Figure 8.10A), if sandy. A high rate of retreat of 1 m hr^{-1} at low stage flow (560 cumecs) was observed for banks along a primary channel near Bhaptiahi. Banks with cohesive, finer sediments in the upper parts slump into the water and break into pieces. This results in vertical banks and vertical sides of mid-channel bars and side bars. This will naturally lead to widening of channels. The limit to which channel widening is carried is not known.

ISLANDS

Between Chatra and Koparia, there are many islands of different sizes. The length and width of these islands range from 1 to 29 km and 0.5 to 4.5 km respectively, giving a length/width ratio between 1.0 and 6.4. The size of islands becomes smaller in the downstream direction in zones 2 and 3. The surface of the islands forms the highest topographic levels of the area. Islands are vegetated—grasses and shrubs are most common and tall trees are abundant in places—especially in the reach upstream of the Bhimnagar barrage. In zone 1, abandoned channels are observed over the island surfaces. Where these channels are cut into gravelly surfaces, they are partially filled with sandy sediments. Thin layers of silt covering the

Figure 8.10. Morphological features of the channel environment. (A) Banks consisting of sand just before failure. Shear plane developed due to undercutting, Zone 2; (B) A linguoid bar eroded at the top by wind activity to expose internal structures, Zone 2; (C) A field of megaripples over a big linguoid bar. Megaripples in turn superposed by ripples. Locality: Bhimnagar; (D) Linear accumulation of gravels over the level 3 surface, covered by sand. Flow towards the observer. Locality: near Chatra; (E) Lobate accumulation of gravel over sand in an abandoned channel bed: 20 km downstream of Chatra; (F) An elongated depression in a shallow part of a subchannel being filled up at the upstream end by advance of two bars with crests (marked by dashed lines) at an angle of 50°, Zone 2

deeper parts of the abandoned channels are very common on level 4 in zones 1 and 2.

Islands and banks probably originate by a complex series of events of deposition and erosion. Vertical accretion by deposition of silts and muds during receding floods is a prominent process. Vegetation on the surface also helps in entrapment of fine material.

UNIT BARS

The term 'bar' is used to designate any exposed or slightly submerged major element of the river channel. Following Smith (1974), two types of bars—unit bars and compound bars are recognized. Unit bars have simple depositional histories and are essentially unmodified forms whereas compound bars have been moulded by a complex succession of depositional and/or erosional events. Various unit bars observed in the area are linguoid, diagonal and longitudinal bars. Linguoid bars (Collinson, 1970) or transverse bars (Smith, 1971, 1972) are common in the entire main braided and transitional reaches (zones 2 and 3) of Kosi River. These have well-defined slip faces which are generally convex in the downstream direction in planform (Figure 8.10B). The crests of the bars at places may also have margins which are straight, sinuous, or lobate in the downstream direction. At Bhaptiahi, a solitary linguoid bar was observed with length and breadth of 35 m and 8 m respectively. Closely related diagonal bars (Cant and Walker, 1978) are observed in the curved parts of channels.

In the abandoned channel beds small longitudinal bars can be recognized. These are composed mostly of gravel with some accumulation of coarse sand at the downstream end. These have rhombic shape and the surface of the bar slopes gently toward both the upstream and downstream ends.

COMPOUND BARS

The surfaces of compound bars correspond to both levels 2 and 3. Four types, mid-channel bars, side bars, alternate bars, and point bars are identified.

Mid-channel Bars

The primary channels are commonly divided by mid-channel bars, as mentioned earlier. Over the bar surface smaller bars are present with lobate slip faces and flat or elongated depressions along the upstream surfaces. This suggests the presence of transverse bars or linguoid bars. Large mid-channel bars in the main braided zone commonly have lengths of 3.3 to 5.5 km and widths of 1.7 to 1.8 km respectively. At places minor islands with development of level 4 (with vegetation) may occur as nucleus of the mid-channel bars. Mid-channel bars at Chatra are composed of gravels, with some sands deposited as pockets in the abandoned chutes and channels on the surface of the bars. Coarse sands may also cover large downstream parts of the bars, which gradually slope into the channels.

Chains of islands separating the primary braided channels in zone 2 and portions of banks far away from the primary channels are more stable than mid-channel islands and bars of the primary braided channels. However, the stability of all these features is of the order of a few years, as brought out by the comparison of cross-sections (Figure 8.8) and of post-flood planforms of the Kosi river bed (see *Comparison of planforms*) for different years. The presence of small areas with trees at level 4 is due to chance preservation of those areas during repeated migration of channels from one side of the river bed to the other. Mid-channel bars may appear independently or may form around the nucleus of the remnant islands.

Side Bars

Side bars are common in the main braided zone of Kosi River. Side bars are attached to the banks or sides of an island (Figure 8.7). These have no slip faces but slope gradually into channel. Their upper surfaces carry superimposed linguoid bars and ripples, which are most common on the upstream portion of the side bars.

Small side bars attached to the banks or mid-channel bars are also observed in the gravelly reach of the Kosi near Chatra. These have lobate to triangular lee side margins. Along the downstream margin, where flow is not strong, sands are deposited. The side bars in this reach may be marked by large chute channels. A slough channel may separate the bars from the bank at the downstream end.

Zone 3 is characterized by extensive development of side bars attached to the banks at regular intervals. The side bars are typically triangular in plan on air-photos, linguoid bars can be recognized in the active channel along the thalweg and exposed parts of side bars. In this zone one of the side bars can be identified to be a scroll bar with a slip face parallel to the bank very similar to ones described by Jackson (1976).

Alternate Bars

Bars attached to the banks of secondary straight channels (with sinuosity of about 1.2) draining level 4 in the main braided zone of the Kosi River are considered to be alternate bars. The size of these bars is fairly uniform and ranges from 0.5 to 1.5 km in length and 200 to 500 m in width. In plan they resemble elongate rectangles with their sides parallel to the stream. Slip faces of a large transverse bar moving into the slough at the downstream end of the alternate bar can be recognized at a few places but these are probably obliterated by low flow at most places. A number of linguoid bars can be recognized on the air-photos along the meandering thalweg of the channel. Field studies during the dry season of an alternate bar in a secondary straight channel west of Supaul revealed two sets of megaripples with amplitudes of about 20 cm—one set migrating parallel to the meandering thalweg and the other moving into the slough separating the bar from the bank at the downstream edge of the bar. Both alternate bars of secondary straight channels and side bars of zone 3 are considered to have originated in a similar way at high stages of flow.

Point Bars

Point bars are formed in the sinuous reaches of both the secondary and primary channels. Most small secondary channels showing the develop-

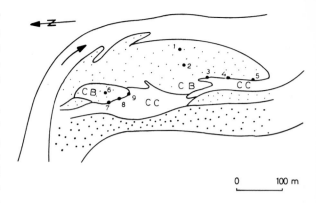

Figure 8.11. Sketch map of a point bar in Zone 4. 1, 2, 3, etc. denote trenching locations. CB indicates a chute bar; CC, a chute channel

ment of point bars have a sinuosity of more than 1.5.

A plan view of a point bar studied from zone 4 is given in Figure 8.11. The downstream portion of the bar is characterized by a small scroll bar with a slip face dipping westward across the flow direction of the main channel. Trenches at locations 1–5 (Figure 8.11) show that the scroll bar is migrating westwards with ripples on its back. Also, chute bars are observed at the upstream end of the point bar and trenching in this part indicates that chute bars are migrating downstream and forming planar cross-beds.

PRIMARY BED FORMS

Mainly two primary bed forms, megaripples and ripples, are observed. Megaripples (dunes and sand waves of Southard, 1975) have been observed as trains on the back of large linguoid bars (Figure 8.10C). Their chord length ranges from 0.8 to 2.5 m and the height varies from 10 to 30 cm. They have sinuous downstream margins with well-defined lee faces. Their upstream faces may be flat or slightly convex upwards. Small ripples are commonly superimposed on the back of the megaripples. The fronts of megaripples are marked by depressions which may be partially filled during low stage flow. Both in-phase and out-of-phase relationships within the trains of megaripples are observed.

Ripples are the smallest bed features. These occur on the back of large bed features (Figure 8.10C). Their amplitude is less than 8 cm and chord length commonly between 10 cm and 20 cm. Ripples may be straight crested, linguoid, lunate, or rhomboid type.

OTHER CHANNEL BED FEATURES

Linear accumulations of gravel surrounded by sand seem to represent edges of former small channels over gravel bars, where these channel have been almost filled with sand by the receding flows.

The top sandy surfaces of some bars in the gravelly zone of the river bed are marked by an accumulation of gravels in a linear fashion parallel to the current flow direction (Figure 8.10D) and by lobate patches (Figure 8.10E) of gravels. The gravels arranged in straight zones may result from linear transport of pebbles over sand by fluid vortices with their horizontal axis near the bottom. The origin of the gravel patches is not clear.

During late stages of receding flows, current parallel, shallow linear depressions (50 cm depth and about 4 m width) are observed. Two bars (diagonal) with bar crests at an angle of 50° and the axes asymmetrically inclined to the axis of the channel advance from the sides and gradually fill the depression from the upstream end. The crests of the two advancing bars do not join but an oval shaped depression is left at the upstream end of the channel (Figure 8.10F). Another feature is observed in a part of the channel with a slight curvature of the channel bank with convexity to the bank side. This leads to the formation of a slough (80 cm depth) and solitary bar with its crest parallel to the main flow direction (about 50 cm amplitude) is seen to migrate into it.

Crabs, lamellibranchs, insects, and earthworms are active in shallow parts of the channels and indeed over the exposed areas of the whole fan. Muddy areas of the river bed and fan are almost completely reworked by earthworms during the rainy season. Many different types of burrows, trails, and pellets are observed in silty and muddy areas.

Bed Features of the Overbank Environment

LEVEES

Levees are observed on the concave erosional banks of the meandering reach of Kosi River near Koparia, and can be recognized on air-photos by their medium grey tones. Well-defined levees similar to those of the Brahmaputra (Coleman, 1969) are not recognized in the braided zones of the Kosi, though parts of level 4 adjoining the major subchannels have a slightly higher elevation than the rest of the level.

SPLAYS

In the main braided zone of the Kosi numerous splays can be recognized on air-photos on level 4 because of textural differences, splay deposits (sands) being coarser than those covering level 4 (silts and muds). Splays are associated with all types of primary and secondary channels and may be confined to one or a network of extremely shallow channels. These shallow channels may be used to convey flood waters from a large subchannel to a small one over level 4 (Figure 8.12). At places, splays may be just spreads of a thin veneer of sediment (50 cm thickness) resembling fox tails in plan view and without any of the well-defined channels that are observed on the outer side of small bends in straight secondary channels. In contrast to braided zone 2, splays are usually confined to well-defined crevasses over the levees in zones 3 and 4.

FLOODPLAIN

As explained earlier, with increases in the clay content of surficial sediments of level 4, it passes slowly into a floodplain in zone 3. Muds are dispersed over large areas of the floodplain by receding floods. Mud ponds and swamps are present in this area. Type 1 secondary channels are most abundant. Abandoned channels with neck cutoffs and chute cutoffs are common.

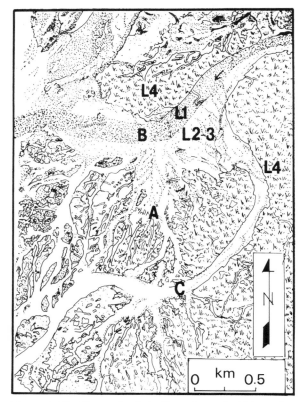

Figure 8.12. Network of shallow channels (A) used to convey water from a large subchannel (B) to a secondary channel (C) on level 4 during peak flood flows, Zone 2. L1, L2, etc. refer to different topographic levels

Modification of River Bed

The Kosi river bed is modified extensively during peak flows of floods. Due to the difficult and unpredictable nature of the river during floods, modifications made by the river during these periods are only inferred. The receding floods are next in importance in shaping the river bed and most of our observations about river bed changes are made on these flows. The other processes sculpturing the river bed are wind activity and wave action.

MODIFICATION BY FALLING STAGE CURRENTS

During the peak flow, larger bed features, like megaripples and linguoid bars are generated and these migrate. During the early recession of the flow, sheet flow can be observed over shoaling bars in which sediment is transported over a plane bed of the upper flow regime. Post-sheet flow is the late recessional flow, when some parts of the channel bed are exposed and are incised and dissected. In some cases, peak flow may pass into post-sheet flow during the recession of a flood, indicating a high rate of falling flow.

The Kosi river bed undergoes extensive modification during post-sheet flows when the water is channelled along favourable gradients. Parts of the channel under shallow water (80 cm depth) show great variation in direction of flow areally and with time. When the peak flow passes directly into post-sheet flow, levels 2 and 3 of mid-channel and side bars, with large areas covered by linguoid bars, are emergent during early falling stages. Delta lobes are built into depressions in front of low, shallow channels cut into the surface of the linguoid bars. Microdeltas form in parts of the bars with low relief. However, linguoid bars exposed during the post-sheet flows are extensively dissected and modified so that these become barely recognizable (Figure 8.13A). Reworking is maximum in the highest parts of the mid-channel bars. Post-sheet flows lead to accumulation downstream and leave behind linguoid bars (amplitude about 25 cm) and small channels filled from the sides over the surfaces of mid-channel and side bars.

During peak floods, large parts of the Kosi river bed in the main braided zone are covered, giving the appearance of one or two large primary channels. As the floods recede, the primary channel divides into two unequal channels. This process is probably continued to even low stage flows, when numerous subchannels appear. This is supported by measurements of discharge by the Kosi Project authorities in various subchannels, which differ vastly from one subchannel to another (Figure 8.14).

Unequal bifurcation of the main channel and lateral abandonment of the smaller channels is accompanied by net aggradation in the smaller channels. This observation is explained by laboratory experiments (Linder (1952) who found that at a stream fork the stream carrying the smaller discharge has the higher concentration of bed

MORPHOLOGY OF THE KOSI MEGAFAN

Figure 8.13. Modifications of channel bed features. (A) A rhomb-shaped bar emerged at low stage flow. The rhombic shape of the bar is due to modification of the bar by falling stage flows. Locality: Bhimnagar; (B) Sands from level 3 surface being deposited by wind on the adjacent, abandoned channel bed. Locality: Mahesi, Zone 3

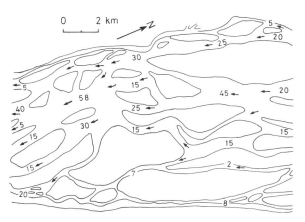

Figure 8.14. Distribution of flow into different sub-channels of Kosi River 8 km upstream of the Bhimnagar Barrage. Numbers indicate flow per cent

load and thus the smaller stream is more likely to become aggraded).

Level 3 surface may receive sediments by rainwash from level 4 on the adjoining islands and banks. Also gradual growth of vegetation helps entrapment of fines from the receding floods and level 3 is built up into level 4.

MODIFICATION DUE TO WAVE ACTION

Waves play no significant role in modifying the river bed, but they do generate some interesting structures. Waves generated by strong winds are active in shallow (20 cm) parts of the channels. Swash marks are formed due to wave activity on gently sloping banks of sand, adjoining the channel. These may be comprised of thread like ridges arranged parallel to the bank or form a net-like pattern of connected ridges. At times crests of current-formed ripples may be planed off or symmetrical ripples with crests parallel to the bank may be formed. Interference ripples consisting of current and wave-formed ripples may be present in shallow water. Sometimes the lee sides of bars show heavy mineral concentration in the form of ridges and terraces.

In the case of linguoid bars and megaripples, decrease in the slope of the lee side or formation of sand wedges is observed due to wave action.

WIND ACTIVITY ON EXPOSED AREAS

The Kosi river bed remains dry from October to May. Winds with velocities up to 40 km hr^{-1}, capable of transporting sand, are prevalent during the dry months of January to April. As a result, vegetation-free surfaces of levels 2 and 3 are subjected to continued modification. Dry sands are eroded away from the exposed areas and carried away until they are stopped by vegetation or some other obstruction. Migrating sands may be deposited as mounds. Slight elongation of these mounds is governed by the prevalent wind direction at a particular time as winds blow from E, NE, W, and SW during the dry months. At the edges of the mound-fields, sands may be carried in the form of a large bar with or without a slip face into the adjoining abandoned channels (Figure 8.13B). Migrating sands take up barchan or transverse dune or elongated ridge shapes. Wind forms ballistic ripples in coarse sand in zone 1 and at places interference ripples in medium sand due to changes in the prevailing wind direction. Aeolian sands may fill up depressions in front of linguoid bars.

Wind velocity is sufficient to erode slightly consolidated sand, thus exposing internal structures. Erosion may be up to 15–20 cm deep over large surfaces (Figure 8.10B).

Channel Pattern Changes and Sedimentation

An enormous literature on channel pattern changes with time is available for meandering streams (see review by Lewin, 1977). However, very few workers (Coleman, 1969; Williams and Rust, 1969) have carried out investigations on channel pattern changes in braided streams. The present study describes the channel pattern changes as deciphered from topographical survey maps for the period 1956 to 1977 over most of the braided and straight reach, 120 km in length, of Kosi River, and compasses them with trends of sedimentation.

CHANNEL PATTERN ANALYSIS

The river course downstream of Chatra has been divided into six blocks of 20 km length each, taking distances along the centre-line of the river bed. The blocks are delineated on the map of the area in Figure 8.2. Since a braided stream is a network of segments, its network can be studied in terms of topological theory as suggested by Howard et al. (1970) and Krumbein and Orme (1972). Following Howard et al. (1970) six topological parameters were measured for various blocks for different years. These are: (1) total number of channel segments including the segments lying entirely within the block and those bisected by the lines bounding the block; (2) total number of bisected channel segments; (3) total number of entire channel segments; (4) total number of unbisected islands within the block; (5) total number of nodes within the block; and (6) mean length of islands in each block. Also in the present study Brice's braiding index (1964) has been calculated from the Kosi Project maps, which do not differentiate non-vegetated bars and vegetated islands.

As expected, all topological parameters and braiding index show similar trends with time for a particular block and from block to block in a particular year, so any of these parameters can be used as an index of braiding. Thus only the plots for the total number of nodes are presented here and changes in its values are discussed in terms of variation in the intensity of braiding (Figure 8.15).

CHANGES IN BRAIDING INTENSITY WITH TIME AND WITH CHANGES IN ANNUAL FLOW

In the pre-barrage period (Figure 8.15A) the intensity of braiding showed various behaviour from block 1 to block 2. Further downstream, it steadily increased up to block 5 and then dropped off. The construction of the barrage seems to have affected the overall trends in the intensity of braiding (Figure 8.15B). The intensity of braiding increased from block 1 and block 2 upstream of the barrage, and blocks 3 and 4 just downstream of the barrage were then marked by a relatively sudden decrease in the intensity. The general downstream trend is that of an increase from blocks 1 to 2, a decrease from block 2 to 4, and then an increase in block 5 in the intensity of braiding. Further downstream the braiding intensity may drop off or increase.

An attempt has also been made to investigate changes in the intensity of braiding with time in the reaches upstream and downstream of the barrage. Figure 8.16 suggests that:
1. In the pre-barrage period, the braiding intensity for the reach upstream of the barrage varied widely with time without any trend and it showed a slight decrease with time for the downstream reaches.
2. In the post-barrage period, both the reaches upstream and downstream of the barrage exhibit a gradual increase in the braiding intensity with time until about 1972. Then there is a drastic fall followed by an increase in the intensity upstream.

MORPHOLOGY OF THE KOSI MEGAFAN 167

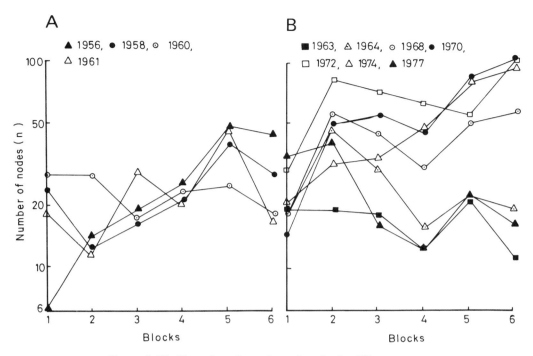

Figure 8.15. Plot of total number of nodes in different years

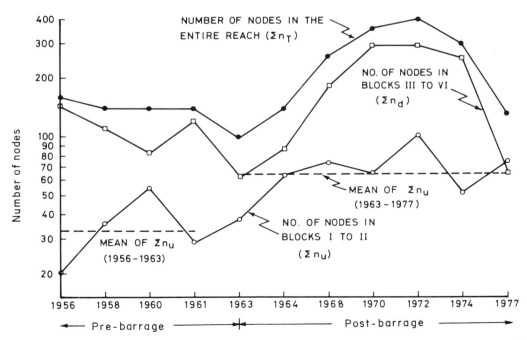

Figure 8.16. Plot of total number of nodes against time for reaches upstream and downstream of the barrage and the whole length under consideration

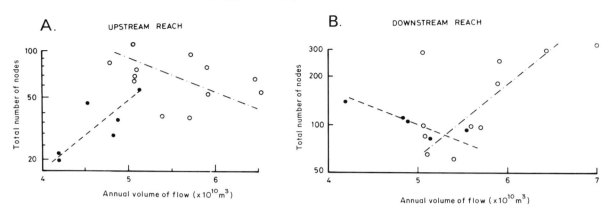

Figure 8.17. Plot of total number of nodes versus annual volume of flow in pre-barrage and post-barrage periods for the reaches upstream (A) and downstream (B) of the barrage. Note the change of ordinal scale

3. The braiding intensities for the post-barrage period for both reaches—upstream and downstream of the barrage—are higher than those for the pre-barrage period.

The reach upstream of the barrage shows an increase in the intensity of braiding with an increase in the volume of flow in the pre-barrage period (Figure 8.17), whereas in the post-barrage period it experiences a decrease in the intensity of braiding with an increase in the volume of flow. Exactly opposite relationships are observed for the reach downstream of the barrage. It should be emphasized that the yearly volume of flow is mainly a function of the volume of flow during monsoon months.

BRAIDING INTENSITY AND RATE OF SEDIMENTATION

Sanyal (1980) presented observations regarding the amount of sedimentation in the Kosi river bed for the period 1955 to 1974 pertaining to almost the same reach of the river as presently investigated. The channel was divided length-wise into six blocks of different lengths. Taking Chatra as the starting point and measuring distances along the centre line of the river bed, the downstream boundaries of the various blocks of his study fall at distances of 26, 40, 60, 80, 104, and 122 km. Results of bed level changes for the periods 1955 to 1963 and 1963 to 1974 for different blocks are shown in Figure 8.18. They indicate that, in the pre-barrage period, almost the whole observed reach of the river was degrading. However, the rate of degradation differed in different blocks.

Construction of the barrage at Bhimnagar in 1963 has led to high aggradation upstream of the barrage, but the area just downstream of the barrage has been marked by a small degradation. Further downstream, there has been an increased aggradation with distance from the barrage. Thus construction of the barrage has changed the predominantly degradational regime of the river to an aggradational one. It seems that the distal part of the fan has been a site of higher aggradation during pre- as well as post-barrage periods as compared to the proximal parts, though aggradation is less than that in the reaches on the Proximal Triangular Alluvial Plain after 1963.

Variations in the intensity of braiding in the downstream direction at a particular time are intimately related to long-term relative rates of sedimentation (Figures 8.15 and 8.19). In the degrading regions of the river bed, the less the degradation the higher the braiding intensity. For aggrading regions, higher aggradation is marked by higher intensity of braiding and in general the aggrading reaches are characterized by higher braiding intensities than degrading reaches. The relationship between the intensity of braiding in

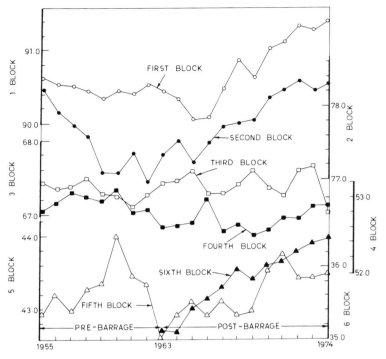

Figure 8.18. Bed level changes with time for different blocks of Sanyal's (1980) study. Note the individual scale for each plot

these reaches of the river bed and their relative degradation and aggradation is summarized schematically in Figure 8.19.

COMPARISON OF PLANFORMS

Comparison of the planforms of the Kosi (Figure 8.20) between the years 1960–61 and 1963–64 (pre- and post-barrage), for the reach upstream of the barrage indicates that there is no consistency in the formation or dissection of braid islands from year to year. However, in the post-barrage period the river width became broader in the reach just upstream of the barrage.

If the discharge exceeds 8400 cumecs, which is fairly common, the whole area between the two embankments is submerged. The river simultaneously silts up some of the pre-existing channels and islands on the one hand and dissects some other islands on the other hand, thus remoulding the whole river bed. It seems that the channel bed with braid bars, a few of which measure several kilometres in length, is in dynamic equilibrium with the discharge of the river and is constantly remoulded by it. Such a conclusion is also suggested by the fairly good relationship observed between the intensity of braiding and yearly volume of flow (Figure 8.17). However, these planform changes brought about by variation in monsoon discharges are considered to be superimposed on those brought about by longer-term aggradation and degradation.

High variability with time in the pre-barrage period, for the reach upstream of the barrage, without a trend, indicates that the Proximal Triangular Alluvial Plain area above the fan was used for temporary storing of the detritus just as the head trench is used in small fans (Bull, 1977). Slow currents deposited a fair amount of their loads in these sections and higher velocity currents associated with floods transferred the loads downstream. The steep slope of the area also helped in the onward transfer of material.

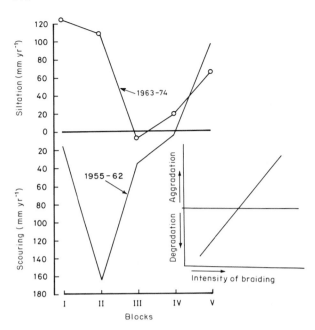

Figure 8.19. Plot of rate of scouring and siltation in blocks I to V (data from Sanyal, 1980). Inset: Generalized relation between intensity of braiding and aggradation/degradation

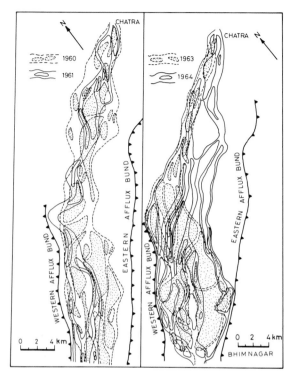

Figure 8.20. Comparison of the planforms of the years 1960–61 and 1963–64

Morphology of the Kosi Fan

The morphological features of the Kosi megafan have been analysed using Landsat images, air-photos, and field studies. About 3600 km² area of the western part of the fan has been mapped using air-photos on 1:25 000 scale (Figure 8.21).

YOUNG ALLUVIAL PLAIN (YAP)

The Young Alluvial Plain underlain by relatively young sediments showing little or no weathering and marked by white to light grey colours. It can be further divided into five subunits: Abandoned channels, Interchannel areas, Low-lying plain, Distal fan plain, and Proximal triangular alluvial plain.

Abandoned Channels

These represent the vestiges of the old Kosi Channels. During its sweep over the fan, Kosi River has left its imprint on the fan surface in the form of numerous abandoned channels, of which at least 16 distinct courses were recognized by Singh and Singh (1971). These channel courses exhibit low sinuosity, though some of the smaller channels show a sinuous pattern. Their widths range between 0.75 and 2.2 km and they are covered mainly by sandy sediments. The old sub-channels and bars of primary braided channels can be identified on air-photos. The abandoned channel courses no longer carry flow from the Kosi River and are presently drained by groundwater-fed streams. Most of these streams are perennial and flow along topographically low areas of the old abandoned courses. These streams have widths consistent with the present discharge characteristics, and have mostly meandering patterns with sinuosity ranging from 1.5 to 4.0. The widths are up to 10 m in the proximal part and increase up to 40 m in the toe

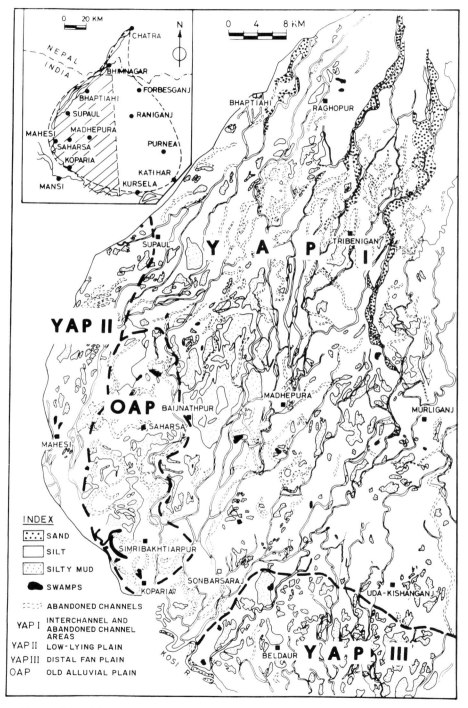

Figure 8.21. Photomorphological map of the western part of the Kosi megafan

Figure 8.22. Features of the Young Alluvial Plain. (A) A meandering, groundwater-fed stream with a point bar developed at the inner side of a bend. The point bar has two tiers, the lower one covered by fine sand and the upper one covered by grasses; (B) Sand mound developed over an old Kosi bar; (C) Hyacinth growth in a swamp of the Low-lying Alluvial Plain

part of the fan. Likewise, water depth ranges between 0.5 m in the proximal part and 1.5 m in the distal part during low flow stages. The width and depth of the channels increase manyfold during monsoon months due to the addition of surface runoff. The streams are incised in the distal part of the fan. Many of the groundwater-fed streams are interconnected and interchannel areas form permanent islands. Thus these streams exhibit an anastamosed pattern. They have well-developed point bars (Figure 8.22A) which generally show two distinct portions. The upper portion is usually covered by grass and is active only at high flows during monsoon months, whereas the lower portion of the bar is mainly comprised of loose sand and silt and it experiences sediment transport for a fair part of the year. Similar stepped profiles of point bars with 'two tiers' corresponding to two dominant discharges have been described earlier by Harms *et al.* (1963) and McGowen and Garner (1970). Migrating diagonal bars with an amplitude of about 30 cm and a distinct slip face, covering almost the whole width of the channel, are observed along the straight portion between the adjacent bends of small streams during low stage flows.

The contemporary streams have dissected valleys into the earlier bars and channels of the Kosi. Reworking of the earlier deposits is a significant process where the convex bends of the streams abut against the valley walls. Even at low stage flows silt to coarse sand is transported as bed load, indicating their substantial capacity to rework the earlier deposits.

The groundwater-fed streams have higher sinuosity in the eastern part than in the western part of the fan. The streams, especially the smaller ones, seem to degenerate with time mainly due to the growth of hyacinth weeds and filling up by fines washed in from the adjoining areas by rain water. Some streams have become swamps or have been almost filled up. This phenomenon is more common in the eastern parts than the western parts of the fan.

Large, old bar surfaces measuring up to 2 km in length and 0.75 km in width and dotted with sand mounds are observed at a number of places along the abandoned channel courses. The mound field is characterized by a relief of up to 2 m near the centre, from where the area slopes away in all directions. The local height of mounds ranges between 0.5 and 2 m. They are roughly circular (Figure 8.22B) and no preferred alignment can be deciphered. Trenches cut into sand mounds indicate that the upper 0.4 to 1 m thick sand of the mound has a distinct break with the underlying fluvial sand. The lithofacies and grain-size distributions of the sand of the upper parts of mounds described by Gohain (1984) suggest deposition by winds. The lower half of the relief of the mounds as observed in trenches is due to scooping out of sand from the fluvial deposits. Thus, the relief of mounds is due to both erosion and deposition. These areas are presently being extensively reclaimed for cultivation and probably they covered much larger areas than seen at present.

The origin of the circular shape of the mounds is thought to be related to the wind pattern in the area. Meteorological data from Purnea (Anonymous, 1953) indicate that winds blow sometimes from NE and E, and sometimes from the opposite directions SW and W in this area. Thus sand, instead of being transported out of the area, is heaped into mounds. In fact, similar sand mounds are forming presently at places along the active river bed.

Interchannel Areas

The interchannel areas constitute the largest morphological unit on the fan area and are highly disturbed due to human settlements and intensive agricultural practices used in the area. As a consequence of this, subenvironments of the old Kosi such as channel bars and banks are difficult to recognize. However, field observations reveal that settlements are concentrated on the high level surfaces. These high lands are underlain by coarser material. Some of the low-lying grounds remain water-logged for most of the year. Aquatic weeds grow and in the long run such areas are filled up by organic matter.

The surficial sediments of the interchannel areas consist of silty sands in higher parts, silts in large flat areas, and silty mud and organic materials in the swampy areas.

The abandoned channel areas constitute slightly higher regions than the interchannel areas and thus in a way are comparable to alluvial ridges (Fisk, 1952) of the meandering stream floodplains.

Low-lying Plain

A north–south trending plain exists in the mid-fan area along the eastern side of the Eastern Embankment of the Kosi (Figure 8.21). It extends from Supaul to the west of Simri–Bakhtiarpur. This area remains water-logged during a substantial part of the year. Many swamps with abundant growth of hyacinth are found in the area (Figure 8.22C). The surficial sediments of the area are silty muds with high organic content. It seems that the low-lying plain was a part of the modern Kosi river bed and has been separated from it by construction of the Eastern Embankment. Seepage of water from the active Kosi River, which is at a higher level as the result of sedimentation following the construction of the embankments, probably is responsible for water-logging in this area.

Distal Fan Plain

The distal fan plain lies in the southernmost part of the fan. This area is drained by southerly flowing, groundwater-fed streams which join the meandering Kosi River flowing eastwards at the foot of the fan. At low flood discharges water from Kosi flows into the groundwater-fed streams which may then flood the adjoining areas on the fan. At peak floods, the Kosi floodwaters cover large parts of the plain. These areas are covered by thick (over 3 m) deposits of silts and muds.

Proximal Triangular Alluvial Plain

This confined area occurs at the top of the circular megafan of the Kosi and is bounded by faults on both the eastern and western sides. A morphological map of the plain based on studies of the Landsat False Colour Composite (scale 1:250 000) and soil map (Anonymous, 1982) is presented in Figure 8.23. Diamond-shaped bars and subchannels can be recognized over most of this plain, except in a small area south of Chatra. Recognition of bars and channels is possible because this plain has been brought under cultivation only recently and features of the Kosi bed have not been obliterated. Bars are covered by coarse to medium sands and channels are overlain by finer sand and silts. A minor part of this plain is being reworked by the Sursar River, a groundwater-fed stream.

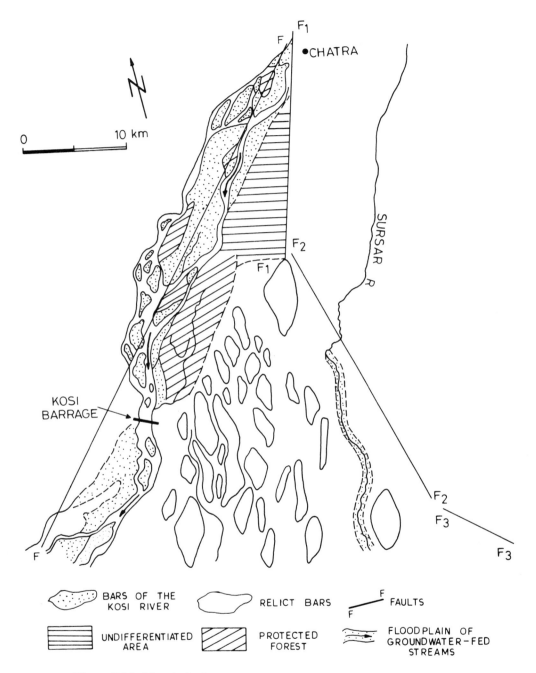

Figure 8.23. Morphological map of the Proximal Triangular Alluvial Plain

OLD ALLUVIAL PLAIN (OAP)

The Old Alluvial Plain is drained by southerly flowing, meandering streams (Figure 8.21). Many channel cut-offs, oxbow lakes, and swamps are found in the area (Figure 8.24) and these are sites of dense vegetation and accumulation of organic rich muds and silts. The surficial sediments, up to 2.5 m depth at most places, are medium to fine sand, yellow to orange coloured because they are thoroughly permeated with iron oxide. They apparently do not show any sedimentary structures. At places, large areas are covered by 15 to 20 cm thick, unweathered, fresh white sands. Groundwater-fed, meandering streams have formed inset sediment ribbons up to 1.5 m thick. Because of the higher weathering, the presence of iron oxide colouration, and even the apparent obscurity of sedimentary structures, sediments (except the fresh top layer and modern sediments) of this plain have been thought to be older than sediments of the YAP. In comparison, the sediments of YAP are usually light grey to white and show little or no weathering, and sedimentary structures can be observed easily in these.

Comparison of the topographic survey maps of the year 1937 with 1973 air-photos of the area indicate most of the oxbow lakes, chute cut-offs, and stream courses were present before the recent sweep of the Kosi past this area. This also indicates the older nature of the sediments underlying the OAP.

Shifting of the Kosi River and Drainage Evolution on the Fan

Kosi River has been thought to have changed its course by slow avulsion during its sweep from Purnea to Saharsa. Deeper parts of the channel were filled gradually and abandoned, whereas shallow parts of the channel were deepened and new areas were brought under the channel processes slowly as described by Gole and Chitale (1966). It is obvious that the sediments of the YAP have been deposited by the latest cycle of deposition as the river swept across the fan. However, the presence of the OAP on the fan suggests that the river has changed its course by discrete steps, also leaving large tracts without significant recent deposition of sediment. Such pattern of avulsion was probably due to neotectonics. The relief of the adjoining areas in the YAP and OAP is apparently similar and no older sediments are observed in trenches up to 3 m deep in the YAP, indicating that the OAP is probably a slightly higher block as compared to the rest of the fan. Though the whole of the fan area is subsiding, it seems that different parts of the fan may subside differentially.

During its recent sweep from Purnea to Saharsa, producing the sediments of the YAP, Kosi River had dominantly a braided character, as indicated by the older maps and sediments. Even during the deposition of sediments of the OAP,

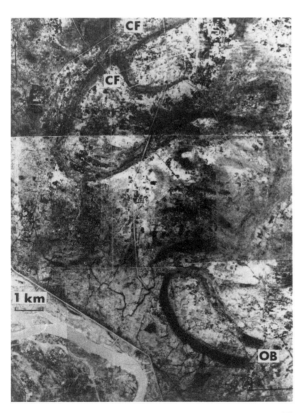

Figure 8.24. Air-photo mosaic showing channel fills (CF) due to chute cut-off and ox-bow lakes (OB) due to neck cut-off formed by groundwater-fed streams on the Old Alluvial Plain

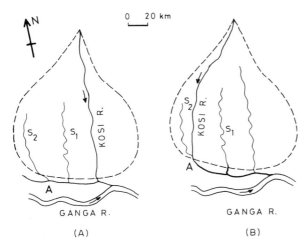

Figure 8.25. Drainage of meandering, groundwater-fed streams in the Distal Fan Plain

Kosi had a braided character as inferred from the mainly sandy character of the sediments. However, since the shifting away of the Kosi, many parts of the fan have been overtaken by meandering, groundwater-fed streams which are reworking the earlier sediments. Some of the groundwater-fed streams and older channels have become swamps or filled up with finer sediments. Meandering streams over the surface of the OAP have a higher sinuosity, and they have wandered over and reworked larger areas leaving behind oxbow lakes and abandoned channels, as compared to streams on the YAP. The OAP streams may have had a longer period to bring about these changes. Thus, in the case of unconfined, braided rivers like the Kosi in a humid region, with the shifting away of the main streams meandering, groundwater-fed streams may rework the older sediments and they may increase in sinuosity with time.

The entrenchment of the meandering, groundwater-fed streams in the distal part of the fan is explained as follows. When the Kosi earlier flowed roughly north–south in the eastern half of the megafan, on leaving the fan it took an easterly turn and flowed for a distance before joining the River Ganga. A groundwater-fed stream in the middle of the fan would join a smaller stream (in area A of Figure 8.25A) at the foot of the fan, which then joined the Kosi. When the Kosi shifted to the western half of the fan, it flowed through the area (area A in Figure 8.25B) at the foot of the fan where earlier a small stream was flowing and cut deeper because of the higher discharge. The old groundwater-fed stream or similar stream in its place meets the Kosi at the foot of the fan at a lower level and thus entrenches its course into the fan.

Conclusions

The following significant points emerge from the present study:
1. The active Kosi river bed between embankments behaves in a way similar to a valley-confined braided stream in that it has developed four topographic levels corresponding to distinct monsoon and lean period discharges.
2. The river is marked by distinct downstream changes in channel patterns *viz.* gravelly–sandy braided, sandy braided, straight, and meandering.
3. High and receding flows and wind activity are the main processes shaping the river bed.
4. Changes in braiding intensity of the river in space and time, as measured from topological parameters, are related to relative aggradation and degradation of different reaches.
5. During the latest sweep over the fan from east to west, the Kosi has shifted in discrete steps in addition to slow shifting.
6. The modern, groundwater-fed streams and wind activity are reworking the fan sediments. The floods of the meandering Kosi at the foot of the fan affect the distal part of the fan.

Acknowledgements

We are highly grateful to Dr. G. V. Middleton (McMaster University), Dr. P. F. Friend (Cambridge University), and Dr. Frank Ethridge (Colorado State University) for reviewing the manuscript and offering many critical comments. These comments have helped us immensely in improving the presentation of the paper. Funds for preparation of the manuscript from the O.I.D.B. Project at Roorkee University are thankfully acknowledged. The Kosi Project authorities provided generous assistance throughout the progress of the work. Sincere thanks are due to Mrs. Ela Gupta for translating some air-photos into line diagrams. Permission of the Survey of India to use some air-photos in publication is duly acknowledged.

References

Anonymous 1953. *Climatological tables of observations in India*. Government Central Press, Bombay. 558 pp.
Anonymous 1982. *Land Systems Maps*. His Majesty's Government of Nepal, Kathmandu.
Akiba, C., Amma, S., and Ohta, Y. 1973. Arun River region. In Ohta, Y. and Akiba, C. (Eds), *Geology of the Nepal Himalayas*. Saikon, Tokyo. 292 pp.
Boothroyd, J. C. and Ashley, G. M. 1975. Processes, bar morphology and sedimentary structures on braided outwash fans, northeastern Gulf of Alaska. In Jopling, A. V. and McDonald, B. C. (Eds), *Glaciofluvial and Glaciolacustrine Sedimentation. Society of Economic Paleontologists and Mineralogists Special Publication*, **23**, 193–222.
Bordet, P. 1961. *Recherches geologiques dans l'Himalaya du Makulu*. Editions Centre Nationale de la Recherche Scientifique, Paris. 275 pp.
Brice, J. C. 1964. Channel patterns and terraces of the Loup Rivers in Nebraska. *United States Geological Survey Professional Paper*, **422–D**, 41 pp.
Bull, W. B. 1977. The alluvial fan environment. *Progress in Physical Geography*, **1**, 222–270.
Cant, D. J. and Walker, R. G. 1978. Fluvial processes and facies sequences in the sandy braided South Saskatchewan River, Canada. *Sedimentology*, **25**, 625–648.
Church, M. and Gilbert, R. 1975. Proglacial fluvial and lacustrine environments. In Jopling, A. V. and McDonald, B. C. (Eds), *Glaciofluvial and Glaciolacustrine Sedimentation. Society of Economic Paleontologists and Mineralogists Special Publication*, **23**, 22–100.
Coleman, J. M. 1969. Brahmaputra River: channel processes and sedimentation. *Sedimentary Geology*, **3**, 129–239.
Collinson, J. D. 1970. Bedforms of the Tana River, Norway. *Goegrafiska Annaler*, **52**, 31–56.
Fisk, H. N. 1952. Mississippi River valley geology in relation to river regime. *American Society of Civil Engineers Transactions*, **117** (Paper 2511), 667–689.

Gansser, A. 1964. *Geology of the Himalayas*. Interscience, London. 289 pp.
Geddes, A. 1960. The alluvial morphology of the Indo-Gangetic plains. *Institute of British Geographers Transactions*, **28**, 253–276.
Gohain, K. 1984. *Sedimentation on the Kosi alluvial mega-fan, North Bihar, India and Nepal*. University of Roorkee, Ph.D. Dissertation (unpublished). 277 pp.
Gole, C. V. and Chitale, S. V. 1966. Inland delta building activity of Kosi River. *American Society of Civil Engineers Proceedings, Journal of the Hydraulics Division*, **92**(HY2), 111–126.
Hagen, T. 1969. Report on the geological survey of Nepal. Volume 1. Preliminary reconnaissance. *Denkschriften der Schweizerische Naturforschungs Gesellschaft*, **86**(Heft 1), 185 pp.
Harms, J. C., Mackenzie, D. B., and McCubbin, D. G. 1963. Stratification in modern sands of the Red River, Louisiana. *Journal of Geology*, **71**, 566–580.
Hilwig, F. W. 1972. Aerial photo-interpretation in relation to the physiography and soils of the Ganges Plain in Uttar Pradesh. *Appreciation Seminar, Indian Photo-interpretation Institute, Dehradun*. 6 pp.
Holmes, A. 1965. *Principles of Physical Geology*, 2nd edition. Ronald Press, New York. 1288 pp.
Howard, A. D. 1967. Drainage analysis in geologic interpretation: a summation. *American Association of Petroleum Geologists Bulletin*, **51**, 2246–2259.
Howard, Alan D., Keetch, M. E., and Vincent, C. L. 1970. Topological and geometrical properties of braided streams. *Water Resources Research*, **6**, 1674–1688.
Jackson, R. G. II 1976. Depositional model of point bars in the lower Wabash River. *Journal of Sedimentary Petrology*, **46**, 579–594.
Karunakaran, C. and Rao, A. R. 1976. Status of exploration of hydrocarbon in the Himalayan region—contribution to stratigraphy and structure. *Himalayan Geology Seminar, New Delhi. Geological Survey of India Miscellaneous Publication*, **41**(5), 1–66.
Krumbein, W. C. and Orme, A. R. 1972. Field mapping and computer simulation of braided-stream networks. *Geological Society of America Bulletin*, **83**, 3369–3380.
Lewin, J. 1977. Channel pattern changes. In Gregory, K. J. (Ed.), *River Channel Changes*. Wiley, Chichester, 167–184.
Linder, C. P. 1952. Diversions from alluvial streams. *American Society of Civil Engineers Transactions*, **118**(Paper 2546), 245–288.
McGowen, J. H. and Garner, L. E. 1970. Physiographic features and stratification types of coarse-grained point bars: modern and ancient examples. *Sedimentology*, **14**, 77–112.
Miall, A. D. 1977. A review of the braided river depositional environment. *Earth Science Reviews*, **13**, 1–62.

Mookerjea, D. 1961. The Kosi—a challenge in river control. *Journal of the Institution of Engineers (India)*, **42**, 117–142.

Mookerjea, D. and Aich, B. N. 1963. Sedimentation in Kosi—a unique problem. *Journal of the Institution of Engineers (India)*, Part CI, **43**(1), 187–198.

Nakata, T. 1982. A photogrammetric study of active faults in the Nepal Himalayas. *Nepal Geological Society Journal*, **2**, 67–80.

Raiverman, V., Kunte, S. V., and Mukherji, A. 1983. Basin geometry, Cenozoic sedimentation and hydrocarbon prospects in northwestern Himalaya and the Indogangetic plains. *Petroleum Asia Journal*, **6**, 67–92.

Sanyal, N. 1980. Effect of embankment on River Kosi. In *International Workshop on Alluvial River Problems. University of Roorkee, India*, 5.55–5.62.

Sastri, V. V., Bhandari, L. L., Raju, A. T. R., and Datta, A. K. 1971. Tectonic framework and subsurface stratigraphy of the Ganga basin. *Geological Society of India Journal*, **12**, 222–233.

Singh, R. L. and Singh, K. N. 1971. Middle Ganga Plain. In Singh, R. L. (Ed.), *India—A Regional Geography*. National Geographical Society of India, Varanasi. pp. 922–946.

Smith, N. D. 1971. Transverse bars and braiding in the lower Platte River, Nebraska. *Geological Society of America Bulletin*, **82**, 3407–3420.

Smith, N. D. 1972. Some sedimentologic aspects of planar cross-stratification in a sandy braided river. *Journal of Sedimentary Petrology*, **42**, 624–634.

Smith, N. D. 1974. Sedimentology and bar formation in the upper Kicking Horse River, a braided outwash stream. *Journal of Geology*, **82**, 205–224.

Southard, J. B. 1975. Bed configuration. In Harms, J. C., Southard, J. B., Spearing, D. R., and Walker, R. G. (Eds), *Depositional Environments as Interpreted from Primary Sedimentary Structures and Stratification Sequences*. Society of Economic Paleontologists and Mineralogists Short Course, **2**, 5–44.

Williams, P. F. and Rust, B. R. 1969. The sedimentology of a braided river. *Journal of Sedimentary Petrology*, **39**, 649–679.

CHAPTER 9

The Portage La Prairie 'Floodplain Fan'

William F. Rannie
University of Winnipeg, Winnipeg

Abstract

At Portage la Prairie, Manitoba, Canada, Assiniboine River emerges from a narrow valley onto the nearly flat floor of former glacial Lake Agassiz. The reduced gradient and loss of confinement have produced an atypical alluvial fan with a radius of 30–45 km, an average gradient of only 0.0005, and a maximum thickness at the apex of about 10 m. In addition to its relatively large size and low gradient, the fan is unusual in that most of the fan sediment is concentrated in alluvial ridges adjacent to palaeochannels which are similar in size and geometry to the modern Assiniboine, a perennial, meandering, suspended load river. Sedimentary facies are more representative of fine-grained floodplains than normal alluvial fans. The fan is the consequence of repeated construction and abandonment of these ridges and corresponding channels by levee-breaching during floods or, possibly, during ice jams. Deposition of the fan began early in the Holocene Epoch and has continued to the present with an average interval of about 1000 years for the formation of major alluvial ridges. In 1970, the opening of a diversion channel at the apex of the fan greatly reduced the likelihood of future avulsions and so fan formation has virtually ceased.

Introduction

In recent years, many traditional concepts of alluvial fan form and process have been revised. For example, the view that fans require an abrupt reduction in gradient has been questioned (Bull, 1977) and the association of alluvial fans with arid or semiarid climates has been weakened by descriptions of fans in humid-temperate and tropical climates (e.g. Kochel and Johnson, 1984; Kesel, 1985). This paper describes a feature deposited by Assiniboine River at Portage la Prairie, Manitoba, Canada (Figure 9.1), for which the term alluvial fan seems most appropriate but which has little in common with other fans and which appears to require further extension of the range of alluvial fan morphogenetic types.

Assiniboine River is one of the longest in western Canada, with a drainage area of 153 000 km^2 and a mean annual discharge of 47.3 m^3 s^{-1} (1914–84) at Headingley, Manitoba (Environment Canada, 1985). From its headwaters above the Manitoba Escarpment to Portage la Prairie, the course of the river is highly confined, first in a large glacial spillway and then, downstream of Brandon, in a deep, narrow valley (average gradient about 0.0006) excavated in glaciodeltaic sediments (Figure 9.2). At Portage la Prairie, Assiniboine River emerges onto the nearly flat floor of former Lake Agassiz and from that point to its junction with Red River in the city of Winnipeg, it is totally unconfined (Figure 9.2). Aggradation after this sudden loss of confinement and sharp reduction in regional gradient has pro-

Alluvial Fans: A Field Approach edited by A. H. Rachocki and M. Church
Copyright © 1990 John Wiley & Sons Ltd.

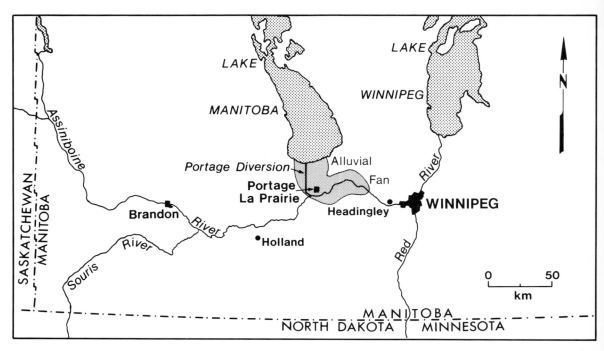

Figure 9.1. General location map

duced a fan-shaped veneer of alluvium over the glaciolacustrine clays of Lake Agassiz Plain. Although the alluvial origin of the materials has been recognized in descriptions of the surficial materials and soils, and aspects of the Holocene history of the area between Portage la Prairie and Lake Manitoba have been studied by Fenton (1970) and by Teller and Last (1981), little attention has been paid to the alluvial fan itself, possibly because its size and atypical characteristics obscure its true nature.

Assiniboine River Hydrology and Sediment Transport

Selected hydrologic, suspended sediment transport, and bed material characteristics of Assiniboine River are given in Table 9.1 and Figure 9.3 for three stations: near Holland 50 km upstream of the fan, Portage la Prairie near the fan apex, and Headingley 30 km downstream of the eastern fan margin (Figure 9.1). Because of the difference in time period, data for Portage la Prairie and Headingley are not strictly comparable with those at Holland. Data at the former two stations are given for the restricted period 1963–69 because the opening of a major diversion and control reservoir at Portage la Prairie in 1970 significantly altered the subsequent hydrologic and sediment regimes at these stations; the period 1963–69 was one of generally below-average mean discharge and the average suspended load for these years may underestimate long-term transport rate. In contrast, the longer time-period for the Holland data was one of above-average maximum and mean discharge and includes the two largest floods (1974 and 1976) observed on the Assiniboine since records began in 1913 (Environment Canada, 1985).

Compared with most other watersheds which terminate in alluvial fans, unit-area suspended sediment yield from the Assiniboine watershed is small, probably because of the small runoff ratios and relatively flat terrain which predominate throughout most of the watershed. At Holland upstream of the fan, the mean annual yield of 569 000 t represents only 3.7 t km^{-2} yr^{-1} aver-

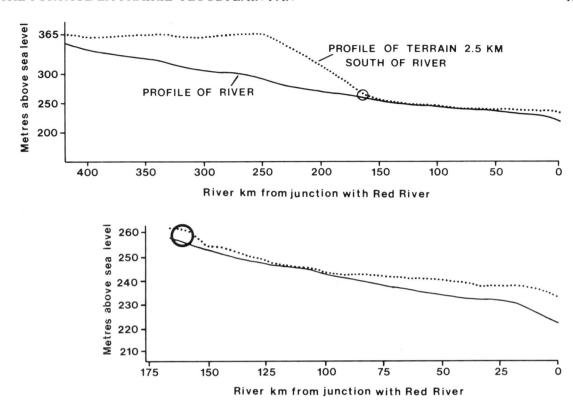

Figure 9.2. Longitudinal profile of Assiniboine River and adjacent terrain. The profiles from the fan apex to Red River are given in greater detail in the lower figure

Table 9.1. Selected hydrologic and sediment transport characteristics, Assiniboine River at Holland, Portage la Prairie, and Headingley

	Years	Holland	Portage la Prairie	Headingley
Drainage area (km^2)	–	152 000	152 000	153 000
Mean annual discharge	1963–69	–	39.5	39.8
(m^3 s^{-1})	1970–85	64.1	–	–
Maximum daily discharge	1963–69	–	626	595
(m^3 s^{-1})	1961–85	1460	–	–
Mean annual suspended	1963–69	–	509 000	452 000
sediment discharge (t)	1970–85	569 000	–	–
Maximum annual suspended	1963–69	–	1 187 000	888 000
sediment discharge (t)	1970–85	1 910 000	–	–
Maximum daily suspended	1963–69	–	103 000	57 000
sediment discharge (t)	1970–85	99 000	–	–
Mean annual maximum daily	1963–69	–	1797	1365
concentration (mg L^{-1})	1970–85	1229	–	–
Maximum daily concentration	1963–69	–	3080	2110
(mg L^{-1})	1970–85	3740	–	–

Data Sources: Environment Canada, 1980, 1985, 1987.

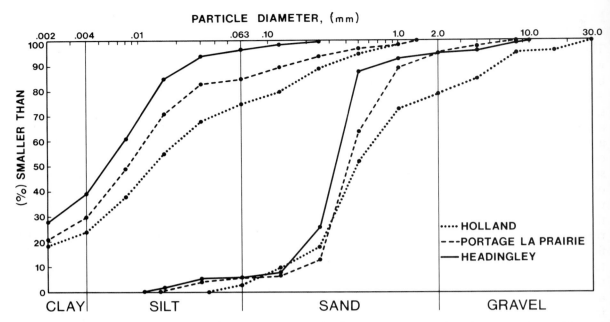

Figure 9.3. Average grain size curves for suspended load (left) and bed material (right) of Assiniboine River at Holland, Portage la Prairie, and Headingley (Source: annual volumes of sediment data for Canada published by Inland Waters Directorate, Environment Canada)

aged over the entire basin. Sediment transport rates during major runoff events are of course much greater, as in 1976 when the suspended load discharge at Holland was 1 910 000 t; from 1970 to 1985, annual yields of more than 1 300 000 t at Holland occurred four times (Environment Canada, 1987). Sediment concentrations and annual yields from the two largest tributaries, Souris and Qu'Appelle Rivers (which drain about 73% of the Assiniboine basin), are normally small (Environment Canada, 1987) and most of the sediment delivered to the fan appears to be acquired along the mainstem reach between Brandon and Portage la Prairie where Assiniboine River has excavated a major valley into glaciodeltaic sediments.

Average grain size curves of river bed material are given in Figure 9.3. At Holland, gravel (>2.0 mm) comprises about 20% of the bed material, but it is insignificant or absent at Portage la Prairie and Headingley where the river bed is almost entirely composed of sand. In 1971, Galay (1974) measured bed-material size at fifteen points in the 15 km reach downstream of Portage Reservoir (located at the fan apex). Gravel percentage ranged from less than 2% of the bed material at eight sample points to 20% at one location near the dam; average percentage over the fifteen locations was 7%. From the size distribution of sediments which accumulated in the Portage Reservoir between 1970 and 1977 (a period which includes the two largest, most competent, floods on record), Vitkin and Penner (1982) concluded that gravel constitutes only a minor proportion of the total sediment transported by the Assiniboine River, a conclusion which is supported by the paucity of gravel in the fan sediments (see below).

The calibre of the suspended load varies considerably both within and between years. On average, however, the proportion of sand declines progressively from about 25% at Holland, to about 15% at Portage la Prairie, to 3% at Headingley (Figure 9.3). At all three stations, more than 50% of the suspended load is in the silt sizes.

Fan Morphology

The topography of the fan, shown on Figure 9.4, was compiled from 1:24 000 maps with a contour interval of one foot[1] (0.3 m) for the area south of Assiniboine River (Manitoba Drainage Commission, 1921) and from 1:50 000 topographic maps augmented by miscellaneous benchmarks for the area north of the river. The convex pattern of contours suggests an alluvial fan with its apex just southwest of Portage la Prairie and a radial extent of 30 to 45 km. Near the apex, the fan is continuous through an arc of nearly 180°, but toward the distal end it consists of a series of alluvial ridges (elevated meander belts) which extend in fingerlike fashion onto Lake Agassiz Plain. Only to the northeast do higher elevations constrain the extension of the fan. The eastern edge of the alluvial sediments is approximated by the 785-foot (240 m) contour; to the north, the fan margin has been drowned by the southward encroachment of Lake Manitoba. Neglecting the drowned portion, the area of alluvium is about 1300 km^2.

The long profile of Assiniboine River is gently concave-up with no distinct breaks in slope except near Winnipeg where the river is entrenched as it approaches Red River (Figure 9.2). Sinuosity is high, generally exceeding 2.0 except in the Winnipeg reach and in a section on the lower fan. On the fan, natural levees as much as 3.5 m higher than the adjacent terrain have produced an alluvial ridge which is broad and dome-like near the fan apex but which becomes narrower and more sharply-defined downfan (Figure 9.5). No alluvial ridge exists between the fan margin and Winnipeg and in that reach the meander belt 'inverts' from an elevated ridge to a shallow depression (Figure 9.5).

The gradient of the fan surface is gentle, averaging only 0.0005 from the apex to the eastern fan margin; nevertheless this slope is significantly greater than that of Lake Agassiz Plain onto which the alluvium has been deposited (Figure 9.2). To the north and south of the present river course, the only topographic expression is provided by numerous palaeochannels and their associated ridges and scroll bars (Figure 9.7). Including the present Assiniboine, eight major courses of the river are identifiable on aerial photographs, soil maps, and topographic maps. These channels are morphologically similar to the modern Assiniboine but are now occupied by misfit streams with discharges at least an order of magnitude smaller than their meander geometries would suggest. All of the major palaeochannels radiate outward from the upper segment of the fan, some entering Lake Manitoba to the north, others trending eastward to Red River. Each is marked by an alluvial ridge with levees as much as 4.5 m higher than the adjacent floodbasin (Figure 9.8). Where they are well preserved, the scroll bars produce an undulating topography with a local relief of about 1 m, indicating active lateral migration where the channels are formed in relatively coarse-grained materials (Figure 9.9A). In contrast, channels on the lower fan and off the fan are bounded by cohesive lacustrine clay and lateral activity is inhibited or absent (Figure 9.9B).

As fan slope and the mobility of the major channels decrease downfan, meander amplitude, and wavelength decrease (Figure 9.9). On Figure 9.10, the mean and range of meander wavelengths are given for all channels south of and including modern Assiniboine River, grouped within 10 foot (3 m) contour intervals on the fan. Both average and maximum wavelengths show a consistent decrease until the channels encounter the underlying clays; thereafter to Winnipeg, no systematic change occurs.

Numerous minor channels with much smaller dimensions occur in the floodbasins between the major alluvial ridges, apparently as distributaries of the main channels. These secondary water courses lack the elevated ridges and scroll bars displayed by the main channels, however, and do not contribute significantly to the overall morphology of the fan.

Fan Sediments

Alluvial sediments overlie lacustrine Lake Agassiz silty clay and clay in a thin wedge which is

---

[1] In this paper, elevations are given in feet, following the source maps. However, distances and local dimensions are given in customary SI units.

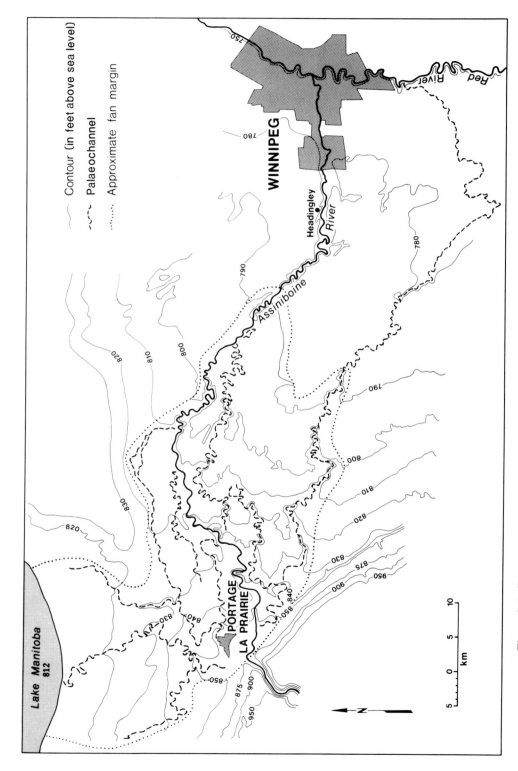

Figure 9.4. Fan topography and locations of major palaeochannels. Note that contours are given in feet

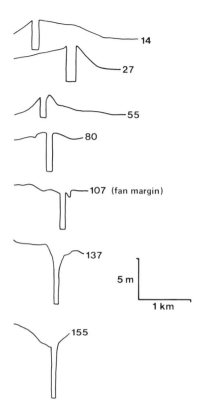

Figure 9.5. Cross-profiles of Assiniboine River channel. Numbers refer to river distance (km) from fan apex and locations of profiles are indicated on Figure 9.6

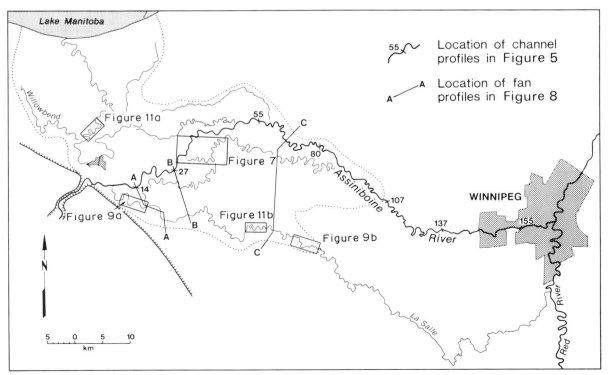

Figure 9.6. Locations of profiles, maps and photographs in Figures 9.5, 9.7, 9.8, 9.9, and 9.11

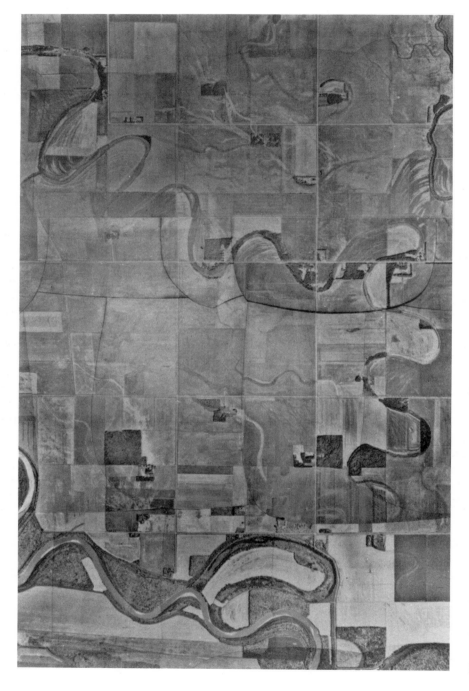

Figure 9.7. Air-photo mosaic showing major palaeochannel south of Assiniboine River (visible on left); note the crevasse splay on right. Location of photo is given on Figure 9.6. (Manitoba Department of Natural Resources Mosaic: Twp. 12, R 4W, 5 W)

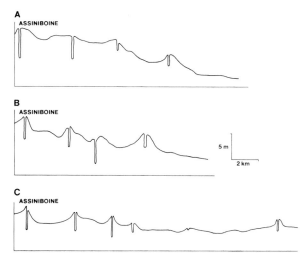

Figure 9.8. Cross-profiles of the fan south of Assiniboine River showing major alluvial ridges. Profile locations are given in Figure 9.6

continuous and 5–10 m thick at the apex (Gilliland, 1965) but which becomes thinner, discontinuous, and increasingly concentrated in the alluvial ridges toward the fan margin. Gravel is rare and virtually all fan sediments are medium sand-sized or smaller, sizes which are well represented in the suspended load of the modern Assiniboine River. Representative spatial patterns in the texture of the surface materials are shown on Figures 9.11A and B (modified from soil maps compiled by Michalyna and Smith, 1972, and Michalyna et al., 1982). The palaeochannels, scroll bars, and inner levees are underlain by well-sorted fine to medium sand and silty sand units in which cross-stratification may be observed. On the upper fan, extensive lateral migration of the channels has distributed these sand units over a wide area and they commonly underlie finer-textured surface sediments of overbank flow origin; lower on the fan, the sand is confined to the vicinity of the palaeochannel-point bar–levee complex. The proportions of silt and clay in the surface sediments increase with distance from the channels and the texture grades laterally through clayey silt on the outer flanks of the levee to silty clay and clay in the floodbasins between the alluvial ridges. These finer-textured materials deposited by overbank flow are also found in the swales of the undulating scroll bars, as lenses (frequently organic rich) within the scroll bar sand, and as a surficial drape over the coarse deposits of the abandoned channels. As the alluvium thins away from the channels and downfan, zones of lacustrine silty clay to clay 'show through', occupying progressively larger and more continuous areas toward the distal end of the fan (compare Figures 9.11A and B).

Fan-building Processes

The Portage la Prairie fan has been formed by lateral and vertical accretion processes typical of actively meandering rivers in floodplain settings. Major alluvial ridges have been repeatedly constructed and abandoned by avulsion but because the river is entirely unconfined downstream of Portage la Prairie, the abandoned ridges radiate outward, in contrast to the subparallel configuration of alluvial ridges in a typical confined floodplain.

No avulsions have occurred on Assiniboine River during historical time and the circumstances which produce such events can only be speculated upon here. Necessary preconditions for levee-breaching and permanent channel abandonment undoubtedly include a distinctly elevated meander belt and river stages near or above bankfull. The upper and middle fan reaches of Assiniboine River and its ancestors may have been particularly susceptible to avulsion because the highly mobile, large amplitude, elevated, meander belts of these channels are situated between the confined reach above the fan apex and the laterally inactive, relatively fixed downstream reaches.

Specific initiating events commonly associated with flooding include bank caving and erosion by concentrated overbank flow at low points in the levee. On the Assiniboine, where most significant floods are caused by snowmelt and spring rainfall, ice jams may play an important role. Ice jams observed on the Assiniboine (Figure 9.12) are normally short-lived events either because the ice jam breaks or because the volume of impounded

Figure 9.9. Contrast in lateral activity between palaeochannels (A) near fan apex (Manitoba Department of Natural Resources Mosaic: Twp. 11, R6W) where the channel is formed in sandy alluvium and (B) on eastern fan margin (Manitoba Department of Natural Resources Mosaic: Twp. 10, R3W) where the channel is bounded by lacustrine clay. Note the large difference in meander wavelength. Photos are located on Figure 9.6

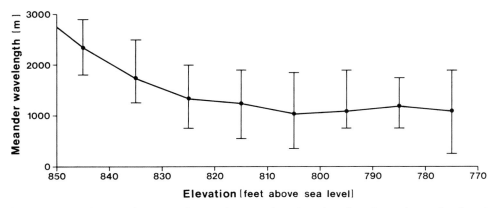

Figure 9.10. Mean and range of meander wavelengths of major channels south of and including Assiniboine River, grouped within 10-foot contour intervals

water cannot sustain protracted overbank flow. However the action of ice against the levee and the localization of overbank flow at the ice jam has the potential to produce a crevasse or breach in the levee which can be exploited by subsequent flood discharges. Ice jams have occurred frequently on the upper to middle fan reaches of the Assiniboine where natural sinuosity is high and numerous sharp bends exist. Since 1882, several of the most pronounced of these bends have been artificially cut off specifically to reduce ice jam formation and without this channel modification, it is likely that the natural incidence of ice jamming would be more frequent.

The sequence of events observed during a moderately-large flood in 1955 may be typical of the circumstances which lead to avulsion in the absence of human intervention.

> Following the break-up, a severe ice jam ... caused the water level at Portage la Prairie to rise on April 9 to its highest crest in the 41-year period of record ... Overflowing water spread south to the (La Salle) River causing flooding of short duration at several points along the river. (Meanwhile) heavy rains continued to increase the discharge ... (Between Portage la Prairie and Winnipeg) dykes defended both sides of the river ... On the north side ... the dyke was maintained except for a break near Poplar Point on May 20. This break was prevented from enlarging but water continued to escape north of the river for some time ... On the south side, a break occurred on May 22 about 10 miles (16 km) east of Portage la Prairie, quickly widening to about 1400 feet (427 m), and allowed about 25 per cent of the Assiniboine's discharge to flow south flooding a large area. (Morris, 1955, p. 7–10.)

Although the radial arrangement of channels gives the appearance of a distributary stream system typical of most alluvial fans and deltas, the modern Assiniboine behaves as a single channel river and the morphology of the abandoned channels suggests that they functioned similarly. Thus, it is hypothesized that the fan has evolved by the addition of successive alluvial ridges, each associated with a single active channel. Nevertheless, overbank flow during Assiniboine River floods does diverge since the ridges and the overall convex form of the fan cause water to spread away from the channel. Where the sites of water loss from the major channels have been sufficiently persistent, secondary 'channels' conveyed the water into and through the floodbasins between the ridges but these 'distributaries' appear to have been relatively unimportant in fan-building. Some of this water eventually returns to the river but a considerable portion may bypass the lower Assiniboine altogether, through channels northward to Lake Manitoba, and/or to Red River via the La Salle palaeochannel (Figure 9.6). The latter route is most common with the present channel configuration but significant flow between the Assiniboine and Lake Manitoba has been reported in several floods.

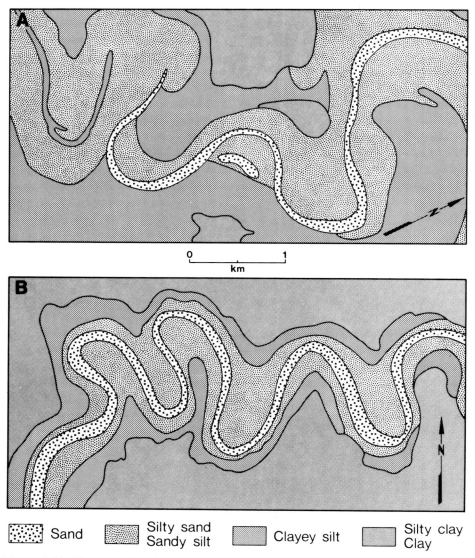

Figure 9.11. Texture of surface sediments (A) on upper fan, and (B) on lower fan. Locations of maps are given on Figure 9.6. (Based on soil series maps prepared by Michalyna and Smith, 1972, and Michalyna et al., 1982)

The evolution of the fan-channel system will be discussed in detail in a separate paper (Rannie et al., in press), but some aspects of this evolution are relevant here. The fan sediments overlie those of late-glacial Lake Agassiz and thus the fan is a Holocene feature which post-dates the final recession of the lake from the region about 9200 B.P. (Teller and Last, 1981). A ^{14}C date of 7030 ± 60 B.P. (TO–243) from the oldest recognizable channel (Willowbend: Figure 9.6) indicates that alluvial ridge formation had begun by that time, along a course which took flow from Assiniboine River northward to Lake Manitoba. Dates from several other channels suggest an average interval of about 1000 years for the formation and abandonment of major alluvial

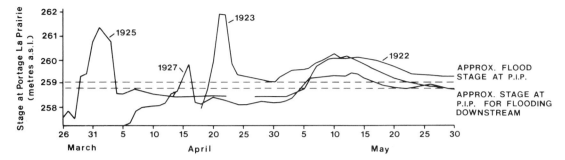

Figure 9.12. Hydrographs of Assiniboine River at Portage la Prairie during selected ice-jam episodes (modified from Craig, 1927)

ridges. Because the fan is situated astride the low divide between Lake Manitoba and Red River, the most notable consequence of channel abandonment has been to cause the ancestral Assiniboine to switch outlets from Lake Manitoba to Red River at least once, and possibly several times (most recently about 3000 yr B.P. via La Salle Channel) with potentially dramatic consequences for the water and sediment inflow to each.

Human modification of the flood regime of the modern Assiniboine has important implications for the continued evolution of the fan. Beginning early in this century, dyking and channel straightening increased the channel capacity in the vicinity of Portage la Prairie from less than 300 m^3 s^{-1} (Riesen, 1961) to about 636 m^3 s^{-1}. More important, however, was the completion of the Portage Diversion in 1970 which permits up to 708 m^3 s^{-1} to be diverted from Assiniboine River to Lake Manitoba. Because the peak discharge recorded on the Assiniboine between 1913 and 1973 was 614 m^3 s^{-1}, this spillway, together with the dyking and Shellmouth Reservoir on the upper Assiniboine, have virtually eliminated flooding on the alluvial fan under all but the most extreme conditions. Exceptional floods in 1974 and 1976 (which had peak discharges of 907 and about 1800 m^3 s^{-1} respectively) were successfully contained with limited emergency measures (Mudry et al., 1981) but without the Diversion, the 1976 event would have inundated much or all of the alluvial fan. Given the elevated state of the Assiniboine alluvial ridge, these floods might have triggered avulsive abandonment of the present channel and initiated a new phase of fan construction. Thus, in addition to fulfilling the intended objective of reducing flooding, the Diversion has had the possibly more important long-term effect of virtually removing the threat of future avulsion and ending the period of active fan formation.

Comparison with Other Fans

The Portage la Prairie fan exhibits few of the characteristics normally associated with alluvial fans. Most fans described in the literature are comparatively small features, seldom exceeding a few hundred square kilometres, with gradients in excess of 0.01 and ratios of basin area to fan area of 1–10, whereas Portage la Prairie fan has an area of *ca.* 1300 km^2, an average gradient of only 0.0005, and a basin–fan area ratio of more than 120. Bull (1977) suggested that the thickness of fan deposits should be more than 1% of fan length while the maximum thickness of Portage la Prairie fan sediments is only 0.02% of fan length. In addition, the literature on alluvial fans has emphasized their episodic sedimentation by braided or multiple distributary streams, their coarse sediments, the importance of debris-flows and increasingly, their active tectonic environment. In contrast, the Portage la Prairie fan has been produced by the repeated construction and abandonment of alluvial ridges by a perennial, meandering, single-channel, suspended load river with a modest sediment yield. The fan sediments contain no debris-flow material and little gravel, and exhibit sedimentary facies more typical of a fine-grained floodplain. The tectonic environment of the fan is stable.

Denny (1967) stated that the juxtaposition of highland and lowland areas is the 'prime requisite of fan formation' (p. 83) and Blissenbach (1954) felt 'bold relief' in the source area to be essential since

> only under such conditions will there be profound erosion and transportation together with a strong tendency for deposition as the mountain streams reach areas of low gradient. (Blissenbach, 1954, p. 177.)

Descriptive terms such as 'highland', 'lowland', and 'bold relief' are subjective and relative to the size of the system. The locus of the Portage la Prairie fan at the transition from the escarpment–glaciodeltaic environment to the nearly flat Lake Agassiz Plain may fit these criteria somewhat but the relief ratio, erosion rate, and unit-area sediment yield of the fan source area are small and as a consequence, the sediment transport characteristics of Assiniboine River are very different from those of most fanhead streams. Only in the loss of confinement of the fanhead stream adjacent to a low-gradient zone over which alluvium can be spread radially does the Portage la Prairie fan conform to the conventional view of alluvial fans.

The closest analogue to the Portage la Prairie feature may be the exceptionally large, low gradient Kosi River fan in northern India, but while the two features have some characteristics in common, the differences between them are substantial. In contrast to the single-channel, meandering pattern of the modern Assiniboine and its ancestors, Kosi River contains braided, anastomosed, and distributary reaches common on other fans (Gole and Chitale, 1966; Wells and Dorr, 1987a,b; Gohain and Parkash, this volume). In part, this difference in channel pattern may be attributable to the differences in the character of the sediments transported by the two rivers. The average sediment concentration of Kosi River is an order of magnitude greater than that of the Assiniboine. In its upper reaches, the Kosi channel sediments are coarse (boulders and cobbles) whereas the Assiniboine channel sediments and alluvium are almost entirely sand-sized or finer and the downfan diminution in particle size reported on the Kosi (Gole and Chitale, 1966) is not present in the Portage la Prairie fan sediments.

In conclusion, the Portage la Prairie feature has had a morphogenesis which has not been reported elsewhere but its radial configuration of channels, overall convexity of form, and locus at the sudden loss of confinement appear to satisfy the criteria for an alluvial fan. The writer proposes the term 'floodplain fan' and suggests that the spectrum of fan types be extended from the classic Death Valley types and their humid counterparts, through the Kosi River form, to encompass features such as the one described here, produced by normal alluviation by a perennial meandering stream which experiences a sudden loss of confinement.

Acknowledgements

The author is grateful to the University of Winnipeg for its financial support of this study and to Weldon Hiebert and Betty Harder for their assistance in the preparation of the figures and manuscript. A particular debt is owed to Mr. Harvey Thorleifson who contributed greatly to the development of this paper and who, but for a whim of his employer's publications policy, would be deservedly listed as co-author.

References

Blissenbach, E. 1954. Geology of alluvial fans in semi-arid regions. *Bulletin, Geological Society of America*, **65**, 175–190.

Bull, W. B. 1977. The alluvial-fan environment. *Progress in Physical Geography*, **1**, 2, 222–270.

Craig, J. 1927. *Report on the Assiniboine River Floods, April and May, 1927*. Canada Department of the Interior, *Dominion Water Power and Reclamation Service*, 5 pp. + App.

Denny, C. S. 1967. Fans and pediments. *American Journal of Science*, **265**, 81–105.

Environment Canada 1980. *Historical Sediment Data Summary, Canadian Rivers, To 1978*. Inland Waters

Directorate, Water Survey of Canada, Ottawa, 133 pp.
Environment Canada 1985. *Historical Streamflow Summary, Manitoba, to 1984*. Inland Waters Directorate, Water Survey of Canada, Ottawa. 257 pp
Environment Canada 1987. *Sediment Data, Manitoba*. Inland Waters Directorate. Water Survey of Canada, Ottawa. 69 pp.
Fenton, M. M. 1970. *The Pleistocene Stratigraphy and Surficial Geology of the Assiniboine River to Lake Manitoba Area, Manitoba*, M.Sc. thesis, University of Manitoba. 121 pp.
Galay, V. J. 1974. *Assiniboine River Degradation Investigation*. Manitoba Department of Mines, Research and Environment Management, Water Resources Branch, unpag.
Gilliland, J. A. 1965. *Geological and Ground Water Investigation for the Portage Diversion*. Manitoba Department of Mines, Research and Environment Management, Water Resources Division. 72 pp.
Gohain, K. and Parkash, B. 1988. Morphology of the Kosi megafan. This volume.
Gole, C. V. and Chitale, S. V. 1966. Inland delta building activity of Kosi River. *Proceedings, American Society of Civil Engineers*, **92**, HY2, 111–126.
Kesel, R. H. 1985. Alluvial fan systems in a wet-tropical environment, Costa Rica. *National Geographic Research*, **1**, 450–469.
Kochel, R. C. and Johnson, R. A. 1984. Geomorphology and sedimentology of humid-temperate alluvial fans, Central Virginia. In Koster, E. H. and Steel, R. J. (Eds), *Sedimentology of Gravels and Conglomerates. Canadian Society of Petroleum Geologists, Memoir*, **10**, 109–122.
Manitoba Drainage Commission 1921. *Topographic Maps of Manitoba for Townships 1–11, Ranges 2E–10W*. Winnipeg.
Michalyna, W. and Smith, R. E. 1972. *Soils of the Portage la Prairie Area*. Manitoba Department of Agriculture, Soil Report No. 17. 100 pp.
Michalyna, W., Langman, M. N., and Aglugub, C. 1982. *Soils of the West Portage and MacGregor Map Areas*. Manitoba Department of Agriculture, Soils Reports No. D20 and D44, 202 pp.
Morris, W. V. 1955. *Report on Assiniboine River Flooding*. Manitoba Department of Mines and Natural Resources, Water Resources Branch. 126 pp.
Mudry, N., Reynolds, P. J., and Rosenberg, H. B. 1981. Post-project evaluation of the Red and Assiniboine River flood control projects in the Province of Manitoba, Canada, *International Commission on Irrigation and Drainage, 11th Congress Grenoble, France, Special Session*, **R9**. 147–148.
Rannie, W. F., Thorleifson, L. H., and Teller, J. T. in press. Holocene evolution and paleohydrology of the Portage la Prairie alluvial fan. *Canadian Journal of Earth Science*.
Riesen, H. G. 1961. *Proposed Shellmouth and Holland Reservoirs and Portage Diversion*. Manitoba Highways Department, Water Control and Conservation Branch, 28 pp.
Teller, J. T. and Last, W. M. 1981. Late Quaternary history of Lake Manitoba, Canada. *Quaternary Research*, **16**, 97–116.
Vitkin, N. and Penner, F. 1982. *Portage Reservoir Sedimentation Study, 1982*. Manitoba Department of Natural Resources, Water Resources Branch, 19 pp. + App.
Wells, N. A. and Dorr, J. A. Jr. 1987a. Shifting of the Kosi River, northern India. *Geology*, **15**, 204–207.
Wells, N. A. and Dorr, J. A. Jr. 1987b. A reconnaissance of sedimentation on the Kosi alluvial fan of India. In Ethridge, F. A., Flores, R. M., and Harvey, M. D. (Eds), *Recent Developments in Fluvial Sedimentology. Society of Economic Paleontologists and Mineralogists Special Publication*, **39**, 51–60.

CHAPTER 10

Fan Deltas—Alluvial Fans in Coastal Settings

William A. Wescott
Amoco Production Company, Houston

and

Frank G. Ethridge
Colorado State University, Fort Collins

Abstract

A fan delta is an alluvial fan that progrades into a standing body of water from an adjacent highland. Modern fan deltas generally occur along narrow coastal plains adjacent to high relief mountains. These conditions frequently are associated with rift basins and the early stages of continental break-up along divergent plate margins, along island arcs on convergent plate boundaries, and in pull-apart basins in strike-slip settings. The classic cone-shaped morphology is best developed along microtidal marine coastlines and in lacustrine settings.

There are three major geomorphic zones on fan deltas: the subaerial fan, which is essentially an alluvial fan; the transitional zone, where fluvial processes and deposits interact with and are modified by littoral processes; and the subaqueous fan delta whose characteristics are least well known and can be quite variable. At least three depositional models are required to classify the known variations in fan-delta deposits. These include models for slope, shelf, and Gilbert-type fan deltas. Studies of fan deltas and ancient fan-delta deposits are important for understanding basin-margin tectonics and basin evolution in a variety of plate-tectonic settings and climates.

Introduction

Alluvial fans that prograde into standing bodies of water have been termed fan deltas (Holmes, 1965; McGowen, 1970). The characteristics necessary for the formation of fan deltas are high-relief terrain adjacent to a shoreline and high gradient, bedload streams flowing into a subaqueous basin. Where these criteria are met a coarse-grained, fan-shaped sedimentary deposit forms comprising subaerial, transitional, and subaqueous components (Wescott and Ethridge, 1980). Fan deltas are unique environments where the fluvial processes characteristic of alluvial fan sedimentation interact with the shoreline processes characteristic of the receiving basin. Therefore, they are sites of active, frequently high-energy sedimentation that straddle the important transition from subaerial to subaqueous processes and resultant facies. Deltas which form under these constraints are different from finer-grained deltas because both the subaerial and subaqueous portions are characterized by relatively high gradients. During the past several years there has been a growing interest in and marked increase in research on fan deltas. It has been recognized that

Alluvial Fans: A Field Approach edited by A. H. Rachocki and M. Church
Copyright © 1990 John Wiley & Sons Ltd.

their deposits are important components of basin-margin architecture and that they form significant hydrocarbon reservoirs.

Fan deltas occur in both lacustrine and marine settings. The first description of a fan delta deposit was by G. K. Gilbert (1890) of the shoreline deposits of Pleistocene Lake Bonneville. Modern lacustrine fan deltas are common on the margins of lakes in mountainous areas and are well developed along the margins of the African rift lakes. Marine fan deltas are constantly modified by waves and tidal currents. Yet they occur and maintain their characteristic fan-shaped morphology in a variety of marine settings from microtidal to macrotidal and under a wide range of wave regimes. Fan deltas may also, under special conditions, prograde into restricted environments such as bays (McGowen, 1970; Hayes and Michel, 1982) and the heads of fjords (Bogen, 1983; Syvitski and Farrow, 1983).

The purposes of this paper are to review the tectonic settings in which fan deltas form, summarize the processes that deposit and modify their sediments, and describe the characteristics of fan-delta deposits.

Tectonic Setting

Holocene fan deltas are located mainly along or near the margins of plate boundaries and along the margins of fault-bounded intracratonic seas and lakes. In terms of plate tectonic settings, fan-delta systems are common along divergent, convergent, and transform plate boundaries where the requisite relief conditions are generated.

Divergent plate margins are conducive to fan-delta sedimentation because of the relief established by the process of rifting. Fan-delta sedimentation in these settings is best developed during the pre-drift phase of basin evolution and shortly after drifting begins, before the basin shoulders collapse to form a passive margin. Excellent examples of fan deltas in pre-drift grabens can be seen along the margins of the large lakes such as Tanganyika and Malawi in the western arm of the East African Rift (Figure 10.1A). Fan deltas in marine settings along rifted margins where sea-floor spreading has occurred include those along the northeast coast of Baja California (Thompson, 1968; Meckel, 1975) and fans along the Red Sea coast (Hayward, 1982). Modern examples of fan deltas in aulocogens are difficult to describe since truly failed rifts, such as those associated with Atlantic rifting (i.e. Benue Trough), are filled in by rivers and have lost the high relief necessary for fan delta development. However, a possible example of a rift basin that may be considered as a proto-aulocogen is the Gulf of Suez. This basin is still underlain by continental to transitional crust and is not extremely active seismically. Fan deltas are present along both margins of this rift.

Convergent margin plate-tectonic settings include forearc, interarc, backarc, and retroarc or foreland basins. Fan deltas in a continental forearc setting have been described by Hayes and Michel (1982) from Cook Inlet, Alaska. Intraoceanic convergent plate settings with fan deltas in fore and backarc basins include the east and west coasts of Japan, eastern New Zealand (Wescott and Ethridge, 1980) and the New Hebrides (Figure 10.1B) (Dubois et al., 1973).

Wrench fault systems are characterized by both compressional and tensional elements in close proximity which may result in uplifted source terrains and subsiding pull-apart basins in close proximity (Mitchell and Reading, 1978). Hempton and others (1983) described modern fan-delta sedimentation in a continental lacustrine setting, the Lake Hazar pull-apart basin along the East Anatolian Transform fault in Turkey. The Dead Sea Rift is actually a left-lateral transform fault that extends from the northern Red Sea north to the Zagros suture zone (Garfunkel et al., 1981). Fan deltas have been described by Sneh (1979) from Late Pleistocene deposits along the western margin of the Dead Sea, a hypersaline lake. Along the southern end of the transform, coarse-grained sediments are fed into the marine waters of the Gulf of Aqaba by fan deltas.

Examples of ancient fan deltas from various tectonic settings, interpreted from their deposits in the stratigraphic record, have been summa-

Figure 10.1. Fan deltas from different tectonic settings. (A) Lacustrine fan delta on the northeastern coast of Lake Malawi, Tanzania; (B) Fan delta off the southern coast of Espiritu Santo, New Hebrides (photo courtesy of R. H. Griffen, from Ethridge and Wescott, 1984, reprint with permission of the Canadian Society of Petroleum Geologists); (C) Lynmouth fan delta on the south coast of the Bristol Channel, North Devon, England (photo courtesy of H. G. Reading); (D) Yallahs fan delta, southeastern Jamaica (photo by J. S. Tyndale-Biscoe from Wescott and Ethridge, 1980, reprinted by permission of the American Association of Petroleum Geologists)

rized by Wescott and Ethridge (1980) and Ethridge and Wescott (1984).

Fan Delta Processes

SUBAERIAL FAN

The subaerial portion of a fan delta is an alluvial fan, in all ways similar to the wholly terrestrial members of the alluvial fan–fan delta continuum. Alluvial fans have been classified as arid (dry) fans or humid (wet) fans (Schumm, 1977). Arid region fans are characterized by ephemeral fluvial discharge and are constructed of debris-flows, sieve deposits, and fluvial channel and overbank sediments. Kochel and Johnson (1984) have classified humid region alluvial fans as humid-glacial, humid-tropical, and humid-temperate fans. Although, humid region fans are dominated by 'normal' fluvial processes, short-term fluctuations in precipitation and rapid changes in discharge can also produce debris-flows on wet fans (Rust and Koster, 1984). The major sedimentary processes on modern humid-temperate fans in Virginia (Kochel and Johnson, 1984) and New Zealand (Pierson, 1980) appear to be debris-flows and avalanches which are triggered by intense rainfall. The resultant debris deposits consist of poorly-sorted, mud-supported gravels with inverse grading and poorly-developed stratification.

The depositional and erosional processes acting on subaerial fan-delta plains are common to all alluvial fans, and are observed on all alluvial and deltaic fans in similar tectonic settings and climates. Therefore, depositional models developed for alluvial fan sedimentation are appropriate and adequate for deciphering the depositional history of the subaerial portions of fan deltas. It is in zones where fluvial processes interact with those of the receiving basin that the truly distinctive facies association develops that allows fan-delta deposits to be distinguished in the stratigraphic record from those of alluvial fans.

LITTORAL FAN (TRANSITION ZONE)

Within this transitional zone, fluvial deposits are modified by littoral processes and interfinger with either lacustrine or marine sediments. Fan deltas in lacustrine settings are generally modified only by relatively low energy waves. However, fans on the margins of large lakes, such as Lake Tanganyika, are modified by very large waves generated by strong winds blowing down narrowly confined valleys over a long fetch. These conditions can also generate longshore currents which modify the deltas and transport sediments into and out of the littoral zone. Ancient lacustrine fan deltas have been described by Stanley and Surdam (1978), Gloppen and Steel (1981), Pollard and others (1982), and Nemec and others (1984).

Fans associated with ephemeral lakes provide a unique opportunity for alluvial fan and fan delta characteristics to become intimately associated with each other. Depositional conditions may vary from subareal to subaqueous over short periods in time and short distances. Fans in these settings are frequently associated with mud flats and salt pans (Hardie *et al.*, 1978). Although these features may be more alluvial fan than fan delta, they do incorporate marginal lacustrine facies and fluvial facies that have been reworked by shoreline processes into their dominantly arid alluvial fan architecture.

In marine settings, the classic fan-shaped morphology can develop in a broad range of hydrographic conditions. Fan deltas occur in the microtidal ranges of the Caribbean Sea (Wescott and Ethridge, 1980) and in macrotidal settings such as the south coast of the Bristol Channel, North Devon, Great Britain (Figure 10.1C) (Holmes, 1965) and along both shores of lower Cook Inlet (Hayes and Michel, 1982).

On wave-dominated fan deltas, such as the Yallahs delta in southeastern Jamaica (Figure 10.1D), the beach-shoreface is the most areally extensive environment of the transitional zone (Wescott and Ethridge, 1980; Wescott, this volume). Constructional beaches are generally broad and sandy with well-developed berms, wide backshore areas, and comparatively gentle foreshore slopes. Beach sediments are interbedded sands, granules, and pebbles. Destructional beaches are narrow, comparatively steep, and are composed of very coarse sands to cobble-size clasts. The very coarse deposits develop as a lag as finer sands are winnowed by constant wave action.

Fans along the southeastern coast of Alaska (Boothroyd and Ashley, 1975; Boothroyd, 1976; Boothroyd and Nummedal, 1978) prograde into a basin where both wave and tidal processes modify their marine transition zone. On these fan deltas the distal margins are characterized by tidal flats, barrier islands, and barrier spits. Galloway (1976) classified the Copper River delta, also in SE Alaska, as a mixed wave and tide dominated, humid-region fan delta. Major environments of this delta-complex (Reimnitz, 1966; Galloway, 1976) include: (1) distal fan, comprising marsh and swamp muds, and braided estuarine distributary channel fills; (2) tidal lagoon, with sand and mud flats incised by tidal channels; and (3) the shore, consisting of marginal barrier islands, breaker bars and middle shoreface sands, and lower shoreface sand and muds.

The facies associations that characterize the littoral (transition) zone of fan deltas can be quite variable and depend upon basin configuration, fluvial controls (both the frequency and magnitude of water and sediment discharge, grain size, slope), and intensity of wave and tidal currents. Hayes and Michel (1982) documented the effects fluvial and basin processes have on the geometry

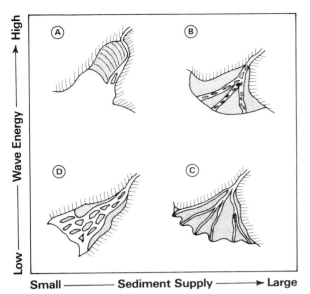

Figure 10.2. General relationship between fluvial input, sediment supply, and wave energy on fan delta morphology (after Hayes and Michel, 1982): (A) bayhead beach-ridge plain; (B) symmetrical, lobate fan delta; (C) fluvially dominated fan delta with elongated distributaries; (D) tide dominated delta

and nature of deposition on fan deltas in Lower Cook Inlet, Alaska. Although it is a macrotidal setting, the interaction of sediment supply and wave action accounts for most of the variation in fan delta morphology (Figure 10.2). High gradient, bed-load laden streams deposit symmetrical fan deltas where large waves impinge upon the delta front. Where wave energy is low more fluvially dominated fan deltas form with elongate distributaries separated by small embayments. In bedrock-bounded embayments lower gradient streams with low sediment discharge form bayhead beach-ridge plains where they are exposed to waves. In protected embayments tide-dominated deltas develop under these conditions.

In areas with similar wave and tidal regimes, the relation between stream gradient and the depth of water into which a fan delta progrades can have a pronounced effect upon sedimentation in the transition zone. These effects are graphically summarized in Figure 10.3. High gradient streams entering deep water build deltas with steep delta front slopes. In that setting (Figure 10.3A) wave energy is expended almost directly on the foreshore resulting in a narrow surf zone and the combined steepness of the delta front and fan surface allows for only a narrow intertidal zone. The Yallahs fan delta in southeastern Jamaica is an example where these conditions are met (Wescott and Ethridge, 1980; 1982).

In shallow water a major break in slope may not develop at the transition zone and wider surf and intertidal zones can develop (Figures 10.3B and 10.3C). The width of these zones depends upon the gradients of the delta front and fan surface; the steeper the gradient, the more narrow the transition zone. Where gradients are quite low, barred shorelines may form and the transition zone may be characterized by tidal flats, lagoons, and marshes. The Copper River fan delta is an example of a low gradient fan prograding into shallow water (Galloway, 1976).

Low gradient streams also form a steep delta front slope where they prograde into deep water (Figure 10.3D). This situation can result in surf zones with variable widths. During high tide there is a broader shallow platform for the waves to traverse and consequently a wider surf zone. However, during periods of low tide, the surf zone is much narrower because waves do not feel bottom across the steeper delta front until they are very close to the foreshore. The width of the intertidal zone depends upon the severity of the break in slope between the fan surface and the delta front. Two bayhead fan deltas with steep offshore slopes and broad intertidal zones have been described from Knight and Bute Inlets in British Columbia by Syvitski and Farrow (1983).

Accurate reconstruction of the palaeography of conglomeratic sequences in the stratigraphic record requires that alluvial fan deposits be distinguished from those of fan deltas, the possibility for which depends upon the establishment of criteria for recognizing the presence or absence of transitional zone facies. One significant aspect to this differentiation of alluvial fan and fan delta deposits is the recognition of beach gravels in the latter. Criteria for distinguishing beach and fluvial gravels have been summarized by Ethridge and

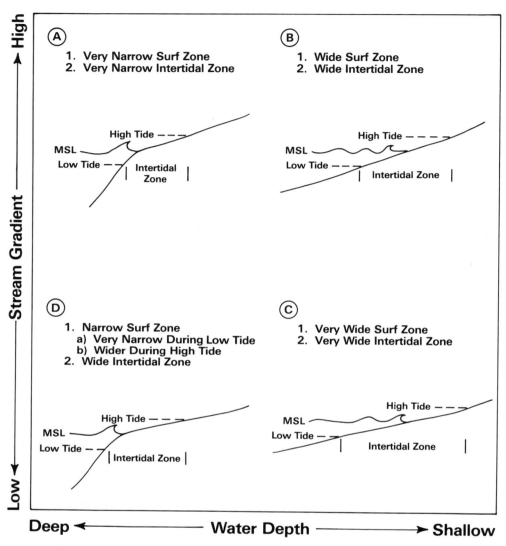

Figure 10.3. Diagram showing the generalized relationship between stream gradient and depth of water into which a fan delta progrades and consequent effects upon transition zone sedimentation. A similar wave and tidal regime are assumed for all four situations

Wescott (1984) and are reproduced here as Table 10.1.

SUBAQUEOUS FAN DELTA

The subaqueous component of fan deltas is the least well studied and most poorly understood portion of this depositional system. Processes that effect sedimentation and erosion in this environment are difficult to observe, especially those related to low frequency, high magnitude storms and other rare events that may result in considerable modification of the delta front. Because fan deltas generally form in tectonically active areas, structural characteristics of the basin margin play a significant role in the development and distribution of the delta front facies. In areas with relatively wide shallow slopes, normal fining-

Table 10.1. Criteria for distinguishing beach and fluvial channel gravels (Reproduced by permission from Ethridge and Wescott, 1984)

Beach gravels	Fluvial gravels
1. Well sorted	Poorly sorted
2. Well-segregated beds	Poorly-segregated beds
3. Continuous beds	Lenticular beds
4. Gravels interbedded with gravelly sandstones—rare	Gravels interbedded with gravelly sandstone—common
5. Erosional basal contacts—rare	Erosional basal contacts—common
6. Repeated small-scale fining-up sequences—rare	Repeated small-scale fining-up sequences—common to rare
7. Maximum clast size smaller than in adjacent channels	Maximum clast size larger than in adjacent beach
8. Clayey or coaly laminae—rare	Clayey or coaly laminae—common to rare
9. Imbrication—seaward dipping	Imbrication—landward dipping
10. Sphericity—low Roundness—high	Sphericity—high Roundness—low
11. Horizontal beds of different size gravels or swash laminations	Horizontal beds or high angle trough cross-beds

basinward delta-front facies are deposited. However, many fan deltas build onto very steep slopes or prograde to the edge of fault scarps. The slopes of fan deltas in these settings are significantly affected by (1) seismic activity, (2) steep gradients, (3) high sedimentation rates which increase pore water pressure, and (4) frequent sediment failure by slumping and mass-gravity sliding (Wescott and Ethridge, 1982).

A sedimentation model for the subaqueous portions of fan deltas dominated by conglomeratic mass flows has been developed by Postma (1984), and Postma and Roep (1985) by studying side-scan sonar images of modern arctic deltas and from ancient examples in southeastern Spain. Postma (1984) subdivided the subaqueous delta into three zones: delta front, delta slope, and prodelta. The delta front is the portion above wave base, the delta slope is characterized by a relatively steep gradient and is transitional between the upper delta front and the prodelta, which dips gently toward the basin floor. Mass flow processes are predominant on the subaqueous zones of conglomeratic fan deltas.

Slumping is frequently initiated by the development of high pore water pressures in sediments during their accumulation. Evidence of high pore water pressures may be preserved in delta-front sediments as water-escape structures (Postma, 1983). In this setting, thin, small debris-flows fill slide scars on the steep delta slope. Thicker, larger-scale flows move farther down the delta and are deposited on the lower slope and prodelta. This results in the following stratigraphic sequence (Postma, 1984): steeply dipping conglomerates in slide scars at the distal end of the delta front; gently dipping conglomerates on the delta slope; and prodelta muds.

Where fan deltas build out from faulted basin margins the location of the first submerged fault scarp limits their distal growth. Two examples of how faults control delta front sedimentation will be discussed here, the first from a lacustrine setting and the second from a marine basin.

Crossley (1984) described two deltas that prograde into Lake Malawi from the adjacent western highlands of the East African Rift. Off Dwangwa River the first fault is approximately 7 km lakeward from the western lake margin. Consequently, a large shallow delta front with a gentle slope has been deposited on the intervening lake floor. However, farther to the north the North Rukura delta has prograded to the edge of a submerged fault scarp. Here the river discharges its coarse sediments directly onto the steep, fault-controlled slope. Bathymetric profiles show the slope to be

Table 10.2. Comparison of submarine fan and fan-delta deposits (Reproduced by permission from Ethridge and Wescott, 1984)

Submarine Fan	Fan-Delta
1. *Slope deposits*—mainly hemipelagic lutites; rare lenticular, massive channel conglomerates and sandstones or olistostromal mudstones and sedimentary breccias. 2. *Inner fan deposits*—one or more discrete channel conglomerates and sandstones with erosional bases and fining and thinning-upward sequences grading into overbank turbidite sandstones and lutites; bulk of deposits consist of hemipelagic lutites interbedded with thin turbidite sandstones. 3. *Mid-fan deposits*—bulk of deposits consists of repeated thinning-upward cycles of turbidite sandstones resulting from deposition in multiple shallow braided channels. 4. *Outer fan deposits*—characteristic deposits consist of hemipelagic lutites and repeated thickening-upward cycles of turbidite sandstone beds which have sheet-like geometries.	1. *Subaerial delta plain deposits* resemble alluvial fan deposits and usually consist of braided stream conglomerates and sandstones and/or coarse-grained meanderbelt deposits. Fine-grained floodplain deposits are rare. In arid regions, debris-flow and/or sieve deposits may constitute a significant fraction of the proximal deposits. 2. *Delta front and shallow submarine fringe deposits*—depending upon the interplay of fluvial and marine processes and climatic factors, deposits may consist of beaches, spits, tidal flats, shallow marine bars, delta foresets and/or carbonate reefs or algal mounds. 3. *Shelf and slope deposits*—in shallow seas shelf deposits consist of muds interbedded with shallow nearshore submarine bar sandstones. Fans that build to the shelf edge overlie slope deposits similar to those described in (1) under submarine fans.

characterized by irregular topography, which suggests that slumping is quite common. Crossley (1984) presumed that much of this coarse sediment is transported into much deeper water by mass flows.

The Yallahs fan delta in southeastern Jamaica progrades into the Caribbean Sea up to the faulted edge of Yallahs Submarine Basin (Burke, 1967). Bottom reflection profiles show that the Yallahs delta front is characterized by two bathymetric zones, a narrow island shelf and a steep slope dissected by the heads of at least three submarine canyons (Wescott and Ethridge, 1980, 1982; Wescott, this volume). The refraction and diffraction of waves across the canyons and shelf control the sediment distribution on the shallow delta front. The narrow shelf zone is approximately 600 m wide with a flat bottom. Mean grain size of sediments decreases from the shoreline seaward across this shelf. In the slope zone fluvial sediments and sediments transported by a westward longshore drift are trapped in the canyon head depressions or temporarily stored on steep, unstable slopes between the canyons. Gravity processes periodically move these sediments downslope and into the deep water Yallahs Basin.

As the above examples both suggest, fan deltas can be intimately associated with deep water facies, including submarine fan and turbidite deposits. Although in modern settings there is no difficulty in separating the two depositional systems, the distinction is not always clear when interpreting conglomeratic sequences in the rock record. Both systems contain poorly-sorted, lenticular conglomerates and sandstones in proximal areas and more sheet-like conglomerates and sandstones in distal areas, and generally display overall coarsening upward sequences. Criteria for differentiating these two depositional systems have been summarized by Ethridge and Wescott (1984) and are listed in Table 10.2.

Depositional Models

Fan-delta deposits generally fit or are variants of one of three depositional models; slope-fan deltas, shelf-fan deltas, and Gilbert-type fan deltas (Ethridge and Wescott, 1984). The sedimentary architecture of fan deltas can be quite variable depending upon the width of the shelf and the interaction of fluvial and marine or lacustrine processes, climate, eustacy, and associated depositional systems.

All fan deltas, regardless of tectonic setting, have certain characteristics in common. Individual fan deltas are relatively small physiographic features with the subaerial fan usually occupying an area covering only tens of square kilometres. The landward margin of fan-delta deposits is almost always fault bounded, with proximal deposits unconformably overlying bedrocks. The overall geometry of a stratigraphic sequence of fan-delta deposits is a wedge or prism of coarse-grained conglomeratic sediments that thins away from a mountain front. Texturally and mineralogically, fan-delta deposits are immature to submature, reflecting the close proximity of bedrock source area. Depositional sequences usually coarsen upward. This relationship is better developed in shelf- and Gilbert-type than in slope-type fan deltas. The thickness and areal extent of fan-delta deposits depend upon the complex interplay of mountain-front uplift, sediment supply, and basin subsidence. The thickness of individual fan-delta deposits may be on the order of tens of metres; however, thick fan-delta clastic wedges developed along the margins of plate boundaries over considerable periods of geologic time may have cumulative thicknesses on the order of several thousands of metres and extend up to several tens of kilometres from the mountain front.

SLOPE MODEL

A generalized depositional model of a slope-type fan delta, based on studies of the Holocene Yallahs fan delta in southeastern Jamaica (Wescott and Ethridge, 1980; 1982) is illustrated in Figure 10.4. This sequence is characteristic of fan deltas that prograde onto an island or continental slope or build out to the faulted margin of a basin. The overall coarsening-upward sequence may not always be as well developed in slope-type fans because: (1) they are truncated by the shelf/slope break; and (2) coarse-grained slump deposits associated with the heads of submarine canyons may be locally developed just seaward of the non-marine to marine transition zone.

Proximal fan deposits are characterized by poorly-sorted, massive conglomerates that result from surge flows in confined channels (McGowen and Groat, 1971). These are interbedded with crudely horizontally bedded, landward-imbricated conglomerates and conglomeratic sandstones that result from deposition on longitudinal bars in shallow braided streams. Distal subaerial fan deposits consist of crudely horizontally bedded, landward-imbricated conglomerates and conglomeratic sandstones, and rare trough cross-bedded conglomeratic sandstones of braided stream origin. Minor amounts of fine-grained, organic deposits may represent deposition in isolated coastal lagoons and/or ponds. The non-marine–marine transition zone is characterized by well-sorted and segregated conglomerates and sandstones that display horizontal bedding, swash laminations, and seaward-dipping clast imbrication. Fossils and bioturbation structures are often abundant in marine sandstones. Slope deposits consist of marine muds and matrix-supported conglomerates which are frequently distorted by slumping. Processes active in these steep slopes include debris-flows, mud flows, liquifaction transport, and submarine sliding (Prior *et al.*, 1981).

SHELF MODEL

An idealized depositional model of a fan delta prograding onto a shelf, based upon numerous studies of fans and fan deltas along the southern Alaska shoreline (Reimnitz, 1966; Boothroyd and Ashley, 1975; Galloway, 1976; Hayes and Michel, 1982) is illustrated by Figure 10.5. A typical proximal to distal succession of sedimentary structures on the subaerial portion of a small glacial outwash fan delta is: (1) poorly-bedded, well-imbricated, poorly-sorted, and coarse-grained gravels; (2) imbricated, fine-grained gravel, and planar laminated and planar cross-bedded sand; and (3) interbedded planar cross-laminated sand and ripple-drift sand (Boothroyd, 1976). Fans that have not built directly onto the coast often

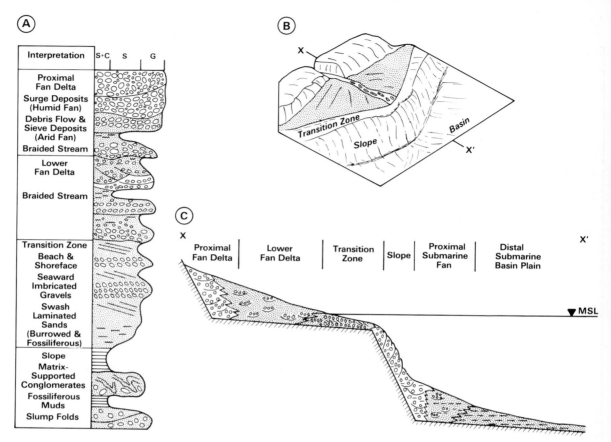

Figure 10.4. (A) Vertical sequence and (B) plan view of an idealized slope-type fan delta based upon the Yallahs fan-delta system, Jamaica (Wescott and Ethridge, 1980; 1982, and Ethridge and Wescott, 1984); (C) Generalized cross-section through a slope-type fan delta associated with a submarine fan (after Wescott and Ethridge, 1983)

have a distal facies consisting of marsh and meandering stream deposits. The distal margins of fan deltas that prograde directly into the Gulf of Alaska are reworked into a series of barrier spits. Data from the Copper River fan delta reveal that the coastal zone is characterized by a broad, sand-rich, tidal lagoon with active tidal channels and subaqueous to intertidal sand flats (Galloway, 1976). Sediments seaward of this tidal lagoon were deposited in marginal island, beach, barred shoreface and mid- to lower delta front environments. The entire sequence is presumably underlain by prodelta and shelf muds (Reimnitz, 1966; Galloway, 1976).

Shelf fan deltas often have better developed coarsening-upward sequences than do slope-type fans (Fortunato and Ward, 1982). Because they are not truncated by the shelf–slope break, they can prograde for longer distances and, therefore, tend to develop sandy braided, coastal lagoon and barrier-spit deposits on their distal margins.

Numerous variants exist on the shelf-model presented above. Holocene and ancient fan deltas along divergent plate margins and intracratonic rifts often show a close association between coarse-grained clastic delta deposits and carbonate reefs or carbonate algal mounds. These fan deltas usually build into lagoons fringed by carbonate deposits. An excellent example of this type of shelf fan delta is given by Dutton (1982) for Pennsylvanian fan-delta deposits in the Palo Duro Basin, Texas.

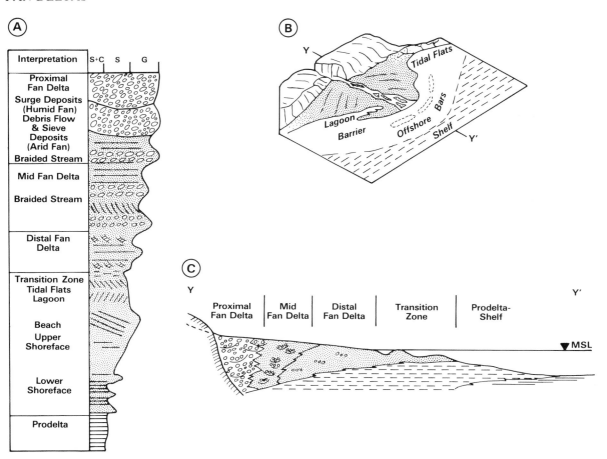

Figure 10.5. (A) Vertical sequence; (B) plan view; (C) idealized cross-section of a shelf-type fan delta based upon fan deltas along the southern Alaskan coast (Galloway, 1976; Boothroyd, 1976; Boothroyd and Ashley, 1975; Boothroyd and Nummedal, 1978) and Pennsylvanian–Permian fan deltas from the mid-continent of the United States (Brown, 1979)

GILBERT-TYPE MODEL

The ideal Gilbert delta consists of topset, foreset, and bottomset beds and in plain view it has a fan-shaped geometry (Figure 10.6). Topset beds are deposited by shifting channels on an alluvial fan. Foresets occur where the traction load is dropped at the river mouth and continues down the delta front under the influence of gravity. Foreset beds, in the absence of external forces, are deposited at the angle of repose. Fine-grained sediment is carried basinward in suspension and settles in front of the delta face forming bottomset beds. In stream-modified Gilbert deltas (Postma and Roep, 1985), the foreset geometry is in part controlled by density underflows of sediment-laden river currents.

The type Gilbert delta was deposited as a fan delta building into Pleistocene Lake Bonneville (Gilbert, 1885; 1890) and it is quite probable that many modern fan deltas prograding into lacustrine basins are also of this type. Modern examples of Gilbert-type deltas in lacustrine settings have been described by Born (1972), Gustavson et al. (1975), Bogen (1983), and Dunne and Hempton (1984). Examples of ancient Gilbert-type fan deltas have been reported by Stanley and

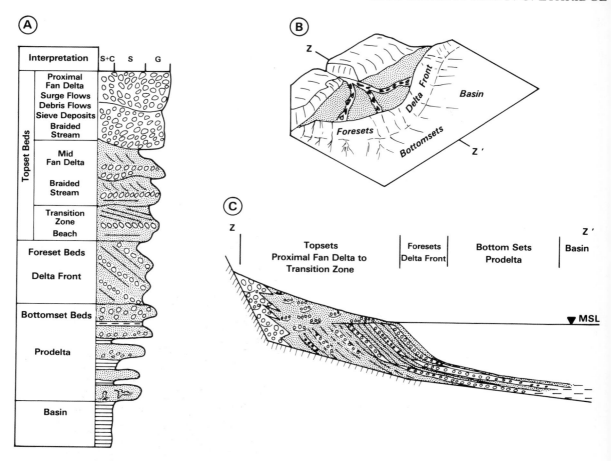

Figure 10.6. (A) Idealized vertical sequence; (B) plan view; (C) cross-section of a Gilbert-type fan delta (after Gilbert, 1885, 1890; Dunne and Hempton, 1984; Postma and Roep, 1985)

Surdam (1978) in Eocene lacustrine rocks in the Washakie Basin, Utah.

Gilbert-type fan deltas in marine settings have been documented from Late Carboniferous, fault-bounded, intracratonic basins in Colorado and New Mexico (Casey and Scott, 1979; Walker and Harms, 1980; and Millberry, 1983) and from the Pliocene of southeastern Spain (Postma and Roep, 1985). Modern Gilbert-type fan deltas have been described from several proglacial settings (Salisbury, 1892; Church, 1972; Bogen, 1983, and Syvitski and Farrow, 1983). These deltas are commonly deposited at the heads of narrow, rock bounded bays and fjords with steep cliffs. In this setting classical fan-shaped geometry is inhibited by the confining rock walls. However, the sedimentary processes are the same as those on other fan deltas.

The principal characteristic that differentiates Gilbert-type fan deltas from the other models is the presence of large-scale gravelly foreset beds in the transitional and shallow subaqueous zones. In lacustrine settings delta front foresets are thought to develop as the result of homopycnal flow. However, Dunne and Hempton (1984) have shown that the slope of the basin floor is a major factor controlling whether or not foresets will develop on the delta fronts in Lake Hazar. Deltas building onto slopes greater than 3 degrees form gravelly foresets while those with lower gradients do not.

Large gravel foresets have been described in marine fan deltas in confined proglacial settings on Baffin Island (Church, 1972), from Norway (Bogen, 1983), and British Columbia (Syvitski and Farrow, 1983). The foresets develop where the deltas prograde into relatively deep water. Delta front slopes in this setting are frequently modified by slumping, which results in rugged nearshore bathymetry (Prior et al., 1981).

Bogen (1983) studied fan deltas in both fjord valley lakes and in fjords. He noted that foreset geometry was related to sediment load and water depth. Where the ratio of bed load to suspended load is large and water depths not very deep, classical Gilbert-type foresets are deposited. In very deep water where a broad range of grain-sizes is available the foreset slope is steep near the top and becomes more gentle near the base. Bogen (1983) also noted that marine fan deltas were more prone to subaqueous sliding than lacustrine fan deltas.

Postma and Roep (1985) described a Pliocene Gilbert-type fan delta in Spain that was deposited in a similar, bedrock enclosed marine embayment that was sheltered from wave and tidal actions. In this setting mass-flow rheology was the major factor controlling the delta-front geometry. A major characteristic of this fan delta sequence is the presence of resedimented conglomerates in the lower delta slope bottomset beds. The conglomerates originated from the transformation of delta-front slides into high-density turbulent flows, debris-flows, or turbidity flows. Postma and Roep (1985) termed this type of mass-flow dominated fan delta a bottomset-modified Gilbert-type delta.

Discussion

During the last decade there has been a marked increase in research on fan deltas and interest in this geomorphic system is continuing to grow. The diversity of settings and complexity of facies of modern fan deltas has necessitated the development of at least three models for their recognition in the rock record. Still, care must be taken when interpreting tectonic setting and reconstructing palaeogeography of conglomeratic fan-delta deposits because our knowledge of these systems is far from complete. For example, along a faulted basin margin fan deltas can occur as isolated features with a point sediment source (Figure 10.7A) or as a laterally coalesced apron of fans with several adjacent sediment sources (Figure 10.7B). How these differences in sediment supply influence sediment and facies distribution, and geometry is still unknown.

Detailed studies of modern fault bounded basins are also contributing to building a better framework for understanding fan-delta settings. Recent seismic studies of the East African Rift Lakes have shown that they are not simple, symmetrical grabens. Rather, they are characterized by asymmetric half grabens with alternating

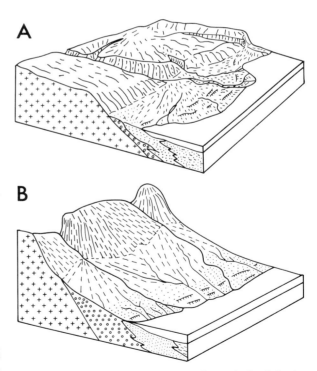

Figure 10.7. Fan-delta patterns along faulted basin margins. (A) Single, isolated fan delta with a single trunk channel; (B) Coalescing apron of fan deltas with multiple, adjacent channels (after Renaut, 1982)

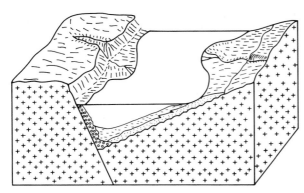

Figure 10.8. Schematic, simplified asymmetric rift-basin model showing how slope-type fan deltas (adjacent to major fault) and shelf-type fan deltas (on gently sloping margin) can develop in relatively close proximity in the same basin. Model is based upon studies of East African rift lakes (see text for references)

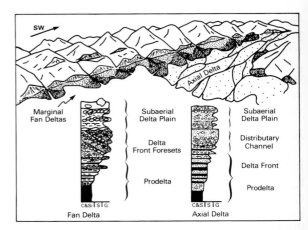

Figure 10.9. Schematic diagram showing the locations and idealized vertical sequences of marginal fan deltas and axial deltas in Lake Hazar, Turkey. The well-developed gravelly forsets are characteristic of Gilbert-type fan delta. (after Dunne and Hempton, 1984)

polarity (Rosendahl and Livingstone, 1983; Ebinger et al., 1984). This basin geometry develops settings in which both slope and shelf-type fan deltas can be deposited in relatively close proximity (Figure 10.8). The previously discussed study of deltas in Lake Malawi (Crossley, 1984) documents a modern basin where this occurs. Likewise, the Gulf of Suez is an asymmetrical rift basin (Chenet and Letouzey, 1983) which provides a similar setting adjacent to a marine basin. Alluvial fan sedimentation is also affected by basin asymmetry. For example, eastward tilting of the Death Valley–Panamint Range block has resulted in major differences in the size and geometry of the fans on opposite sides of Death Valley, California (Denny, 1965).

As pointed out by Miall (1981, his figure 6) and demonstrated in Lake Hazar by Dunne and Hempton (1984), high gradient, relatively small fan deltas can be closely associated with lower gradient, larger deltas in fault bounded basins. Fan deltas are proximal, transverse elements prograding from the basin margins, whereas the lower gradient deltas with finer-grained, more distal facies are deposited along the basin axis (Figure 10.9). If only one delta type were recognized in a stratigraphic sequence, an erroneous tectonic reconstruction could possibly result; however, an awareness that both delta types can occur together in a suspected fault-bounded basin could result in a model for predicting the other facies and an accurate reconstruction of the basin geometry.

Restraint must be exercised not to interpret every coarse-grained, conglomeratic deposit with fluvial and marine elements as a fan delta. A braided stream flowing to the sea does not constitute a fan delta. During the entire pre-Devonian, prior to the advent of land vegetation, deltas characterized by braided fluvial and deltaic plains were the norm. However, in order to be recognized as a fan delta the following criteria must be established: (1) proximity to an adjacent highland, (2) a fan-shaped morphology of the sedimentary body, unless confined by bedrock as in a fjord or rock-bounded embayment, and (3) deposition marginal to a subaqueous receiving basin.

Additional studies of fan-delta systems, both modern and ancient, will result in refinement of fan-delta facies models permitting a better understanding of deposition in tectonically active basins. In areas of known plate tectonic setting, these models provide a tool for predicting the probable basin-fill style. Conversely the recognition of fan-delta deposits in the stratigraphic sequence can be an important clue in interpreting the palaeotectonic setting. Although fan-delta deposits represent only a minor proportion of the stratigraphic column, they record important palaeoclimatic and tectonic influences on sedimentation and basin evolution.

Acknowledgements

Reviews of an earlier version of this manuscript by Robert Raynolds and Lorie Dunne are greatly appreciated. Photographic reproduction, drafting, and typing services were supplied by Amoco Production Company.

References

Bogen, J. 1983. Morphology and Sedimentology of deltas in fjord and fjord valley lakes. *Sedimentary Geology*, **36**, 245–267.

Boothroyd, J. C. 1976. A model for alluvial fan–fan delta sedimentation in cold temperate environments. In Miller, T. P. (Ed.), *Recent and Ancient Sedimentary Environments in Alaska. Proceedings of the Alaska Geological Society Symposium, Anchorage, Alaska*, N1–N13.

Boothroyd, J. C. and Ashley, G. M. 1975. Processes, bar morphology and sedimentary structures on braided outwash fans, northeastern Gulf of Alaska. In Jopling, A. V. and McDonald, B. C. (Eds), *Glaciofluvial and Glaciolacustrine Sedimentation. Society of Economic Paleontologists and Mineralogists Special Publication*, **23**, 193–222.

Boothroyd, J. C. and Nummedal, D. 1978. Proglacial braided outwash: a model for humid alluvial fan deposits. In Miall, A. D. (Ed.), *Fluvial Sedimentology. Canadian Society of Petroleum Geology Memoir*, **5**, 641–668.

Born, S. M. 1972. *Late Quaternary history, deltaic sedimentation, and mudlump formation at Pyramid Lake, Nevada*. University of Nevada, Reno. Center for Water Resources Research, 97 pp.

Brown, L. F. Jr. 1979. Deltaic sandstone facies of the mid-continent. In Hyne, N. J. (Ed.), *Pennsylvanian Sandstones of the Mid-continent. Tulsa Geological Society Special Publication*, **1**, 35–63.

Burke, K. 1967. The Yallahs basin: a sedimentary basin southwest of Kingston, Jamaica. *Marine Geology*, **5**, 45–60.

Casey, J. M. and Scott, A. J. 1979. Pennsylvanian coarse-grained fan deltas associated with the Uncompahgre uplift, Talpa, New Mexico. In *New Mexico Geological Society, 30th Field Conference, Santa Fe County Guidebook*, 211–218.

Church, M. 1972. Baffin Island sandurs: a study of arctic fluvial processes. *Geological Survey of Canada Bulletin*, **216**, 208 pp.

Chenet, P. Y. and Letouzey, J. 1983. Tectonics of the area between Abu Durba and Gebel Mezzazat (Sinai, Egypt) in the context of the evolution of the Suez Rift. *Bulletin des Centres de Recherche Exploration-Production Elf-Aquitaine*, **7**, 201–215.

Crossley, R. 1984. Controls of sedimentation in the Malawi rift valley, central Africa. *Sedimentary Geology*, **40**, 33–50.

Denny, C. S. 1965. Alluvial fans in the Death Valley region, California and Nevada. *United States Geological Survey Professional Paper*, **466**, 62 pp.

Dubois, J., Larue, B., Pascal, G., and Reichenfeld, C. 1973. Seismology and structure of the New Hebrides. In Coleman, P. J. (Ed.), *The Western Pacific, Island Arcs, Marginal Seas, Geochemistry*. University of Western Australia Press. 213–222.

Dunne, L. A. and Hempton, M. R. 1984. Deltaic sedimentation in the Lake Hazar pull-apart basin, southeastern Turkey. *Sedimentology*, **31**, 401–412.

Dutton, S. P. 1982. Facies control of cementation and porosity, Pennsylvanian fan-delta sandstones, Texas Panhandle. *American Association of Petroleum Geologists Bulletin*, **66**, 565 (abstract only).

Ebinger, C. J., Crow, M. J., Rosendahl, B. R., Livingstone, D. A., and LeFournier, J. 1984. Structural evolution of Lake Malawi, Africa. *Nature*, **308**, 627–629.

Ethridge, F. G. and Wescott, W. A. 1984. Tectonic setting, recognition and hydrocarbon reservoir potential of fan-delta deposits. In Koster, E. H. and Steel, R. J. (Eds), *Sedimentology of Gravels and Conglomerates. Canadian Society of Petroleum Geologists Memoir*, **10**, 217–235.

Fortunato, K. S. and Ward, W. C. 1982. Upper Jurassic–Lower Cretaceous fan-delta complex—La Casita Formation of Saltillo area, Coahuila, Mexico. *American Association of Petroleum Geologists Bulletin*, **66**, 1429 (abstract only).

Galloway, W. E. 1976. Sediments and stratigraphic framework of the Copper River fan delta, Alaska. *Journal of Sedimentary Petrology*, **46**, 726–737.

Garfunkel, Z., Zak, I., and Freund, R. 1981. Active faulting in the Dead Sea Rift. *Tectonophysics*, **80**, 1–26.

Gilbert, G. K. 1885. The topographic features of lake shores. *United States Geological Survey Annual Report*, **5**, 104–108.

Gilbert, G. K. 1890. Lake Bonneville. *United States Geological Survey Monograph*, **1**, 438 pp.

Gloppen, T. G. and Steel, R. J. 1981. The deposits, internal structure and geometry of six alluvial fan-fan delta bodies (Devonian, Norway)—a study in the significance of bedding sequences in conglomerates.

In Ethridge, F. G. and Flores, R. M. (Eds), *Recent and Ancient Nonmarine Depositional Environments: Models for Exploration*. Society of Economic Paleontologists and Mineralogists Special Publication, **31**, 49–69.

Gustavson, T. C., Ashley, G. M., and Boothroyd, J. C. 1975. Depositional sequences in glaciolacustrine sedimentation. In Jopling, A. V. and McDonald, B. C. (Eds), *Glaciofluvial and Glaciolacustrine Sedimentation*. Society of Economic Paleontologists and Mineralogists Special Publication, **23**, 264–280.

Hardie, L. A., Smoot, J. P., and Eugster, H. P. 1978. Saline lakes and their deposits: a sedimentological approach. In Matter, A. and Tucker, M. E. (Eds), *Modern and Ancient Lake Sediments*. International Association of Sedimentologists Special Publication, **2**, 7–41.

Hayes, M. O. and Michel, J. 1982. Shoreline sedimentation within a forearc embayment, lower Cook Inlet, Alaska. *Journal of Sedimentary Petrology*, **52**, 251–263.

Hayward, A. B. 1982. Coral reefs in a clastic sedimentary environment: fossil (Miocene, S.W. Turkey) and modern (Recent, Red Sea) analogues. *Coral Reefs*, **1**, 109–114.

Hempton, M. R., Dunne, L. A., and Dewey, J. F. 1983. Sedimentation in an active strike-slip basin, southeastern Turkey. *Journal of Geology*, **91**, 401–412.

Holmes, A. 1965. *Principles of Physical Geology*. New York, The Ronald Press Co. 1288 pp.

Kochel, R. C. and Johnson, R. A. 1984. Geomorphology and sedimentology of humid-temperate alluvial fans, central Virginia. In Koster, E. H. and Steel, R. J. (Eds), *Sedimentology of Gravels and Conglomerates*. Canadian Society of Petroleum Geologists Memoir, **10**, 109–122.

McGowen, J. H. 1970. Gum Hollow Fan Delta, Nueces Bay, Texas. *University of Texas at Austin, Bureau of Economic Geology. Report of Investigations*, **69**, 91 pp.

McGowen, J. H. and Groat, C. G. 1971. Van Horn Sandstone, West Texas: an alluvial fan model for mineral exploration. *University of Texas at Austin, Bureau of Economic Geology Report of Investigations*, **72**, 57 pp.

Meckel, L. D. 1975. Holocene sand bodies in the Colorado Delta area, northern Gulf of California. In Broussard, M. L. (Ed.), *Deltas—Models for Exploration*. Houston Geological Society. 239–265.

Miall, A. D. 1981. Alluvial sedimentary basins: tectonic setting and basin architecture. In Miall, A. D. (Ed.), *Sedimentation and Tectonics in Alluvial Basins*. Geological Association of Canada Special Paper, **23**, 1–33.

Millberry, K. W. 1983. Tectonic control of Pennsylvanian fan-delta deposition, southwestern Colorado (abstract). *American Association of Petroleum Geologists Bulletin*, **67**, 514.

Mitchell. A. H. G. and Reading, H. G. 1978. Sedimentation of tectonics. In Reading, H. G. (Ed.), *Sedimentary Environments and Facies*. N.Y., Elsevier. 439–476.

Nemec, W., Steel, R. J., Porebski, S. J., and Spinnangr, A. 1984. Domba Conglomerate, Devonian, Norway: process and lateral variability in a mass flow-dominated, lacustrine fan delta. In Koster, E. H. and Steel, R. J. (Eds), *Sedimentology of Gravels and Conglomerates*. Canadian Society of Petroleum Geologists Memoir, **10**, 295–320.

Pierson, T. C. 1980. Erosion and deposition by debris flows at Mt. Thomas, New Zealand. *Earth Surface Processes*, **5**, 1952–1984.

Pollard, J. E., Steel, R. J., and Undersrud, E. 1982. Facies sequences and trace fossils in lacustrine/fan delta deposits, Hornelen Basin (M. Devonian), western Norway. *Sedimentary Geology*, **32**, 63–87.

Postma, G. 1983. Water escape structures in the context of a depositional model of a mass flow dominated conglomeratic fan-delta (Abrioja Formation, Pliocene, Almeria Basin, SE Spain). *Sedimentology*, **30**, 91–103.

Postma, G. 1984. Slumps and their deposits in fan delta front and slope. *Geology*, **12**, 27–30.

Postma, G. and Roep, T. B. 1985. Resedimented conglomerates in the bottomsets of Gilbert-type gravel deltas. *Journal of Sedimentary Petrology*, **55**, 874–885.

Prior, D. B., Wiseman, Wm. J., and Gilbert, R. 1981. Submarine slope processes on a fan delta, Howe Sound, British Columbia. *Geo-Marine Letters*, **1**, 85–90.

Reimnitz, E. 1966. *Late Quaternary history and sedimentation of the Copper River Delta and vicinity, Alaska*. Unpublished Ph.D. Dissertation, University of California at San Diego. 160 pp.

Renaut, W. 1982. *Lake Quaternary Geology of the Lake Bogoria Fault-Trough, Kenya Rift Valley*. Unpublished Ph.D. Thesis, Queen Mary College, University of London. 498 pp.

Rosendahl, B. R. and Livingstone, D. A. 1983. Rift Lakes of East Africa—new seismic data and implications for future research. *Episodes*, **1983**, 14–19.

Rust, B. R. and Koster, E. H. 1984. Coarse alluvial deposits. In Walker, R. G. (Ed.), *Facies Models*. Geoscience Canada Reprint Series, **1**, 53–69.

Salisbury, R. D. 1892. Outwash plains and valley trains. *Geological Survey of New Jersey, Annual Report of the State Geologist, Section 7*, 96–114.

Schumm, S. A. 1977. *The Fluvial System*. New York: Wiley–Interscience. 335 pp.

Sneh, A. 1979. Late Pleistocene fan deltas along the Dead Sea Rift. *Journal of Sedimentary Petrology*, **49**, 541–552.

Stanley, K. O. and Surdam, R. C. 1978. Sedimentation on the front of Eocene Gilbert-type deltas, Washakie Basin, Wyoming. *Journal of Sedimentary Petrology*, **48**, 557–573.

Syvitski, J. P. M. and Farrow, G. E. 1983. Structures and processes in bayhead deltas: Knight and Bute Inlets, British Columbia. *Sedimentary Geology*, **36**, 217–244.

Thompson, R. W. 1968. Tidal flat sedimentation on the Colorado River delta, northwestern Gulf of California. *Geological Society of America Memoir*, **107**, 133 pp.

Walker, T. R. and Harms, J. C. 1980. *Fan-delta deposition, Minturn Formation, McCoy area, Colorado*. Unpublished field trip notes, Rocky Mountain Section, Society of Economic Paleontologists and Mineralogists Fall Field Conference.

Wescott, W. A. (this volume) The Yallahs fan delta: a coastal fan in a humid tropical climate. In Rachocki, A. H. and Church, M. (Eds), *Alluvial Fans— A Field Approach*. Chichester, John Wiley and Sons.

Wescott, W. A. and Ethridge, F. G. 1980. Fan-delta sedimentology and tectonic setting—Yallahs fan delta, southeast Jamaica. *American Asociation of Petroleum Geologists Bulletin*, **64**, 374–399.

Wescott, W. A. and Ethridge, F. G. 1982. Bathymetry and sediment dispersal dynamics along the Yallahs fan-delta front, Jamaica. *Marine Geology*, **46**, 245–260.

Wescott, W. A. and Ethridge, F. G. 1983. Eocene fan delta–submarine fan deposition in the Wagwater Trough, east central Jamaica. *Sedimentology*, **30**, 235–247.

CHAPTER 11

The Yallahs Fan Delta: A Coastal Fan in a Humid Tropical Climate

William A. Wescott
Amoco Production Company, Houston

Abstract

The Yallahs fan delta on the southeastern coast of Jamaica is a classical example of a wave-dominated, slope-type fan delta complex which has developed in a humid tropical climate. Yallahs River debouches from the Blue Mountains and has built a 10.5 km^2 lobate fan delta composed of sand to boulder size detritus. The morphology of the fan delta is in part controlled by the bedrock of the mountain front which bounds the fan delta on three sides and the steep offshore which causes waves to break and expend most of their energy directly on the fan delta front. Three distinctive morphological zones make up the fan delta; the subaerial fan, the transitional zone, and the submarine delta front.

Environments composing the subaerial fan include active and abandoned braided fluvial channels, floodplains, and saltwater ponds. High magnitude, low frequency flooding is the dominant process that deposits coarse sediments and modifies the surface. The distribution of sediments and alternation of erosional and depositional beaches in the coastal transition zone are controlled by the refraction and diffraction of waves across the shallows offshore. The submarine portion of the fan delta is characterized by two provinces, a narrow island shelf, and a steep offshore slope incised by the heads of submarine canyons. A dominantly westward longshore drift transports coarse sediments until they are captured in the heads of submarine canyons or deposited on steep, unstable slopes between the canyons. Mass flow processes periodically move these sediments downslope and out into Yallahs Basin.

Introduction

In areas where high coastal relief occurs adjacent to lakes or seas, coarse alluvium is shed into the water resulting in the formation of alluvial fans that prograde into the subaqueous basin. This type of alluvial fan is called a fan delta (Holmes, 1965; McGowen, 1970; Wescott and Ethridge, 1980). Fan deltas are unique in that they represent an environment in which alluvial fan forming processes interact with and are modified by littoral processes. In recent years there has been a large number of studies on fan deltas, particularly concerning the recognition of their deposits in the stratigraphic record (Ethridge and Wescott, 1984). To date, very few modern fan deltas have been studied in great detail. Perhaps the best documented modern fan delta is the Yallahs fan delta in southeastern Jamaica (Wescott and Ethridge, 1980, 1982; and Hendry, 1982).

Alluvial Fans: A Field Approach edited by A. H. Rachocki and M. Church
Copyright © 1990 John Wiley & Sons Ltd.

The purpose of this paper is to review and summarize the setting, processes, and deposits of the Yallahs fan delta within the context of fan deltas as members of the alluvial fan continuum.

Setting

REGIONAL GEOLOGY

Jamaica is located in the northwestern Caribbean Sea (Figure 11.1) on the eastern end of the Nicaraguan Rise at the northern margin of the Caribbean plate where it meets the North American plate (Robinson et al., 1970). The plate boundary is the Cayman Trough, located north of Jamaica, a sinistral transform fault (Molnar and Sykes, 1969).

Yallahs River originates near the western end of the Blue Mountains at an elevation of 1448 m. The river flows southeastward past the Port Royal Mountains and across a belt of low hills underlain by Tertiary limestones before it reaches the sea 37 km from its source. The drainage basin encompasses approximately 163 km^2 and is characterized by local relief of 450 to 600 m with steep slopes at angles of 20 to 30 degrees. Stream frequency is 12.6 km^{-2}. The bedrock is extensively but not deeply weathered due to the steep slopes, frequent mass wastage, and slope wash during intense rainfall.

The Yallahs drainage basin is underlain by

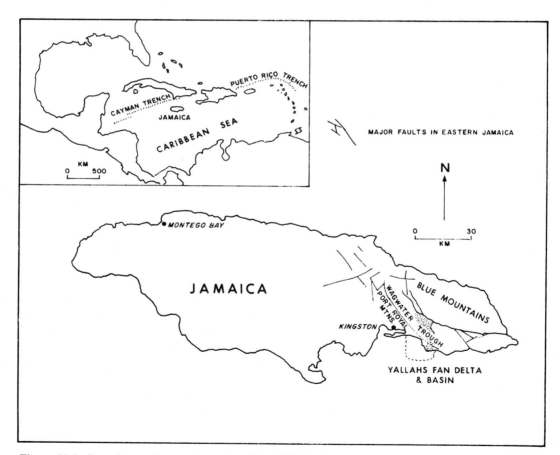

Figure 11.1. Location and tectonic setting of the Yallahs fan delta and Yallahs Basin in southeastern Jamaica. The stippled pattern indicates the location of the Yallahs drainage basin and fan delta. The dashes outline submarine Yallahs Basin (reproduced by permission from Wescott and Ethridge, 1982)

igneous, metamorphic, and sedimentary rocks which constitute 35%, 5%, and 60% of the surface area of the drainage basin respectively. Cretaceous rocks in the northeastern portion of the basin include serpentinites, granodiorites, metavolcanics and marbles of the Blue Mountain Group, quartz–feldspar schists of the Westphalia Schist Group, and sedimentary rocks of the Main Ridge Group (Kemp, 1971). Tributaries entering Yallahs River from the southwestern side of the basin drain an area underlain by conglomerates, sandstones, limestones, and mudstones of the Eocene Richmond Formation, and a large, sill-like body of porphyritic andesite known as the Newcastle Volcanics. In the southern part of the valley Eocene rocks are overlain by the Tertiary Yellow Limestone and White Limestone formations.

Where the river leaves the mountain front it has built a subaerial, lobate fan delta of approximately 10.5 km². The fan delta forms an alluvial headland which extends 2 km from the mountains into the sea. The fan progrades to the edge of a submarine scarp which is probably fault controlled, and the submarine portion of the delta system drops to a depth of 1100 m below sea level over a horizontal distance of 4 km. This slope forms the northeastern boundary of the Yallahs Submarine Basin (Figure 11.2) which covers an area of 100 km² and lies at an average depth of 1350 m below sea level. The basin is filled with approximately 500 m of resedimented siliciclastic sediment derived from the Liguanea–Hope and Yallahs fan deltas on the northwest and northeast, respectively, and carbonate sediments derived from the island shelf to the west (Burke, 1967).

CLIMATE AND HYDROLOGY

Jamaica lies in the belt of northeast trade winds and has a mild tropical climate. In the Yallahs drainage basin, mean annual temperature ranges from 13°C at Blue Mountain Peak to 26°C on the coast. Seasonal variation is slight and fluctuates by 3°C on the coast. Mean annual rainfall values range from 3050 mm per year in the mountains to 1500 mm per year on the coast (Gupta, 1975).

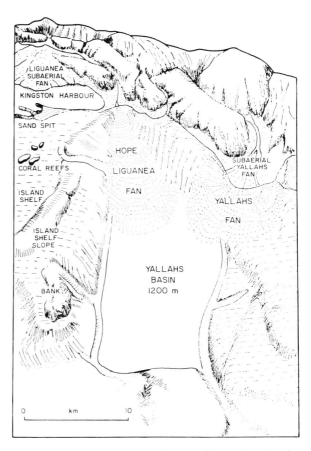

Figure 11.2. Physiographic diagram illustrating the depositional setting of the Yallahs fan delta with respect to Yallahs Basin (after Burke, 1967)

Rainfall in the Yallahs drainage basin is seasonal with the main rainy season occurring from September until the middle of December. Generally, the skies clear in mid-December and there is a dry period until April. Rains begin again in May, although they are not as heavy as during the main rainy season. The upper reaches of the drainage basin receive the greatest amount of rainfall. The dry period is most pronounced on the subaerial fan, where 85% of the annual rainfall occurs during only two months, May and December.

A large portion of the annual rainfall of eastern Jamaica occurs as intermittent downpours of high magnitude and intensity. These storms are associated with southward moving cold fronts, waves in the easterly trade winds, and tropical storms. Gupta (1983) analysed rainfall records from eastern Jamaica and calculated that the 24-hour 5-year recurrence interval rainfall is at least 250 mm and the 10-year recurrence interval rainfall is 300–550 mm in 24 hours. This suggests that high magnitude floods occur on average at least once in every 5-year period in the Yallahs Valley. In fact, hurricanes and tropical storms, which have a pronounced effect on the geomorphic character of the Yallahs drainage basin, occur on average once in 4.75 years (Gupta, 1983). Intense rainfall accompanying these storms may trigger major landslides, induce catastrophic flooding, and generate storm waves that cause severe coastal erosion. Because of the frequent intense rainfall, steep slopes, and intense fracturing of bedrock, mass wasting is the major process responsible for supplying coarse detritus to Yallahs River.

Gupta (1975) mapped the location of major landslides in the Yallahs drainage basin and superposition of these sites on the geologic map shows that most of the landslides occur in the Richmond and Wagwater formations, an Eocene sedimentary sequence of conglomerates, sandstones, and mudstones. One of these slides was documented in October, 1963, when flooding in response to precipitation from hurricane Flora caused Yallahs River to undercut its left bank near Mahogany Vale. This initiated a landslide which delivered approximately 40 000 m^3 of sandstone and shale to the valley floor (Gupta, 1975, 1983).

Sediment supplied to the valley floor in this manner is reworked and transported to the fan delta by normal fluvial processes. Short transport distances, high gradients, and episodic introduction of coarse detritus into the fluvial system by mass wasting result in texturally immature sediments and preclude the development of significant downstream and downfan grain size trends.

Quantitative information on the magnitude of floods in the Yallahs valley is lacking. There are no long-term discharge records for Yallahs River, but limited data from the upper part of the drainage basin at Mahogany Vale (1960–61) and from the apex of the fan at Easington (1973–77) indicate that Yallahs River discharge is seasonal. A maximum discharge of 17.5 m^3 s^{-1} was recorded during the month of November and a minimum discharge of 0.0004 m^3 s^{-1} in June at Easington. Hurricane flows of up to 2330 m^3 s^{-1} have been calculated for Yallahs River.

From the apex of the Yallahs fan delta to the sea, Yallahs River is characterized by decreasing flow as surface waters are lost to subsurface flow and are diverted into irrigation canals. At Yallahs Ford (on the lower fan) the river is completely dry during periods of low flow, and during periods of high flow discharge is less than at Mahogany Vale, in the upper part of the drainage basin (Gupta, 1975). Only periodically is there sufficient water in the channels at the distal fan margin for surface flow to reach the sea. During most of the year the channel is blocked by a wave-built berm. However, after storms or during the rainy season the river breaches the berm and flows to the sea (Figure 11.3).

LITTORAL PROCESSES

Jamaica is located in the belt of northeast trade winds. Summaries of synoptic weather observations in the region of southeast Jamaica from 1958 to 1973 show that winds blow from the east 57% of the time (Figure 11.4). Average speed of these easterly trade winds is 7 m s^{-1}. Winds without an easterly component blow, on the average, only 7% of the year. Consequently, deep-water wind-generated waves approach southeast Jamaica from the east (U.S. Naval Weather Service Command, 1974).

Deep-water waves in this part of the Caribbean Sea have significant wave heights of 0.7 to 1.6 m with periods of less than 6 s (DeLeonibus and Sheil, 1973). Waves in the littoral zone off Cow Bay have average significant wave heights of 1.2 m with periods of 5 to 9 s (Jamaica Industrial Development Corporation, 1974). Waves approaching the Yallahs delta are refracted around its lobate shape and the impinging waves generate a dominantly westward longshore drift. The westward elongated asymmetry of the delta reflects the dominance of these currents.

Jamaica's coastline is classified as microtidal (Davies, 1964) and is characterized by a mixed

Figure 11.3. Yallahs River in flood in October 1963 after Hurricane Flora (photo by J. S. Tyndale-Biscoe)

tidal regime with a mean high tide of +0.14 m and mean low tide of −0.11 m. The limited tidal ranges have little modifying effect on the coarse coastal sediments.

Depositional Environments

The Yallahs fan delta complex is a depositional system comprising environments that can be grouped into three geomorphic settings, subaerial fan, transitional coastal zone, and submarine fan delta (Figure 11.5). Environments of the subaerial fan are braided alluvial channels, abandoned channels, floodplains, marshes, and ponds. Transitional environments are erosional and depositional beaches and coastal dunes. The submarine fan delta includes slope and island shelf environments, including isolated carbonate reefs.

SUBAERIAL FAN

The subaerial fan is the fluvially dominated alluvial fan portion of the fan delta system. It sits on an eroded bed rock surface and is up to 32 m thick in the mid-fan area (Andrew, 1965). The average gradient of the fan, down the Yallahs River axis is 15 m km^{-1} (Figure 11.6).

The active Yallahs River channel is the dominant environment on the subaerial fan. The most common bed forms in the active part of Yallahs River from the fanhead to the coast are longitu-

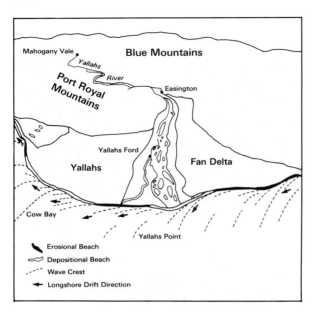

Figure 11.4. Sketch map of the Yallahs fan delta showing distribution of depositional and erosional beaches, wave refraction pattern, and dominant longshore drift direction. Localities mentioned in the text are also indicated on the diagram (after Wescott and Ethridge, 1980)

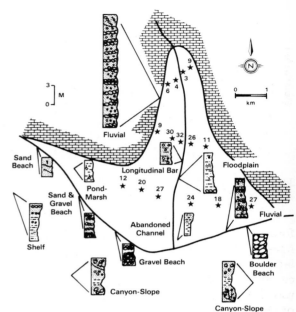

Figure 11.5. Diagram showing the general vertical sequences and lateral associations of subaerial, transitional, and submarine deposits on the Yallahs fan delta. Numbers with stars indicate depth to bedrock, in metres (data from Andrew, 1965). Thicknesses of submarine vertical sequences are inferred (after Wescott and Ethridge, 1980)

dinal gravel bars (Figure 11.7A). They comprise poorly-sorted, well-imbricated gravels with the coarsest sediment concentrated along the bar central axes, and their relief is not much greater than the thickness of the largest clast. They separate individual low-stage channels, which have a flat bottom and are armoured by coarse gravels. Under and between the coarse gravel clasts is coarse sand. The channels have a low silt and clay content. Filamentous algae cover the gravel bed during periods of low discharge.

The principal deposits formed by aggradation in this environment are crudely horizontally bedded, imbricated sandy gravels (Figure 11.7B), which can be seen in eroded high terraces along the river (Wescott and Ethridge, 1980; Gupta, 1983). The floodplain is eroded during periods of very high discharge (5- to 10-year floods) and is rebuilt and modified during the falling stage of the floods (Gupta, 1983). During these floods, abundant coarse-grained material is transported from the river valley and deposited on the downstream fan. Lower magnitude, higher frequency flows reduce the channel width, rebuild the floodplain, and deposit the mid-channel bars.

Abandoned channels are a common, although not an aerially extensive, feature of the subaerial Yallahs fan. On the floodplain abandoned channels form linear topographic lows that are partly filled with water during much of the year. A core taken in an abandoned channel east of the Yallahs river mouth and behind the beach ridge is characterized by muddy sand with isolated pebbles overlying homogeneous muds. The coarser sediment at the top of this core is probably introduced by storm overwash from the beach. Older abandoned channel deposits are exposed in the erosional beaches and in the cutbanks of Yallahs River.

Floodplain deposits are exposed in the cut-bank along the eastern side of Yallahs River below the

THE YALLAHS FAN DELTA

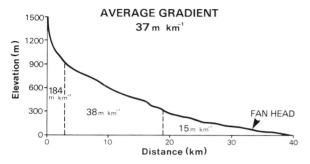

Figure 11.6. Gradient of Yallahs River from its source in the Blue Mountains to the Caribbean Sea

ford. They are predominantly grey, silty, very fine-grained sands characterized by ripple cross-lamination, ripple drift lamination, small-scale cut-and-fill structures, and abundant burrows and roots. A few palaeochannel gravels and clay plugs of abandoned channels are encased in the floodplain sediments. The aerial extent of these fine-grained floodplain deposits, although unknown, is probably not extensive.

The westernmost part of the lower fan is a relatively low area characterized by marshes, swamps, and open ponds. Maps show that in 1804 it was the site of a relatively large pond. Subsequent maps reveal progressive infilling. Today only three small ponds remain, surrounded by mangrove swamps and grassy marshes. Short cores through the marshes show that the sediments consist of muddy sand and organic muds, with isolated pebbles or pebbly layers.

The largest area of the Yallahs subaerial fan is the relatively flat fan delta plain. The fan has been an agricultural site since the 16th century and a secondary growth of grasses or thick shrubs and trees covers large areas not presently farmed. Near the coast, road networks for an undeveloped housing project were built on both sides of the river. Open areas on the fan reveal isolated boulders left behind as a type of lag deposit. Their presence in abandoned channel depressions serves to explain the presence of boulder-sized clasts in fine-grained floodplain deposits. Wave-cut scarps along the eastern coast of the delta reveal that the Yallahs plain is underlain by coarse-grained fluvial deposits. They consist of

Figure 11.7. (A) Cobble armored channel of Yallahs River on the Yallahs fan looking upstream. Note hammer for scale; (B) Subaerial fan deposits exposed in cut-bank of Yallahs River. Finer-grained sandy deposits overlie a coarse gravel channel deposit. Stadia rod is 2 m high

flat, tabular, cobble to boulder conglomerates with a sand matrix interbedded with finer-grained sandy units, and are similar to the modern channel deposits.

In the classification system of Kochel and Johnson (1984) the subaerial Yallahs fan is a humid tropical fan. Fans forming in this setting are dominated by fluvial processes and composed of braided alluvium and debris-flow deposits. Humid region fans can be constructed entirely of fluvial channel and overbank deposits of perennial streams. However, short-term fluctuations in precipitation and rapid changes in discharge can also trigger debris flows on 'wet' fans (Rust and Koster, 1984). At the other end of the spectrum, humid tropical fans, such as those in Papua, are built almost entirely of layered, boulder debris-flows (Ruxton, 1970). The Yallahs fan delta is fluvially dominated and there is no evidence for debris-flows contributing sediment to the upper fan. The reasons for this are unclear, especially since abundant mud is provided to the system by erosion of shales in the drainage basin. It may reflect the seasonality of the climate and short transport distances that allow much of the fine-grained sediment to be flushed from the system during frequent high discharge events.

TRANSITIONAL ZONE

Within this zone, fluvial deposits are modified by marine processes and interfinger with marine deposits. The most extensive environment in this zone is the beach. The beaches of the Yallahs fan delta are classified as depositional or erosional on the basis of their morphology and textural characteristics.

Depositional beaches are generally broad and sandy with well-developed berms, a broad backshore area, and comparatively gentle foreshore slopes (Figure 11.8A). The width of the beaches ranges from 15 to 50 m and averages 30 m. Foreshore gradients range from 0.05 to 0.15 and average 0.10. Grain size generally increases from the berm to the plunge point and tends to decrease from the berm to the backshore. Depositional beaches normally have a high sand/gravel ratio. Near the mouth of Yallahs River the backshore is composed of washover deposits that partly fill the river channels. Just east of this area the backshore grades into a small aeolian dune field. Sedimentary structures of depositional beaches include: (1) seaward-dipping, low-angle, thinly bedded units of sand in the foreshore zone; (2) horizontally bedded sand units in the berm and on flat backshore zones; (3) landward-dipping, low-angle, thinly bedded units of sand in the back-

Figure 11.8. (A) Sand and gravel depositional beach near Grants Pen at the western edge of the fan delta. Depositional beaches are relatively broad and have gentle gradients; (B) Erosional beach composed of cobble and boulder size sediments, east of Yallahs River mouth. The very coarse sediments are eroded from the subaerial fan by storm waves and remain on the beach as a lag deposit. Note that the erosional beach is relatively narrow and steep. Scale rod is 2 m long

shore zone; and (4) horizontally bedded gravel units with imbricated pebbles interbedded with sand units in all three beach zones. These gravel units are particularly common in storm berm and lower foreshore deposits. Generally, vertical to subhorizontal burrows of crabs are common on broad, sandy depositional beaches. Small dune fields are composed of well-sorted, poorly-stratified to massive fine-grained sands which are winnowed from the foreshore by sea breezes and deposited behind the berm. Areally and volu-

THE YALLAHS FAN DELTA

metrically, however, these aeolian deposits constitute an insignificant part of the Yallahs fan delta.

Erosional beaches along the margin of the Yallahs subaerial fan have comparatively steep, narrow, foreshore profiles, little or no backshore, and are composed of very coarse sand to boulder-size clasts (Figure 11.8B). Beach scarps from a few centimetres to 3 m high occur, associated with the beaches. Erosional beaches range from 10 to 20 m wide and average 15 m. Gradients of the foreshore range from 0.12 to 0.19 and average 0.15. Sediments on erosional beaches are essentially a lag or talus deposit resulting from the wave erosion of fluvial deposits along beach scarps and the winnowing of finer grain sizes. Therefore, grain size on these beaches is largely determined by the size of material being eroded.

Along the eastern margin of the delta erosional beaches are composed of wedge-shaped, thick-bedded, imbricate boulder to cobble deposits. The deposits are clast supported and voids are filled with poorly-sorted sand. These units probably have a relatively high preservation potential. Beachrock is commonly associated with erosional beaches in the intertidal zone along the eastern margin of the delta. The beachrock is composed of carbonate-cemented sand to cobble-size clasts. Farther west along the Yallahs fan delta, erosional beaches are composed of coarse sand to cobble-size detritus.

The distribution of erosional and depositional beaches on the fan delta is directly related to the distribution of wave energies and to the presence of coarse-grained fluvial deposits at the coastline. Wave energy impinging upon the beach is controlled by refraction and diffraction processes (Figure 11.9).

The area subject to the greatest wave erosion is between Station 7 and Station 6 (Figure 11.9) on the eastern coast of the delta. On this section of

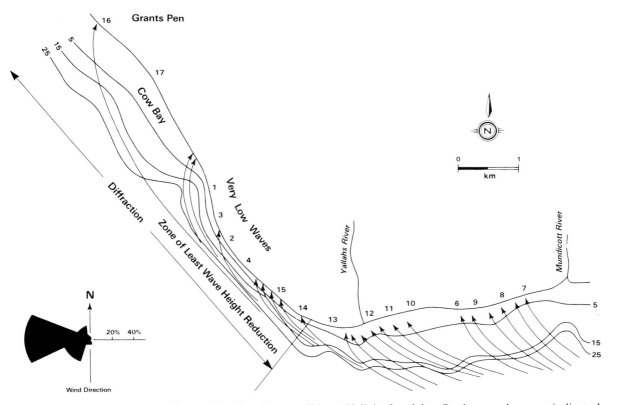

Figure 11.9. Wave refraction diagram for the shallow offshore Yallahs fan delta. Station numbers are indicated along the beach. Bathymetric contours have been simplified (after Wescott and Ethridge, 1982)

the beach, waves from the east are refracted around Yallahs Point and converge slightly, concentrating their attack. The waves approach the shore at nearly right angles and expend their energy directly on the beachface. The section of beach immediately east of the river mouth, Stations 6 to 12, is depositional because of two factors: (1) the westward longshore transport of sediment; and (2) the wave energy striking the beach is reduced in this zone.

The general character of the coastline west of the river mouth is progradational (Wood, 1976). Waves refracted around the river mouth bulge on the delta tend to parallel the offshore contours. Therefore, wave refraction is minimized and wave diffraction is the more important process. Diffraction results in a significant reduction in wave height. The lower wave energy, coupled with the continuous supply of longshore transported sediment, determines the overall depositional character of the beaches on the western side of the delta. There are, however, small areas of beach that are being actively eroded on the western coast of the delta. Erosional beaches located between Stations 15 and 4 may owe their existence to the very steep offshore profile along this segment of the coast. This is also the area where wave heights are reduced the least by diffraction. Here even the reduced waves expend most of their energy directly on the shoreface. Erosion is further enhanced by the loss of longshore transported sediment into the offshore slope and Cow Bay Canyon system.

The second zone of erosion of this side of the delta occurs just west of Cow Bay Point. Here erosion may be due to convergence of waves on

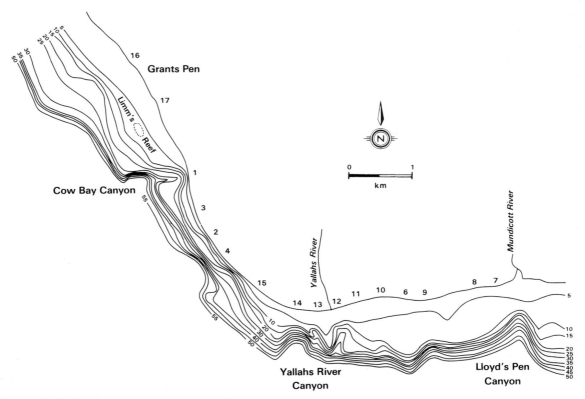

Figure 11.10. Bathymetric chart of the shallow (<55 m) offshore Yallahs fan delta showing the locations of the heads of submarine canyons on the east and central parts of the fan delta front and the narrow shelf west of Cow Bay Canyon (after Wescott and Ethridge, 1980, 1982). Contour interval is 5 m

THE YALLAHS FAN DELTA

the western side of Cow Bay Canyon as they diverge across the canyon. West of this point beaches are depositional to the end of the delta. Waves reduced in height by diffraction cross the relatively broad, shallow shelf in this area and reach the beach with less energy than those along the eastern coast of the delta.

SUBMARINE FAN DELTA

The shoreface and shallow offshore bathymetry in front of the Yallahs delta was determined using a Kelven-Hughes depth recorder; 49 profiles were recorded across the subaqueous delta plain from the Mundicott River to Grants Pen. Navigation was controlled by range poles and sextant. A bathymetric map of the area indicates that the offshore topography can be divided into two morphologic zones (Figure 11.10): (1) a steep island slope with heads of submarine canyons; and (2) a narrow island shelf with patch reefs.

The shoreface and shallow offshore zones from the Mundicott River on the east to Cow Bay Point are characterized by a steep offshore profile and the headward region of three submarine canyons: Lloyd's Pen Canyon on the east, and Yallahs River Canyon (Figure 11.10) and Cow Bay Canyon on the west. A detailed investigation of the nearshore environment off Cow Bay Point reveals that the bottom drops off steeply at approximately 20 to 30°. Sediment-size distribution on the bottom is patchy and ranges from muds, to sand, granules, and cobbles. In relatively stable areas the bottom is covered by a veneer of silt and mud. Marine grasses and algae are common.

Slumping occurs frequently in the heads of the submarine canyons and it is probably the principal process initiating the movement of sediment from the nearshore zone to Yallahs Basin. The bottom topography in the canyons is characterized by low gravel and/or cobble ridges and intervening silt-veneered swales paralleling the coastline. A downdip profile off Cow Bay Point shows a sequence of textures and bedforms characteristic of this nearshore zone. At the plunge point, pebbles and cobbles are concentrated. From just beyond the plunge point to a depth of approximately 5 m symmetric, straight-crested sand waves, composed of coarse sand to granule-size particles, form subparallel with the coast line. At depths of up to 7.5 m, asymmetric longshore current ripples are common. These may be straight crested, bifurcating, or linguoid in form. Wave ripples commonly are superimposed on the longshore ripples. Below 7.5 m bedforms of current origin are rare. At depths of 15 m and greater, grasses and algae are bent by water currents running parallel with the shore. Sediment at these depths is bioturbated by polychaete worms and eels.

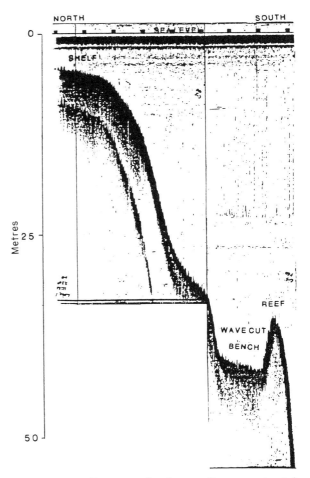

Figure 11.11. Bottom reflection profile approximately 300 m long perpendicular to shore showing the steep fault scarp with a possible wave-cut bench and reef at a depth of approximatly 40 m (reproduced by permission from Wescott and Ethridge, 1982)

In summary, the island's submarine slope in front of the Yallahs fan delta system is characterised by steep gradients (up to 20 to 30° slopes) and submarine valleys and canyon heads. Sediment distribution is patchy and characterized by an apparently random distribution of muds, sands, and gravels. Seaward dispersion of fine-grained, terrigenous detritus in the water column accounts for much of the fine-grained sediments that veneer the slope and comprise a large portion of the slope deposits. The fine-grained material is derived from the suspended load of Yallahs Rivers, suspended load in longshore currents, and winnowing of beach deposits by wave action.

From Cow Bay Point west to Grants Pen the nearshore and shallow offshore zones are characterized by a relatively flat, gently dipping island shelf. The shelf is approximately 600 m wide and is characterized by sand with local veneers of silt and mud. Gravels are uncommon on the shelf. Current-wave ripples are common and grasses and algae are abundant on the shelf.

Near the western margin of the shelf in 12 to 18 m of water is a small, living, patch reef approximately 30 by 60 m in area with its long axis parallel to the shore. Bottom sediment around the reef is composed largely of carbonate-skeletal sands with only minor amounts of terrigenous material. Current ripples are present on the bottom adjacent to the reef, and grasses and algae are common. A series of reefs paralleling the seaward edge of the shelf at depths of 45 m is suggested by the fathometer records (Figure 11.11).

Conclusions

Sediments derived from the Jamaican Blue Mountains have built a fan delta where Yallahs River flows from the mountain front and enters the Caribbean Sea. The subaerial portion of this delta complex is a humid tropical alluvial fan. The very coarse-grained fluvial sediments that characterize the bulk of this sedimentary pile are deposited by relatively high magnitude, low frequency events (i.e. floods with recurrence intervals of 5 to 10 yr). Moreover, the surface morphology of the present fan is probably created and maintained by these episodic events.

Fluvial sediments that reach the adjacent marine environment are modified by littoral and marine processes. The distribution of these sediments and the erosional or depositional nature of the deltaic coastline is most strongly influenced by the wave climate and wave-generated longshore currents. The intensity of wave energy impinging upon the Yallahs coast is controlled by refraction and diffraction processes which in turn reflect the shallow delta-front bathymetry.

The submarine Yallahs fan delta is characterized by two distinctive bathymetric zones. The westernmost offshore zone is a narrow, flat, shallow island shelf terminated seaward by an abrupt scarp. To the east the delta front is dissected by the heads of three submarine canyon systems.

Coarse sediments are introduced onto the fan delta front by Yallahs River when it is in flood, and by wave erosion of beaches. Sediments transported by longshore currents are temporarily stored on the delta-front slope or trapped in the heads of the canyons. Sediments continue to accumulate until a critical stress value is exceeded, at which time they fail and move down slope by gravity driven processes. Turbidity currents generated in front of the Yallahs fan delta have, in part, filled Yallahs Basin, eroded Yallahs Canyon and may ultimately deposit sediments on the Columbian Abyssal Plain.

References

Andrew, E. M. 1965. Seismic surveys over river gravels in Jamaica. *United Kingdom Overseas Geological Surveys Geophysical Report*, **28**, 64–78.

Burke, K. 1967. The Yallahs Basin: a sedimentary basin southwest of Kingston, Jamaica. *Marine Geology*, **5**, 45–60.

Davies, J. L. 1964. A morphogenic approach to world shorelines. *Zeitschrift für Geomorphologie*, **8**, 27–42.

DeLeonibus, P. S. and Sheil, R. J. 1973. Observations of wave spectra in the Caribbean Sea with an airborne laser. *Naval Oceanographic Office, Navoceano Technical Note*, **6110–6–73**, 15 pp.

Ethridge, F. G. and Wescott, W. A. 1984. Tectonic setting, recognition and hydrocarbon reservoir potential of fan-delta deposits. In Koster, E. H. and Steel, R. J. (Eds), *Sedimentology of Gravels and Conglomerates. Canadian Society of Petroleum Geologists Memoir*, **10**, 217–235.

Gupta, A. 1975. Stream characteristics in eastern Jamaica, an environment of seasonal flow and large floods. *American Journal of Science*, **275**, 825–847.

Gupta, A. 1983. High-magnitude floods and stream channel response. In Collinson, J. D. and Lewin, J. (Eds), *Modern and Ancient Fluvial Systems. International Association of Sedimentologists Special Publication*, **6**, 219–227.

Hendry, M. 1982. *Field Guide to the Yallahs Fan-delta Complex*. The Geological Society of Jamaica. 10 pp.

Holmes, A. 1965. *Principles of Physical Geology*. New York, Ronald Press Co. 1288 pp.

Jamaica Industrial Development Corp. 1974. *Cow Bay Wave Recordings*. Unpublished Report, Wimpey Laboratories, Hayes, Middlesex. 11 pp.

Kemp, A. 1971. *The Geology of the Southwestern Flank of the Blue Mountains, Jamaica*. Unpublished Ph.D. Thesis, University of the West Indies, Kingston. 306 pp.

Kochel, R. C. and Johnson, R, A. 1984. Geomorphology and sedimentology of humid-temperate alluvial fans, central Virginia. In Koster, E. H. and Steel, R. J. (Eds), *Sedimentology of Gravels and Conglomerates. Canadian Society of Petroleum Geologists Memoir*, **10**, 109–122.

McGowen, J. H. 1970. Gum Hollow fan delta, Nueces Bay, Texas. *University of Texas, Bureau of Economic Geology, Report of Investigations*, **69**, 91 pp.

Molnar, P. H. and Sykes, L. R. 1969. Tectonics of the Caribbean and middle American regions from focal mechanisms and seismicity. *Geological Society of America Bulletin*, **80**, 1639–1684.

Robinson, E., Lewis, J. F., and Cant, R. V. 1970. *Field Guide to Aspects of the Geology of Jamaica*. American Geological Institute—International Institute.

Rust, B. R. and Koster, E. H. 1984. Coarse alluvial deposits. In Walker, R. G. (Ed.), *Facies Models. Geoscience Canada Reprint Series*, **1**, 53–69.

Ruxton, B. P. 1970. Labile quartz-poor sediments from young mountain ranges in northeast Papua. *Journal of Sedimentary Petrology*, **40**, 1262–1270.

U.S. Naval Weather Service Command 1974. Summary of synoptic meteorological observations: Caribbean and nearby island coastal marine areas. *Area 9, Jamaica South*, **2**, 158–236.

Wescott, W. A. and Ethridge, F. G. 1980. Fan-delta sedimentology and tectonic setting—Yallahs Fan Delta, southeast Jamaica. *American Association Petroleum Geologists Bulletin*, **64**, 374–399.

Wescott, W. A. and Ethridge, F. G. 1982. Bathymetry and sediment dispersal dynamics along the Yallahs Fan Delta front, Jamaica. *Marine Geology*, **46**, 245–260.

Wood, P. A. 1976. Beaches of accretion and progradation in Jamaica. *Journal of the Geological Society of Jamaica*, **15**, 24–31.

SECTION II

INTERPRETATION OF THE ENVIRONMENT OF ALLUVIAL FANS

PART A

Alluvial Fans in Studies of Palaeogeomorphology

CHAPTER 12

Evolution of the Alluvial Fans of the Alföld

Z. Borsy
Lajos Kossuth University, Debrecen

Abstract

The Pannonian Lake which formerly occupied the area of the Alföld was infilled during the Pliocene Epoch. During this process, a system of watercourses evolved which persisted until Würmian time and built extensive alluvial fans into the subsiding basin. One of the largest fans was formed by the Danube River. A large alluvial fan plain evolved in the southern foreland of the Inner Carpathian Mountains and in the northeastern part of the Alföld, where the most active fan building rivers were the Tisza and Szamos. In the southeastern part of the Plain a large alluvial fan was deposited by the Maros.

Examination of material from boreholes has shown that the evolution of the fans was not continuous. The rivers abandoned parts of them for longer or shorter periods and there also occurred instances of slight entrenching. Considerable areas became flood free at times, facilitating soil development.

During the last third of the Quaternary Period, surface evolution differed between the eastern and western parts of the Alföld. In the east, the deposition of fluvial sediments continued, whereas on the Danube–Tisza Interfluve there were several periods of blown sand movement and, at places, of loess formation.

During the early Würm glaciation, the network of waterways began to change, primarily because of continued tectonic foundering. With the exception of the Maros, the rivers abandoned part or the whole of their alluvial fans. Large areas became free of floods. In the cold, dry upper pleniglacial climate, blown sand movement and, at places, loess formation occurred.

After the rivers of the Alföld had abandoned their alluvial fans or become entrenched, their action was restricted to lower lying floodplains. The lowest, sediment collecting depressions between the levees have been transformed during the second half of the Holocene Epoch into poorly drained, waterlogged swamps.

Introduction

In areal extent (52 000 km^2), the Alföld is the largest of the regions of Hungary (Figure 12.1). This region of the Carpathian Basin extends beyond the political frontiers, into Czechoslovakia, the Soviet Union, Romania and Yugoslavia. The highest point of the region can be found at Mezőföld (277 m above sea level). No point east of the Danube reaches a height of 200 m above sea level. The larger part of the Alföld is a classic plain of fluvial accretion. However, at places, eolian processes have also contributed significantly to its surface evolution.

There had been marked tectonic movements in the area of the Alföld even by the end of the Quaternary Era. This factor, however, was not realized by researchers for a long time. This fact mainly accounts, in more than one respect, for their erroneous views on the surface evolution of the Alföld.

It was J. Sümeghy (1944) who for the first time

in the Hungarian geological literature emphasized that, during the course of the Pleistocene Epoch, extensive alluvial fans evolved in the Alföld. In his opinion, the Upper Pleistocene surface of the Alföld was characterized by alluvial fans with low-lying areas between them.

During the decades since Sümeghy's work a large step has been taken towards a fuller understanding of the Alföld, and new features of alluvial fan evolution have been revealed. This is primarily due to the series of deep core drillings sunk by the Lowland Department of the Hungarian Geological Institute since the beginning of the 1960s. Multidisciplinary analysis of the material from these drillings has yielded a great deal of information on several sedimentological properties of the drilled layers, the cycles of sediment formation, the extent of tectonic movements, and the thickness of the Pliocene and Quaternary strata (Rónai, 1985). A great number of data also have been obtained on the evolution of the alluvial fans of the Alföld. Parallel with this research, detailed geomorphological and geological explorations were performed in various parts of the Plain.

Although elucidation of the properties of the Alföld alluvial fans in more detail requires further investigations, results obtained so far entitle us to propound a novel explanation of alluvial fan evolution. The publication of the present study is all the more important because the alluvial fans of the Alföld represent a special type. They evolved in a basin some parts of which subsided as much as 300 to 700 m during the Quaternary Era. Thus, the formation and complete evolution of the alluvial fans were considerably influenced by tectonic movements, as well as by the discharge of rivers, transport of sediments and climatic changes.

Schematic Evolution of the Pannonian Basin in Neogene Time

At the beginning of the Miocene Epoch, the greater part of the present-day territory of Hungary was dry terrain with surface formations char-

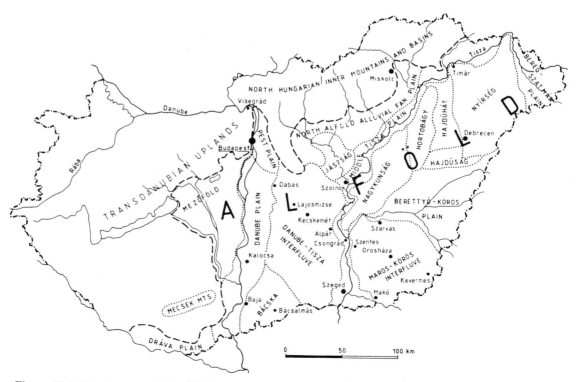

Figure 12.1. Landscapes of the Alföld (Great Hungarian Plain) and principal places mentioned in text

acteristic of subtropical regions. This phase ended in the Carpathian substage when, due to crustal movements, a transgression of the sea began from the southwest. In the Badenian substage the expansion of the sea continued, thus, the dry land area became smaller than the area inundated by this Parathethys Sea.

As a consequence of tectonic movements at the end of the Sarmatian substage the territory of Hungary, or more exactly of the larger Pannonian Basin, began to sink at a faster rate. Hence the water of the Pannonian Lake, now cut off from the Parathethys Sea, covered a considerably larger area in the Lower Pannonian substage than the Sarmatian inland sea. On the other hand, the larger part of the Hungarian Inner Mountains was not inundated. In the Lower Pannonian substage there were, in both Transdanubia and the Alföld, numerous islands in the Pannonian Lake, onto most of which the water of the lake transgressed only in the Upper Pannonian substage (Figure 12.2).

Because of the rise of the Carpathians and the Dinaric Mountains the Pannonian Lake became more and more isolated from the remains of the fragmented Parathethys Sea, and the only contact with even the Wallachian Lake was maintained through the 'Lower Danube'.

It is well-known that every lake is filled up sooner or later, especially when it is surrounded by mountains as high as the Alps, Carpathians and Dinaric ranges. The filling up of the Pannonian Lake took a relatively long time, since its bottom sank very deep, especially in the Alföld. Filling up took place mostly through river deltas. The very fine grained sediments (clay and silt) which were not deposited at the deltas, were distributed by the lacustrine currents over the whole area of the lake, including the deeper parts of the basin.

In the Upper Pannonian (Pontian) substage, in the last phase of the infilling, the Pannonian Lake was completely shallow and on the margins of the Pannonian Basin the evolution of alluvial fan

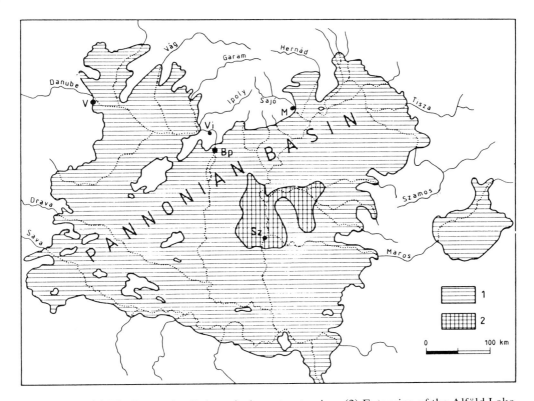

Figure 12.2. (1) The Pannonian Lake at its largest extension; (2) Extension of the Alföld Lake in the Upper Pliocene (so far explored). Constructed by Z. Borsy and F. Franyó

plains began. The rivers, with their large amounts of bedload, penetrated right into the central regions of the basin. Thus, transport by lacustrine currents, which was considerable at the beginning of the Pannonian substage, gradually lost its importance. Step by step the whole sediment collecting basin was transformed into a fluviolacustrine system. The rivers of this special network of watercourses deposited very large quantities of sand; 30 to 100 m thick cross-beds covering nearly the whole of Transdanubia.

The Pannonian Lake was first filled up in Transdanubia and the northeastern part of the Alföld. K/Ar dating (Borsy et al., 1986) shows that 3.5–3.2 million years ago the greater part of Transdanubia was already an infilled plain, and the Pannonian Lake had shrunk back to the area of the Alföld (Figure 12.2). The continuously diminishing lake was relatively quickly filled up by the adjoining watercourses, rich in bedload by the end of the Pliocene Epoch, so the Alföld, too, was transformed into an emergent plain. However, Transdanubia lay, at the end of the Pliocene, somewhat higher than the Alföld, which continued to sink. There were differences between the two regions in the character of the deposits too. As we have noted, in Transdanubia the surface was covered by a great deal of cross-bedded sand, whereas on the Alföld very thick strata of varicoloured clays characteristic of a semiarid climate were formed.

In the course of 8 million years very large quantities of sediments were deposited in the Pannonian Lake; in the whole area of Hungary about 50 000 km^3. On the margins of the Hungarian Inner Mountains 100 to 600 m, and in the middle of the Pannonian Basin 600 to 4500 m thick successions evolved. The thickest deposits were detected in the depressions on the Alföld in drillings carried out during hydrocarbon prospecting.

As the Pannonian Lake was shrinking and became smaller and smaller, a new system of watercourses began to develop in the Pannonian Basin. Nevertheless, the system of watercourses at the end of Pliocene Epoch considerably differed from the present one. The Danube already flowed through the Visegrád Gate (Pécsi, 1959, 1980), but on reaching the Pest Plain it did not take a turn to the south but southeast, towards the Upper Pliocene Alföld Lake (Figure 12.2). When this latter was filled up, it continued to flow along the line Csongrád–Szentes–Szeged–Makó, since at that time this was the deepest and most intensely sinking part of the Alföld. The Tisza established its course across the Nyírség, together with its tributaries, in the direction of Szarvas–Csongrád, and flowed into the Danube somewhere round Szentes–Csongrád.

Filling of the Alföld in the Quaternary Period; Evolution of the Alluvial Fans

In the Quaternary Period the greater part of the Alföld continued to sink. This process was especially remarkable in the region between Csongrád, Szentes, Makó and Szeged, as well as in the Körös and South Jászság basins. The eastern wing of the Maros Alluvial Fan and the northeastern part of the Alföld subsided to a lesser degree, whereas the western belt of the Danube–Tisza Interfluve was still a degradational surface at the beginning of the Quaternary. The Danube, arriving in the Alföld, continued to flow towards the Csongrád–Szeged line (Figure 12.3). From the direction of the Nyírség the Tisza, touching the Körös region, also flowed towards Csongrád. At that time the watercourses of the North Hungarian Inner Mountains penetrated deep into the central region of the Alföld and joined the Tisza south of Hortobágy and the Nagykunság.

The watercourses arriving in the Alföld built up extensive alluvial fans. As a concomitant process of alluvial fan evolution and partly due to tectonic effects the rivers frequently changed their beds. However, the main directions did not essentially change until the Würm glaciation (Borsy et al., 1969).

In the western half of the Alföld two alluvial fans were built up. A smaller one, the alluvial fan of the ancient Sárviz, advanced from the direction of Mezőföld southeastwards to the region of Bácska. North and northeast of this was formed the immense alluvial fan of the Danube, the proximal end of which lies on the Pest Plain.

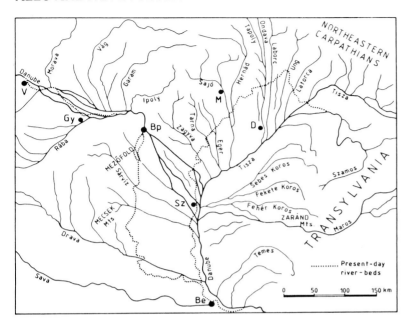

Figure 12.3. Network of watercourses at the beginning of the Quaternary Era

From laboratory experiments it is clear that the alluvial fan of a watercourse arriving on a plain has a relatively large angle of slope in the initial phase (Kádár, 1957; Borsy, Kádár and Pinczés, 1972). The river, deepening its valley in the reach above its alluvial fan, later degrades the proximal end of the fan and, if this process continues, a series of terraces is brought about. At the same time the gradient of the accumulating part of the fan shows a gradual decrease.

Similar processes took place on the alluvial fan of the Danube. At the beginning of the Quaternary Period the Danube deposited large amounts of gravel on its alluvial fan (Franyó, 1980). After depositing Lower Pleistocene sediments the river cut into the head part of the fan several times and formed four terraces. The surface of the Lower Pleistocene alluvial fan and terraces III and IV on the south and southeast margins of the Pest Plain were buried abruptly below the younger Pleistocene formations of the alluvial fan (Pécsi, 1959, 1967, 1970). The coarser-grained Lower Pleistocene fluvial sediments, which are still on the surface on the Pest Plain, are found at a depth of 200 to 250 m near Kecskemét and 500 m at Kiskunfélegyháza (Rónai, 1985). The gravel layer becomes finer in several cycles upwards and contains more and more sand and silt (Franyó, 1980). In later periods mostly sand was deposited on the Danube alluvial fan. The finer-grained floodplain sediments are of minor importance.

As a consequence of the more intensive subsidence of the areas round Csongrád and Szeged the successive sediment layers of the Danube alluvial fan necessarily became thickest (500 to 600 m) to the southeast (Figure 12.4).

The watercourses arriving from the North Hungarian Inner Mountains formed a wide range of alluvial fans running from the Zagyva to the Sajó and Hernád which, crossing the present-day course of the Tisza, advanced deeply into the central part of the Alföld. The smaller watercourses, the Zagyva, Tarna and Eger deposited large amounts of clay and silt, on their alluvial fans too. One of the special features of the range of the North Alföld alluvial fans is that, as a result of tectonic movements, the Quaternary sequence of layers is the thickest (300 to 400 m) in the southern part of the Jászság and the present-day middle reaches of the Tisza. South of this area, in the territory of the Nagykunság, in a belt running northeast–southwest, the thickness of the sequence is 150 to 200 m.

An extensive, compound alluvial fan was built

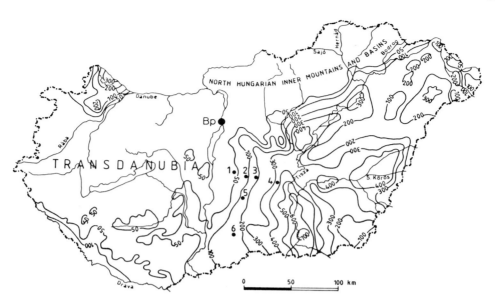

Figure 12.4. Thickness of Quaternary sediments in Hungary (m). (Courtesy of A. Rónai and F. Franyó). Sites of core drillings illustrated in Figure 12.5: 1. Kunadacs, 2. Kerekegyháza, 3. Kecskemét, 4. Nyárlőrinc, 5. Kaskantyu, 6. Jánoshalma

in the northeastern part of the Alföld where all the watercourses coming from the northeastern Carpathians and North Transylvania participated in building the alluvial fan. The largest amount of work was done by the two largest rivers, the Tisza and the Szamos. In the northern and northeastern parts of the alluvial fan (present-day Bodrogköz and Bereg–Szatmár Plain) the rivers deposited a great deal of gravel. The gravel layers gradually become thin and pass into sand layers. In the successive sediment layers in the middle and southern parts of the Nyírség, gravel occurs only in the Lower Pleistocene strata.

The Körös region, which was intensively sinking, remained low-lying during the entire Quaternary Era. The rivers were scarcely capable of filling this area, even though the Tisza transported a great deal of sediment.

In the course of the Quaternary, the Maros, too, built up an extensive alluvial fan. In spite of the fact that the river is rich in bedload, its alluvial fan remained relatively low-lying (Borsy *et al.*, 1969). This condition is characteristic of the Maros alluvial fan even today. On arriving in the Alföld, the Maros deposited large amounts of gravel in the western piedmont area of the Zaránd Mountains. However, the greater part of the alluvial fan was built up of sand, silt and clay. At the proximal end of the fan the Quaternary sequence is thinner than 100 m, but it thickens to 500 to 700 m westwards.

The building of the Alföld alluvial fans was not continuous. As a consequence of their evolution the rivers abandoned smaller or larger parts of their alluvial fans for longer or shorter periods. To a lesser extent they may even have degraded. Thus, large areas may have become free from inundation and soil formation could start on them. Exploratory boreholes in the Alföld have revealed fossil soils in a great number of cases even in the most intensely sinking basins (Rónai, 1972, 1985; Franyó, 1980). At places where chernozem or forest soils were formed aggradation was stopped at least for some time.

Within the last third of the Quaternary the history of the eastern and western parts of the Alföld has been different. Electron microscopic examination of sand from core drillings near Komádi in the Körös Plain (Borsy *et al.*, 1983) and at Tótkomlós on the Maros Alluvial Fan

ALLUVIAL FAN EVOLUTION

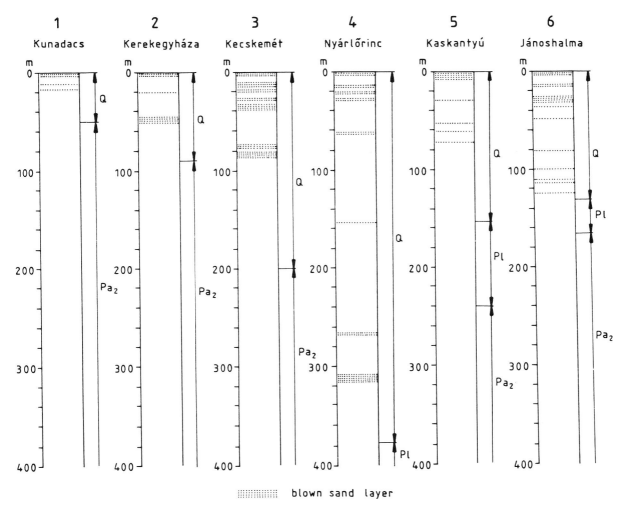

Figure 12.5. Blown sand layers in the core drillings in the Danube–Tisza Interfluve, core locations given in Figure 12.4 (Q—Quaternary, Pl—Pliocene, Pa$_2$—Upper Pannonian)

(Borsy *et al.*, 1985) demonstrated that the sequence of deposits in both boreholes is completely fluvial. At the places mentioned blown sand cannot have been formed from alluvium even during the more arid glaciations since, due to the groundwater level, the surface remained more compact. However, on the Danube alluvial fan, as well as on the alluvial fan of the ancient Sárvíz in the southwestern part of the Danube–Tisza Interfluve (Borsy *et al.*, 1982) deeper lying, blown sand layers occur in each core drilling (Figure 12.5).

The electron microscopic analyses testify that on the Danube alluvial fan the largest amount of subsurface blown sand is found in the vicinity of Kecskemét. This area was under the influence of aeolian effects as early as the first half of the Würm Glaciation. This is reflected in aeolian layers which lie down to the depth of 38 m. The deepest layer of blown sand in the Kecskemét borehole was found between 83.5 and 84.5 m.

Analysis of the sand from the core drilling sunk near Nyárlörinc is very instructive. Altogether seven blown sand strata were found in the borehole. Blown sand conspicuously occurs even at the depths 264 to 266 m and 307.4 to 313 m in typically fluvial transport layers.

In the 130.3 m thick Pleistocene sequence of the core at Jánoshalma (ancient Sárviz alluvial fan) more than 10 blown sand strata were revealed by electron microscopic examinations. Another interesting feature of the drilling at Jánoshalma is that down to 116 m the drill traversed 26 loess, sandy loess, and loessy sand layers. The fluvial material between the aeolian layers shows that these beds were laid down by smaller watercourses, which drained to the southwestern part of the Danube–Tisza Interfluve from the Mecsek Mountains or the Mezőföld (Borsy et al., 1982). The great number of aeolian layers at Jánoshalma shows that in the evolution and formation of the ancient Sárviz alluvial fan, aeolian processes played a significant role.

The sediment deposited by the smaller watercourses was moved and transformed into blown sand by the wind during arid climatic periods. As a rule a loessy blanket was formed on the blown sand, onto which, due to the slow sinking of the area, fluvial successive layers were deposited.

The subsurface blown sand layers of the Danube alluvial fan prove that some parts of that fan were also free from inundation at times. If the climate was arid as, for example, during the glaciations, the fluvial material on the surface was moved by the strong northerly and northwesterly winds, and thinner or thicker blown sand layers were produced. During the later evolution of the alluvial fan further fluvial sequences settled over the aeolian strata.

During the evolution of the blown sand a lot of dust was winnowed out of the material of the alluvial fans and in more arid climatic periods also from the floodplains of the rivers. After the subsidence of storms in the periglacial climate the airborne dust settled down as loess and sandy loess (Borsy, 1973). If this falling dust was carried to the floodplain it got mixed with the finer-grained sediments of floods. In the periods of periglacial climate these gave rise to the formation of 'floodplain loess', floodplain loess-like sediments which occur under the surface at a number of places in Upper Pleistocene formations.

Changes in the Hydrography of the Alföld during the Würm Glaciation

At the beginning of the Würm glaciation considerable changes took place in the Alföld. As a result of the more intensive subsidence of the region round Szeged the Danube left the part of its alluvial fan that lay north of the line Alpár–Kecskemét–Lajosmizse–Dabas, gradually took a southward turn and made its way from the Pest Plain directly towards Szeged. Wandering along a relatively wide band, the river cut slightly into this part of its alluvial fan.

Due to the southward shift of the Danube, the northern flank of the alluvial fan gradually became free from floods and so in the first half of the Würm glaciation aeolian evolution of its surface started.

In the middle of the Würm the Danube abandoned its alluvial fan (Figures 12.6, 12.7), since the depression forming along the Kalocsa–Baja–Zombor line gradually attracted the river to its present-day course. In this way further alluvial fan-building activity by the ancient Sárviz ceased, since this watercourse was now incapable of flowing through to the Bácska territory.

With this change an important phase in the history of the Danube alluvial fan was brought to a close, since now the aggradational activity of the river ceased also on the southern flank of the alluvial fan. Since the upper pleniglacial period, when a cold, arid climate dominated, the wind has been the prevailing surface forming factor.

At the beginning of the Würm the Tisza and the Szamos also abandoned the Nyírség alluvial fan (Figures 12.8a and b) and wandered to near the present-day Ér valley (Borsy and Félegyházi, 1983). The rivers Tapoly, Ondava and Laborc, arriving from the northeastern Carpathians, continued their flow across the Bodrogköz towards the Nyírség and, corresponding to the modest amount of their bedload, continued modest building of the alluvial fan for some time. The Tisza–

ALLUVIAL FAN EVOLUTION

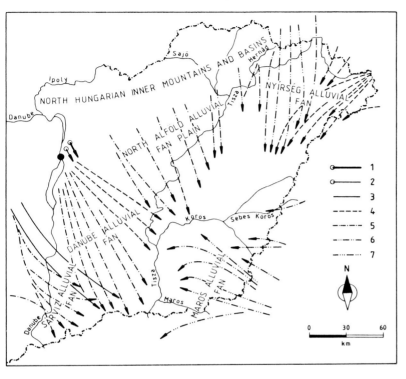

Figure 12.6. Tendencies of alluvial fan evolution in the Alföld. 1. The part of the Lower Pleistocene alluvial fan exposed on the surface. 2. The part of the Middle Pleistocene alluvial fan exposed on the surface. 3. The alluvial fan built until the Upper Pleistocene. 4. The alluvial fan built until the beginning of the Würm. 5. Alluvial fans built until the middle of the Würm. 6. The alluvial fan built until the beginning of the upper pleniglacial period. 7. The alluvial fan built until late glacial time

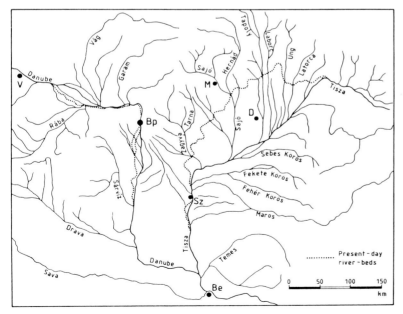

Figure 12.7. The network of rivers in the middle of the Würm glaciation

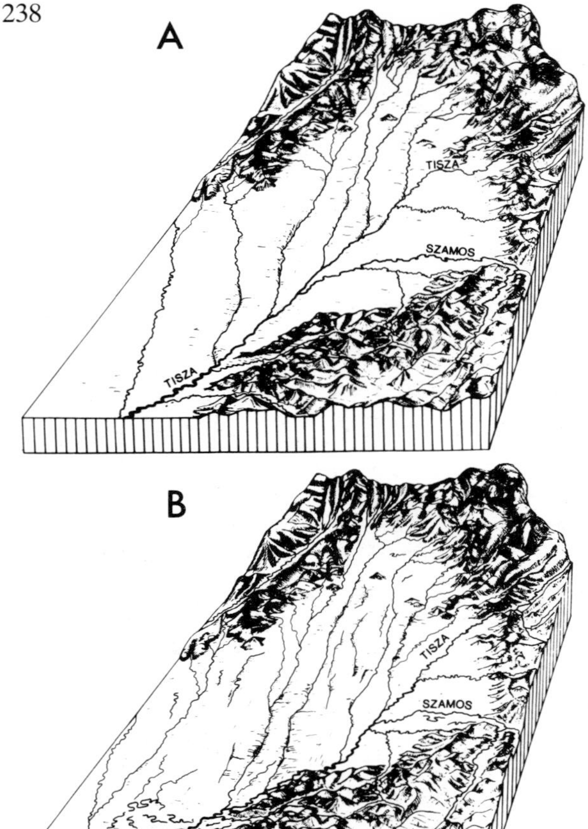

Szamos first developed a wide erosional plain on the southern margin of the Nyírség, then cutting into it because of the subsidence of the Körös region, brought into being the Ér valley which, in Alföld terms, is considered deep. The degradation of the Tisza–Szamos influenced the evolution of the Szatmár plain which, in this way, became somewhat lower than the Nyírség.

At the beginning of the upper pleniglacial period the territories surrounding the Nyírség in the east and north began to sink. Since in Bodrogköz and on the Bereg plain subsidence was more intense than in the Szatmár area, the Tisza abandoned the Ér valley and turned northwest, then took direction towards the Bodrogköz. This shift started the transformation of the Bodrogköz part of the alluvial fan. The Szamos remained for several thousand years in the Ér valley and abandoned it only 16 000 to 14 000 years ago to join the Tisza on the Bereg plain (Figure 12.8c).

The arrival of the Tisza in the Bodrogköz completely changed the earlier hydrography of the region (Figure 12.8c), since the rivers previously flowing southwards towards the Nyírség joined the Tisza. This meant that the depositional activity of the watercourses stopped in the territory of the Nyírség too. On leaving the Bodrogköz the Tisza did not yet flow in its present-day direction, but turned south and meandered on the western margin of Hajdúhát towards the south, then southwest.

The greater part of the North Alföld alluvial fan was low lying as compared to the Nyírség or the Danube alluvial fan even at the beginning of the Würm glacial period. This is partly due to the fact that the rivers building the alluvial fan, had, apart from the Sajó, small water discharge and transported relatively small amounts of bedload. On the other hand, in some parts of the alluvial fan plain, as we have noted, subsidence was intense. The filling-up of such areas required a great deal of sediment.

In the middle of the Würm significant changes took place on this alluvial fan too. Subsidence became more intense in the southern part of the Jászság, and the streams flowing from the Cserhát and Mátra were attracted there. The lower reach of the Zagyva gradually shifted eastwards; the

Figure 12.8. The network of rivers in the northeastern part of the Alföld as constructed by Z. Borsy and E. Félegyházi (1983): (A) at the beginning of the Würm glaciation; (B) at the beginning of the upper pleniglacial period; (C) 16 000 to 14 000 years ago

Tarna, together with its tributary brooks, turned southwest and abandoned the Nagykunság where it had exerted a significant filling activity.

Because of the subsidence of the South Jászság the Zagyva and Tarna cut into their alluvial fans, and these surfaces became free from inundation. The slightly higher-lying areas of Nagykunság also became free from floods, since, due to the subsidence of the Jászság the rivers no longer reached them.

During the Quaternary era the Sajó, leaving the Miskolc Gate, assumed, in general, a south to southeasterly direction towards Hortobágy. At times, however, it touched the northeastern part of the Nagykunság as well. In the middle of the Würm, probably for tectonic reasons, it abandoned this direction and turned from the Miskolc Gate southeast, east, then in the upper pleniglacial period cut into this part of its alluvial fan. Thus, areas free from inundation came into being on the Sajó alluvial fan, too.

At the beginning of the Würm the river Maros also cut into the proximal end of its alluvial fan. On the other hand, in the foreland it continued to build its fan-like alluvium. At the beginning of the upper pleniglacial period the river abandoned the part of the fan situated along the line Kevermes–Orosháza. On the parts of the alluvial fan where evolution continued, it has, since that time, deposited primarily fine-grained sediments.

A few metres deeper than the low-lying Maros alluvial fan lay the Körös Plain, where aggradation continued during the whole of the Würm. The branches of the Körös cut into their alluvial fan only in the second half of the Würm, directly at the foot of the mountains. The meandering, levee-building rivers of the Körös Plain similarly deposited, during the upper pleniglacial period, fine-grained sediments (silt and clay).

Transformation of the Alluvial Fan Surfaces in the Upper Pleniglacial Period

In the foregoing we have shown that by the upper pleniglacial period the Danube had abandoned its alluvial fan. Nor did the Tisza and Szamos flow through the Nyírség, and thus, large inundation-free areas were formed on the North Alföld alluvial fans too. A smaller belt of the Maros alluvial fan also became free of inundation. On the alluvial fans left without flowing watercourses and on the inundation-free parts of the alluvial fans large amounts of fluvial sand were deposited on the surface. This is especially clear on the Danube alluvial fan and the large alluvial fan formed in the northeast part of the Alföld.

In the upper pleniglacial period (28 000 to 13 000 years B.P.) which followed the Stillfried interstadial (35 000 to 28 000 years B.P.), the climate became colder and more arid again. At the maximum of the cooling the mean annual temperature may have been s3°C; s12 to s13°C in January, and 10 to 12°C in July. The annual amount of precipitation was 180 to 250 mm (Borsy et al., 1982). Under these cold, arid climatic conditions the fluvial deposit on the surface was covered only by the scanty vegetation characteristic of cold steppe regions. This vegetation could not provide protection against the strong northwesterly, northerly, and northeasterly winds, thus, in the unprotected areas blown sand movement took place and the evolution of various sand formations could begin. The greatest amount of sand movement took place at the time of the first cold maximum of the upper pleniglacial period, 27 000 to 22 000 years ago. The intensive sand movement transformed the surface of the alluvial fans considerably not only by producing varied blown sand surfaces, but also by covering or deflating the abandoned riverbeds at many places on the alluvial fans.

After the first cold maximum the climate became slightly milder and more humid. The blown sand areas with lower relief energy, the deflation flats, were now better protected by the steppe vegetation; thus, the accumulation of a falling dust blanket could begin on the surface.

In the course of diagenesis under periglacial climatic conditions the deposited dust was transformed into loess and loessy sand. The loessy mantle was formed over large areas particularly in the Nyírség and in the Danube–Tisza Interfluve (Figure 12.9). In addition, remarkable loessy blankets evolved on the blown sand surfaces of the Hajdúhát and Nagykunság. Smaller spots

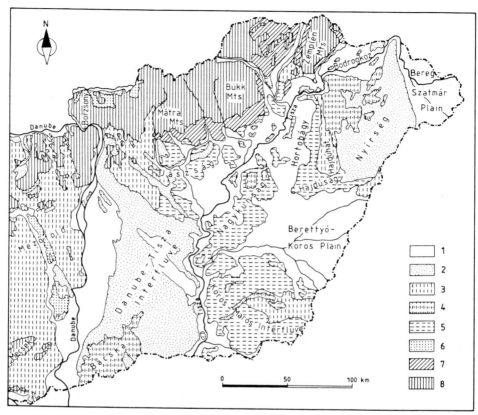

Figure 12.9. Distribution of blown sand, loess, sandy loess, loessy sand, and floodplain loess in the Alföld. 1. Holocene floodplain formations; 2. blown sand primarily of upper pleniglacial age; 3. loess, sandy loess; 4. sandy loess, loessy sand covering dunes; 5. floodplain loesses, floodplain loess-like sediments; 6. other Pleistocene sediments: sand, gravel; 7. brown soil; 8. Pre-Pleistocene formations

were formed near the proximal end of the Zagyva and Tarna alluvial fans and on the Maros alluvial fan.

The thickness of the upper pleniglacial loessy mantles on the alluvial fans varies between 50 and 520 cm today (Borsy, 1973). On studying several hundreds of exposures the thickest loess blanket (Figure 12.10a) was found in the northwestern part of the Nyírség (520 cm). Loesses of similar thickness, 4 to 5 m, were also found in the southwestern part of the Danube–Tisza Interfluve (Figure 12.10b). The loess blanket is of considerable thickness in the northern part of Hajdúhát as well.

The loess or loessy blanket covering the dunes has preserved the sand formations of upper pleniglacial age. Except for a small number of places the blanket is always thinnest on the tops of the dunes; the greatest thickness is observed on the sides of the dunes and in the depressions between them. On the parts of the Alföld alluvial fans where no loess blanket was formed blown sand could move, with longer or shorter interruptions, until the late glacial period. The strike direction of the loess-covered dunes shows very well the wind conditions prevailing on the Alföld in the upper pleniglacial period. It is striking that, even today, the prevailing winds in the Nyírség and in the Danube–Tisza Interfluve have the same direction as those that produced the various sand formations on the Upper Pleistocene alluvial fans.

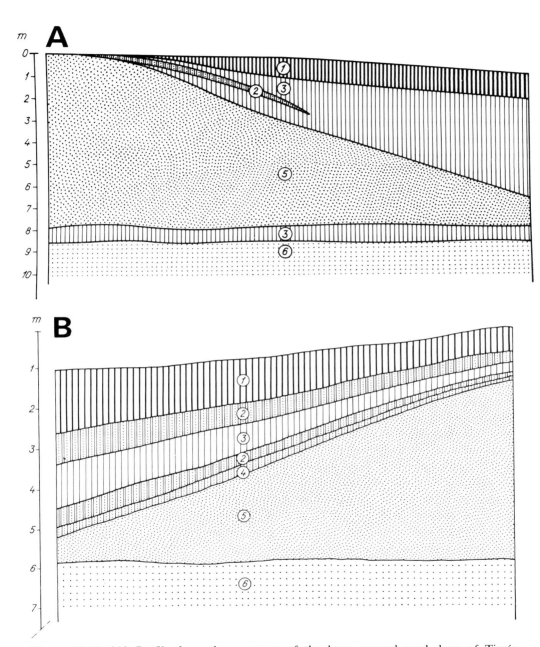

Figure 12.10. (A) Profile from the exposure of the loess-covered sand dune of Timár (northwestern part of the Nyírség); (B) Profile of an exposure 2 km southwest of Bácsalmás (southwestern part of the Danube–Tisza Interfluve). 1. Chernozem or brown forest soil; 2. sandy loess; 3. loess; 4. loessy sand; 5. blown sand; 6. silty fluvial sand

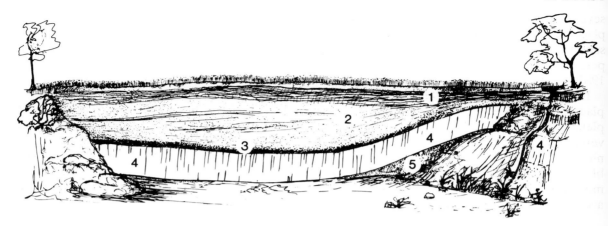

Figure 12.11. Sketch of exposure west of Aranyosapáti (Nyírség). 1. brown forest soil with iron-pan layers; 2. blown sand evolved in the Older Dryas; 3. fossil soil evolved in the Bölling interstadial; 4. loess; 5. blown sand evolved in the first half of the upper pleniglacial period

On the lower-lying parts of the alluvial fans typical loess evolved at only a few places, on inundation-free land. On areas temporarily inundated by the rivers more fine-grained fluvial sediment was deposited than falling dust. From this material, in the course of diagenesis under periglacial climate, floodplain loesses and floodplain loessy sediments were formed. The greater part of the Maros alluvial fan, significant areas in the Jászság, Nagykunság and the margin of the large North Alföld alluvial fan are covered by floodplain loessy sediments (Figure 12.9). In such sediments the proportion of the clay and silt fractions is at places very high.

Blown sand movement decreased considerably on the Alföld alluvial fans at the end of the upper pleniglacial period. In fact, in the Bölling interstadial, when the climate turned somewhat milder and more humid than during, the previous one, the dunes were gradually overgrown by steppe vegetation. At places brush vegetation gained ground and forest groves also appeared. Under the vegetation cover characteristic of cold continental steppes, steppe-like soil formation started. In the Older Dryas this process was interrupted at several places in the higher lying areas of the Danube–Tisza Interfluve, Nyírség, Bodrogköz, and Tarna alluvial fan when blown sand movement started again. This is best demonstrated by the soil profiles from the Bölling interstadial buried in the dunes, which have been exposed recently (Figure 12.11) at several places. The age of the soils was determined by C^{14} dating of charcoal remains found in them to be 12 900 ± 360, 12 680 ± 400, 12 420 ± 360, 12 330 ± 320 years B.P. (Borsey et al., 1982; Borsy et al., 1985; Csongor and Borsy, 1987).

The short-term sand movement in the Older Dryas ended in the Alleröd period. In the warmest stage of the Alleröd the mean temperature in July may have been only about 4°C lower than it is today. As a result of the climate, more humid than that of the Older Dryas, the steppe vegetation became more and more contiguous on the sand surfaces. The areas covered by lower dunes were occupied by forest steppe, whilst the higher and somewhat more arid sand dune surfaces were overgrown by contiguous steppe vegetation, fairly well protecting the blown sand blanket. On the dunes covered by loessy blanket, soil of steppe character evolved under the vegetation, whereas on the blown sand, steppe-like sand soil was formed. These fossil soils have been exposed in recent years at several places under the thinner or thicker sand blankets. On the basis of C^{14} dating their age proved to be clearly Allerödian (11 590 ± 240, 11 400 ± 250, 11 350 ± 360, 11 250 ± 350, 11 000 ± 300 years B.P.). These soils show very well that in the Younger Dryas, following the Alleröd period, blown sand began to move at several places on the Alföld alluvial fans, at places accumulating as much as 3 to 10 m over the sand formations evolved in the upper pleniglacial period (Borsy et al., 1982; Csongor and Borsy, 1987).

In the late glacial period sand movement took place in much smaller areas than in the upper pleniglacial. The greater part of the dunes, covered by loessy mantle, remained motionless. Significant sand movement occurred, even in the blown sand regions, primarily in the higher-lying, more arid dune regions, in particular on the larger accumulated sand fields.

Aeolian Landforms on Alluvial Fans

At the time of the cold, arid climate of the end of the Würm (upper pleniglacial, late glacial), landforms characteristic of partly overgrown or semi-overgrown sand regions evolved on the Alföld alluvial fans (Borsy, 1961, 1968a, 1968b, 1977, 1978, 1986). It was the Nyírség surface that became the most varied. Everywhere in the northern part of the Nyírség wind furrows, oval-shaped sand hummocks and residual ridges evolved, whereas in the southern territories parabolic dunes with undeveloped western wings were produced. In the northern half of the Nyírség one can observe everywhere smaller or larger depressions of deflational origin and larger deflational flats as well. The sand material blown out of these was rearranged in fields of accumulation. However, the majority of the dunes are no higher than 10 to 12 m.

The Danube–Tisza Interfluve did not become so varied as the Nyírség. It is surprising how many flat, featureless sand surfaces and aeolian landforms with low relief are found on these vast alluvial fans. This is primarily due to the fact that in the upper pleniglacial period the groundwater level may have been near the surface at many places on the Danube alluvial fan so the wind was unable to build deeper wind furrows or larger, higher dunes. On the greater part of the fan the height of the dunes does not exceed 5 m. Dunes higher than 10 m only occur in the western part of the region, where they stand out like islands.

Impressively large sand dunes can be seen in high-rising sandfields at places in the Danube–Tisza Interfluve. The vast oval-shaped dunes or sand accumulations virtually attached to their wind furrows and resembling a parabola with short wings can sometimes reach the height of 20 m (Borsy 1968b, 1977). In view of these forms the southwestern part of the Danube–Tisza Interfluve must be viewed as a special area. Here the majority of dunes are covered by a loessy mantle. At some places the characteristic formations are surfaces delicately featured with oval wind furrows and oval-shaped dunes; at other places the most common features are long, narrow wind furrows or smaller flat surfaces of deflational origin. The oval-shaped dunes are frequently pushed into one another to form a longitudinal range. One of the characteristic forms of the region is the hairpin-like parabolic dune. This formation is very similar to the corresponding forms in the Thar desert and on the Californian coastline (Melton, 1940).

The blown sand forms on the North Alföld alluvial fan are also worth attention. This region is today divided into two parts by the Tisza. This is why it was not recognised for a long time that the dunes southeast of the Tisza evolved essentially on an alluvial fan. On the basis of Cholnoky's work (Cholnoky, 1907) the predominant view was that the dunes along the river are riverbank dunes of the Tisza. However, at the beginning of the 1960s it was pointed out that, just as on the immediate foreland of the Inner Carpathian Mountains, in the Nagykunság the predominant forms were wind furrows, oval-shaped sand hummocks and residual ridges. At places, parabola-like sand accumulations occur too (Borsy, 1968a).

Surface Forming Activity of the Alföld Rivers Since the Late Glacial Period

While blown sand movement and evolution of a loessy mantle on the dunes was taking place on the North Hungarian alluvial fans and the Maros alluvial fan, the deposition of fine-grained sediments continued on the frequently inundated

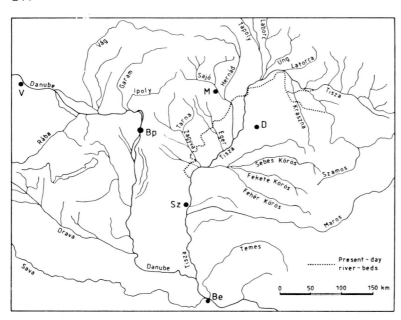

Figure 12.12. Network of watercourses in the Alföld in the late glacial period

areas of the two regions. The same process took place in the lower-lying parts of the Hajduság. From the fine-grained deposits, floodplain loesses and loess-like sediments were formed under the periglacial climate.

In the late glacial period the water output of the rivers increased. The Tisza and its tributaries accordingly increased the intensity of surface forming activity on the subsiding territories around the Nyírség (Bodrogköz, Bereg Plain). The greatest degree of transformation took place in the Bodrogköz where the meandering Tisza eroded more and more blown sand by lateral erosion and deposited its own bedload. The Nyírség was now surrounded from the north, too, by an extensive alluvial plain. The further increase of this plain only enhanced the 'relief island' character of the region. This may be the reason it was not recognised for a long time that the Nyírség is the remainder of a vast alluvial fan.

It has been mentioned already that the Tisza, eventually abandoning the Bodrogköz, still did not flow along its present-day line, but passed west of Hajduhát towards the south, then southwest (Figure 12.12). By lateral, meander forming activity it migrated over the whole of Hortobágy (as can be well seen in satellite photos) and occupied its present-day course only in the Subboreal phase.

To the southwest, as well, an extensive alluvial plain developed by the lateral erosional effect of the river, rising out of which with a break of 2 to 6 m height is the Nagykunság part of the alluvial fan. Because of the wide floodplain of the Tisza and the adjacent South Jászság Plain which, due to subsidence, was becoming more and more extensive, for a long time researchers did not suppose that the Nagykunság is actually the southern wing of an alluvial fan plain.

Along the reach between Szentes and Szeged the Tisza has cut slightly into its alluvial plain since the late glacial period. After the river occupied its present-day course across the Alföld this trenching activity extended upstream to the reach between Szolnok and Szentes as well.

The slight degradation of the Tisza was naturally followed by its tributary, the Maros. Thus, the further building of the Maros alluvial fan also ceased.

In the second part of the Holocene there was no alluvial fan in the Alföld in the process of evolution. On the contrary, due to the lateral erosional activity of the rivers, the existing alluvial fans had become somewhat smaller. The rivers have continued to exert their activities on the lower-lying floodplains, though (because of river regulation works) only a few of them can be regarded as real floodplains today.

The floodplains, in general, have two levels; a lower and an upper floodplain. The upper ones are generally levees. The levees, rising above the lower level sometimes by 2 or 3 m, are inundated only on the occasions of larger floodings. Levees evolved in the Berettyó and Körös Plains as early as the upper pleniglacial period. However, the majority evolved only during the Holocene. Because of the changing course of the riverbeds more than one levee was built even in the same area. Typical levees evolved in the Bereg–Szatmár Plain, in the Bodrogköz and on the Körös Plain. It is in the last region where the greatest number of levees was built on the Alföld. Between the levees, mostly during the younger part of the Holocene, moist, boggy, marshy areas evolved with poor drainage.

At several places on the flood plains of the Alföld we can see regions where no levees evolved, or if they occur, they hardly rise above the general level. At such places some variety is provided in the landscape by the shallow depressions of the oxbow lakes. This floodplain category also covers the majority of Hortobágy and the southwestern part of the Bodrogköz. In the latter territory, when one views the landscape from an elevated point after a major flood one can obtain a very interesting view of a well-developed system of point bars.

Since the end of the last century there have been great changes in the activity of the rivers on their own floodplains. To promote the run-off of floods, a number of meanders have been cut through and the rivers confined between dykes. Today the rivers exert their erosional–accumulational activities only on the areas between their dykes.

The Surface of Alluvial Fans in the Holocene Epoch

At the beginning of the Holocene Epoch, as a result of the gradual warming and the increased amount of precipitation, even the sand surfaces on the alluvial fans became more and more overgrown by vegetation, so that sand movement practically ceased. As our recent investigations prove, there was no significant sand movement in the northeastern part of the Alföld and on the North Alföld alluvial slope during the Holocene (Borsy et al., 1982). This statement is also valid for the major part of the Danube–Tisza Interfluve. Marked sand movement has occurred mostly since the 18th century as a result of deforestation or overgrazing. In the Danube–Tisza Interfluve and in the Nyírség, particularly in the sand fields of accumulational origin, one can clearly see how the evolution of newer forms resulted in the dissection of the dunes evolved in the upper pleniglacial or late glacial periods. On the other hand, these sand movements have not significantly altered the features of the surface of the two alluvial fans. In order to prevent further sand movement afforestation and the plantation of protective forest belts were begun at the end of the last century.

In the course of the Holocene the filling of the abandoned riverbeds of the alluvial fans with finer grained sediments from the surrounding, somewhat higher-lying areas continued. During the past 10 000 years deposits of 2 to 3.5 m thickness have accumulated in some abandoned riverbeds in the Nyírség and in the Danube–Tisza Interfluve (Borsy and Borsy, 1955).

Since the alluvial fans have not accumulated further during the Holocene, soil formation has taken place on their surfaces. In the areas covered by loess or loessy blanket, chernozem or chernozem-like soils developed; while on the blown sand, rust-brown forest soils or humic sand soils developed.

The alluvial fan surfaces today belong to the most important agricultural production areas of Hungary. On the chernozem soils mainly wheat, maize, sugar beet and rough fodder are produced. On the sandy soils characteristic crops include potatoes, rye, sunflowers, and tobacco, and there is significant production of fruit, grapes and, in suitable areas, garden vegetables. People living on the alluvial fans and pursuing their various jobs do not even think that not so long ago in these areas the most important factor of surface evolution was the activity of flowing water.

References

Borsy, Z. 1961. A Nyírség természeti földrajza (The physical geography of the Nyírség). *Földr. Monogr.* V., 1–227.
Borsy, Z. 1968a. Geomorfológiai megfigyelések a Nagykunságban. (Geomorphological investigations in the Nagykunság). *Földr. Közl.*, 129–131.
Borsy, Z. 1968b. The surface of the region between the Danube and Tisza. *Acta Geogr.*, VII, Debrecen, 45–57.
Borsy, Z. 1973. Loess, sandy loess, and loessy sand blankets in Hungarian wind-blown sand regions. *Földr. Közl.*, 172–184.
Borsy, Z. 1977. A Duna–Tisza közi hátság homokformái és a homokmozgás szakaszai (Sand formations of the Danube–Tisza interfluve and their periodic movement). *Alföldi Tanulmányok*, 1, 43–58.
Borsy, Z. 1978. Evolution of relief forms in Hungarian wind-blown sand areas. *Földr. Közl.*, 3–16.
Borsy, Z. 1986. Research in areas of blown sand. In Pécsi, M. and Lóczy, D. (Eds), *Physical Geography and Geomorphology in Hungary*. Budapest, Geographical Research Institute, Hungarian Academy of Sciences, pp. 77–82.
Borsy, Z. and Borsy, Z. 1955. Pollenanalitikai vizsgálatok a Nyírség É-i részében (Palynological investigations in the northern part of the Nyírség). *KLTE Actája*, 1–10.
Borsy, Z., Molnár, B., and Somogyi, S. 1969. Az alluviális medence síkságok fejlödéstörténete Magyarországon (Morphological evolution of alluvial basin-plains in Hungary). *Földr. Közl.*, 237–250.
Borsy, Z., Kádár, L., and Pinczés, Z. 1972. The new physico-geographical laboratory of the Kossuth University, Debrecen (Hungary). *Acta Geographica Debrecina*, 10, 233–243.
Borsy, Z., Csongor, É., Sárkány, S., and Szabó, I. 1982. Phases of blown-sand movements in the northeast part of the Great Hungarian Plain. *Acta Geogr. Debrecina*, XX, 5–33.
Borsy, Z., Félszerfalvi, J., and Lóki, J. 1982. A jánoshalmi MÁFI alapfúrás homoküledékeinek elektronmikroszkópos vizsgálata (Electron microscopic investigation of the sand sediments in the core drilling of MÁFI at Jánoshalma). *Acta Georgr. Debrecina*, XX, 35–50.
Borsy, Z. and Félegyházi, E. 1983. Evolution of the network of water courses in the north-eastern part of the Great Hungarian Plain from the end of the Pleistocene to our days. *Quaternary Studies in Poland*, 115–124.
Borsy, Z., Félszerfalvi, J., and Lóki, J. 1983. A komádi alapfúrás negyedidőszaki homokrétegeinek elektronmikroszkópos vizsgálata (Electron microscopic analysis of Quaternary sand layers in the base borehole at Komádi). *Alföldi tanulmányok*, VII, 31–58.
Borsy, Z., Félszerfalvi, J., Franyó, F., and Lóki, J. 1985. A Tótkomlós III./P. jelü magfúrás homokrétegeinek elektronmikroszkópos vizsgálata (Electron microscopic investigation of the sand profiles in the core drilling III/P. Tótkomlós). *Acta Georgr. Debrecina*, XXII, 47–63.
Borsy, Z., Balogh, K., Kozák, M., and Pécskay, Z. 1986. Recent contributions concerning the evaluation of the Tapolca-basin. *Acta Geogr. Debrecina*, XXIII, 79–104.
Cholnoky, J. 1907. A Tisza meder helyváltozásai (Migration of the riverbed of the Tisza). *Földr. Közl.*, 381–405, 426–445.
Csongor, É. and Borsy, Z. 1987. Radiocarbon dates and Lateglacial blown sand movement in N.E. Hungary. In *Low-level Counting and Spectrometry*. Bratislava, Veda Publishing House, pp. 113–117.
Franyó, F. 1980. Ujabb felszinfejlödéstörténeti és vizföldtani eredmények a Duna–Tisza közi kutatófurások alapján (Recent results in studying the geomorphic history and hydrogeology of the Danube–Tisza Interfluve based on test boreholes). *Földr. Ért.*, 409–443.
Kádár, J. 1957. Die Entwicklung der Schwemmkegel. *Petermans Geographische Mitteilungen*, 101, 241–244.
Melton, F. A. 1940. Tentative classification of sand dunes. *Journal of Geology*, 48, 113–174.
Pécsi, M. 1959. A magyarországi Duna-völgy kialakulása és felszinalaktana (The evolution and geomorphology of the Hungarian section of the Danube Valley). *Akad. Kiadó, Budapest*, 1–346.
Pécsi, M. 1967. A dunai Alföld (The Danubian Plain). *Akad. Kiadó, Budapest*, 1–358.
Pécsi, M. 1970. Geomorphological regions of Hungary. *Akad. Kiadó, Budapest*, 1–45.
Pécsi, M. 1980. A Pannóniai-medence morfogenetikája (Morphogenetics of the Pannonian Basin). *Földr. Ért.*, 105–117.
Rónai, A. 1972. Negyedkori üledékképződés és éghajlattörténet az Alföld medencéjében (Quaternary sedimentation and history of climate in the basin of Great Hungarian Plain). *MÁFI Évkönyv*, LVI.1, 1–416.
Rónai, A. 1985. Az Alföld negyedidőszaki földtana (The Quaternary of the Great Hungarian Plain). *Series Geologica*, 21, 1–446.
Sümeghy, J. 1944. A Tiszántul (The Transtibiscian Region). *Magyar tájak földtani leirása*, 6, 1–208.

CHAPTER 13

Factors Influencing Quaternary Alluvial Fan Development in Southeast Spain

A. M. Harvey
The University of Liverpool, Liverpool

Abstract

Three groups of factors are recognized influencing the morphological development of Quaternary alluvial fans in semiarid southeast Spain, over differing temporal and spatial scales. Tectonic factors influence the gross morphology and location of the fans at the regional scale but exert little direct influence in fan morphology or development sequences. Tectonically induced patterns of regional dissection influence fan location in a negative way, whereby in areas of the deepest dissection, mountain front fan accumulation has been prevented. The fan sedimentary sequences suggest a long-term trend from fines-rich to gravel-rich deposition and from net aggradation to net dissection. These trends relate to a long-term 'ageing' tendency or to a progressive climatic aridification over the Quaternary Period as a whole. Superimposed onto these trends are shorter-term, episodic erosional and depositional sequences reflecting the climatic fluctuations during the Quaternary. The major aggradational episodes appear to correlate with dry Quaternary glacials, and stable or degradational phases with interglacials. Internal morphological factors control short-term fan dissection behaviour, with mid-fan intersection-point trenching thresholds related to fan slope : channel slope relationships and channel cross-sectional geometry. The sensitivity of the threshold is expressed by contrasts between the northern and southern parts of the study area. Fans in the less arid northern part of the area show more complex sedimentary sequences, greater calcrete crust development, and more complex, dissection-prone fan morphology, whereas those in the more arid south show simpler sedimentary sequences and simpler morphological development.

Introduction

Alluvial fans are important zones within fluvial systems, especially in dry regions, in that they influence the strength of coupling between mountain sediment-source areas and axial drainage lines. During periods of fan aggradation, sediment, especially coarse sediment, may be stored within the fan, with little reaching the main river system. However, if the fan becomes dissected and trenched throughout, there may be channel continuity between mountain sources and axial drainages. Hence alluvial fan sequences may preserve a sedimentary record more sensitive to variations in source-area sediment supply than do alluvial sequences of the main rivers. Periods of fan aggradation reflect periods of excess sediment generation, whereas reduced sediment supply or increased transport capacity result in a reduction in the rate of fan sedimentation, or even in fan dissection.

Alluvial Fans: A Field Approach edited by A. H. Rachocki and M. Church
Copyright © 1990 John Wiley & Sons Ltd.

Fans are located where sediment supply from upstream exceeds, or exceeded, the capacity of a stream system for sediment transport, often at a major valley junction or at a mountain front. As the result of variations not only of sediment supply but also of fan morphology, the alluvial fan zone is especially sensitive to variations about the threshold of critical stream power (Bull, 1979). Consequently fan development may show both progressive and threshold-controlled (Schumm, 1977) changes between aggradation and dissection.

The factors which influence alluvial fan development can be identified as those which influence fan location in relation to regional topography, those which influence sediment production from the source area, and those which influence sediment transport through the fan environment. They may be grouped into geologic/tectonic factors, climatic or climatic sequence fac-

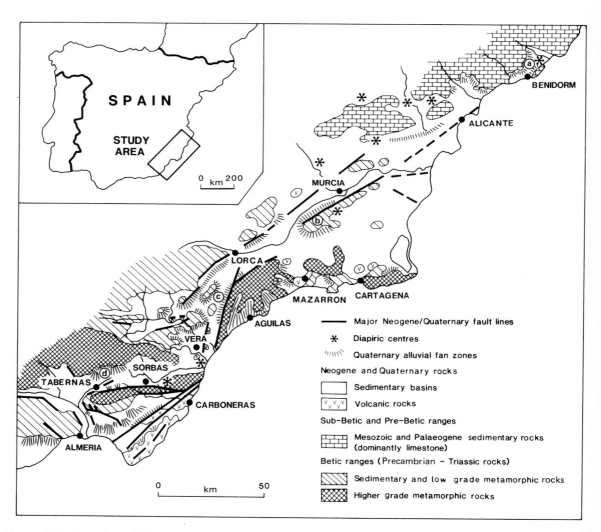

Figure 13.1. Location of Quaternary alluvial fan zones in southeast Spain in relation to generalized geology. Major Neogene–Quaternary fault systems are indicated. Study fan groups are indicated by circled letters: (a) Pre-Betic group; (b) Carrascoy group; (c) Almenara group; (d) Tabernas group

tors which may be expressed by sediment sequences, and morphological factors related to the geometry of the fan environment itself. This paper deals with the influence of these three sets of factors on the development of Quaternary alluvial fans in southeast Spain. Its objectives are to review the relationships between these groups of factors and their relevance at differing temporal and spatial scales to the morphological development of the alluvial fans.

The study fans issue from mountain catchments of the Betic Cordillera. Source-area geology includes high to low grade, Precambrian to Triassic, metamorphic rocks of the Betic nappes, Mesozoic and Palaeogene sedimentary rocks, dominantly limestones, of the Pre-Betic and Sub-Betic ranges, Neogene sedimentary rocks, and Neogene volcanic rocks (Figure 13.1). The fans accumulated in intermontane basins, at faulted mountain fronts and over mountain-front pediment surfaces during the Quaternary Period. Today the climate of the region is semiarid with precipitation totals ranging from *ca.* 350 mm in the north to *ca.* 170 mm in the driest parts of the south (Geiger, 1970). Pleistocene climates too were dry, especially during northern European glacials, when there was insufficient moisture to support a tree cover (Amor and Florschutz, 1964) but, as elsewhere in the drier parts of the Mediterranean region, with intense, presumably seasonal, geomorphic activity (Rohdenburg and Sabelberg, 1980).

Throughout the study area the alluvial fans show a broad similarity of sedimentary and morphological development (Harvey, 1978, 1984b). Early phases, dominantly of aggradation, culminated in the formation of the upper fan surfaces on which, in many cases, calcrete crusts developed. Later phases, dominantly of dissection, resulted in trenching below these surfaces, often through several cut and fill episodes, producing inset terraces within fanhead trenches. Modern behaviour is degradational, in some cases in fanhead trenches but with distal aggradation, and in others by complete through-trenching (Harvey, 1987a). The overall picture is of a progressive change in fan behaviour from dominantly aggradational to dominantly dissectional.

The Tectonic Setting

The Betic Cordillera evolved in response to the relative movements of the European and African plates from Jurrassic to Miocene times (Smith and Woodcock, 1982; Bourrouilh and Gorsline, 1979). Major nappe emplacement had taken place by mid-Miocene times with thrusting from the south of Precambrian to Triassic high to low grade metamorphic rocks of the Betic nappes (*sensu stricto*) towards the thrusted and folded Mesozoic and Palaeogene sedimentary rocks comprising the Sub-Betic and Pre-Betic ranges (Figure 13.1) (Rios, 1978; Alvarado, 1980). By mid-Miocene times sedimentation was taking place in basins between uplifted mountain ranges. Differential and intermittent epeirogenic uplift continued through Pliocene times and the present landscape has developed since Pliocene emergence of the lowland areas. Tectonic activity of several kinds has continued during the Quaternary Period. Regional epeirogenic uplift has continued, with Pliocene marine deposits elevated over 600 m above modern sea levels in parts of the south (Birot and Sole Sabaris, 1959; Harvey and Wells, 1987). Folding and faulting of Pliocene and early Quaternary sediments occurred with the Quaternary tectonic patterns dominated by a system of major strike-slip faults (Bousquet, 1977). Movement has also occurred along many mountain-front faults, especially of the Sierras de Alhamilla and Cabrera in Almeria (Figure 13.2), and Carrascoy in Murcia. Neogene basin-fill sequences have been folded and diapiric uplift has taken place along many major fault lines as well as in larger centres, especially in the Pre-Betic and Sub-Betic ranges (Moseley *et al.*, 1981).

TECTONIC CONTROLS OF ALLUVIAL FAN DEVELOPMENT

During late Cenozoic emergence, pediment surfaces were cut across earlier rocks (Dumas, 1977). Late Pliocene to early Quaternary terrestrial sediments accumulated, both as thin pediment veneers and thicker basin-fill deposits (IGME; Weijermars *et al.*, 1985), in some cases

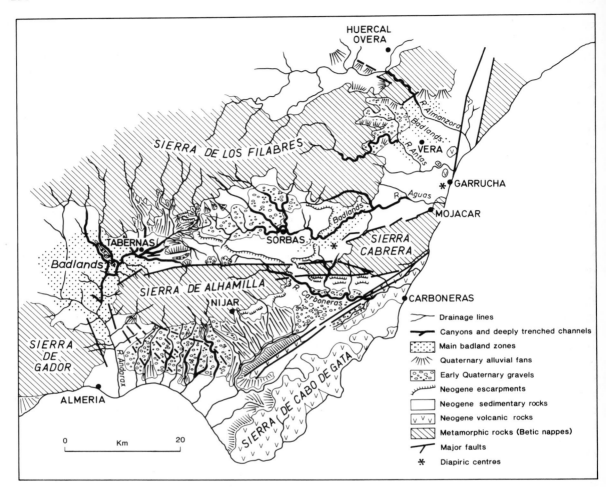

Figure 13.2. Location of fan groups in the Almeria region in relation to geology and Neogene–Quaternary tectonic/geomorphic development

forming alluvial fan-like accumulations. However, the main alluvial fans post-date these early Quaternary sediments, accumulating during mid-Quaternary times.

The relationships between tectonics, fan accumulation, and regional dissection may be illustrated in the Almeria region (Figure 13.2), including the Neogene Vera, Sorbas, Tabernas, and Almeria/Carboneras basins. The area is bounded to the north by the Betic Sierra de los Filabres. In the centre the Sierras de Alhamilla and Cabrera separate the Almeria/Carboneras basin from the other basins. This basin is itself bounded to the southeast, along the Caba de Gata strike-slip fault, by Neogene volcanic rocks. Late Pliocene to early Quaternary emergence created a land surface with a regional slope towards the south, from the Sorbas basin to the Carboneras basin, across the Alhamilla/Cabrera axis of uplift (Harvey and Wells, 1987; Harvey, 1987b).

The earliest, Upper Pliocene, terrestrial sediments in the Sorbas basin are low energy, coastal plain sediments, the Zorreras group of Roep et al. (1979), followed by coarse Plio-Pleistocene fluviatile conglomerates, the Gochar Formation of Roep et al. (1979). These form the final basin-filling unit in the Sorbas basin, and although they resemble fan deposits, accumulated in the basin centre rather than in mountain-front positions. Similar Plio-Pleistocene fluviatile conglomerates occur as a basin fill in the Carboneras lowland, where the former main drainage issued from the Sorbas basin across the axis of uplift. In the Vera basin, in the Almeria basin west of Nijar, and along the margins of the Rioja corridor between the Almeria and Tabernas basins similar conglomerates occur primarily as pediment mantles. Throughout the area these deposits have been folded and faulted (Harvey, 1987b).

During the Quaternary, uplift continued and the river systems became incised into the underlying rocks. Subsequent streams developed rapidly along the outcrop of weak marls of Upper Miocene age. The original southward drainage of the Sorbas basin was captured by the subsequent Rio de Aguas (Harvey and Wells, 1987) cutting back and dissecting the early Quaternary depositional terrain (Figure 13.2). The Andarax valley developed along the Rioja graben (Postma, 1984) and rapidly dissected the western 'badland' part of the Tabernas basin. Small streams on the southwest margin of the Sierra de Alhamilla, a zone of maximum Quaternary uplift, have dissected pediments formed during early Quaternary time.

Alluvial fans accumulated in the mid-Quaternary only in areas away from tectonically induced zones of maximum incision, identified by the occurrence of canyons in resistant rocks and badlands in weak rocks (Figure 13.2). Around the Vera basin the mountain relief is limited and the Neogene marls of the basin itself provide little coarse sediment for fan formation. The drainage is by major transverse rivers, presumably of sufficient power to prevent major fan accumulation. The Quaternary landscape is dominated by dissected pediments (Volk, 1979) and only in the northern part of the basin are there a few small mountain-front alluvial fans. In the Sorbas basin the deep incision of the Aguas system, in response to the regional uplift, has prevented large-scale alluvial fan formation, except along the northern margin of the basin away from zones of tectonically-induced dissection.

The major fan groups are restricted to interior areas of the Tabernas and Almeria/Carboneras basins. A major fan group occurs in the upper part of the Tabernas basin, near the Aguas/Andarax watershed, away from the incision of the badland area. This part of the basin is currently aggrading. In the Almeria/Carboneras basin fans issue from the Sierra de Alhamilla near Nijar, and to the south from La Serrata, low hills of volcanic rocks along the strike-slip faults. Both areas are away from zones of maximum incision along the Carboneras valley, or of uplift and incision at the southwest end of the Sierra de Alhamilla. In the Tabernas and Nijar areas the major alluvial fans occur where substantial mountain drainages issue into the Neogene basins, but away from the zones of tectonically-induced dissection where stream power would have been sufficient to prevent the accumulation of alluvial fan sediments.

At a more detailed scale the geologic and long-term geomorphic controls can be illustrated by considering the geomorphology of fans in four areas, representative of the study area as a whole (Figure 13.3; locations on Figure 13.1).

Fans in the Pre-Betic region (Figure 13.3A) issue from mountains of folded Cretaceous limestones onto pediment surfaces cut across softer Palaeogene sediments, and into lowland basins. The lowland areas are structurally controlled, and include a synclinal lowland near Benidorm, downfaulted terrain at Robelles, and the Altea lowland developed in weak, diapiric, Triassic gypsiferous marls (Moseley et al., 1981). The major fold structures are mid-Miocene in age: Neogene–Quaternary tectonics are largely associated with uplift of the Altea and Finistrat diapirs (Figure 13.3A), and faulting along their margins. Although there is some diapiric disturbance of the fan deposits (see below) the major zones of accumulation are away from the tectonically-induced dissection of the major rivers, Algar, Anchero, and Amadorio. There is only limited dissection at the seaward end of the Tapia fan (Figure 13.3A; see profile on Figure 13.6).

The Carrascoy fans (Figure 13.3B) issue from the Sierra de Carrascoy, an uplifted Betic nappe dominantly of complex, Triassic sedimentary rocks and low grade metamorphic rocks. The Carrascoy range has been uplifted during the Neogene–Quaternary, differentially tilting

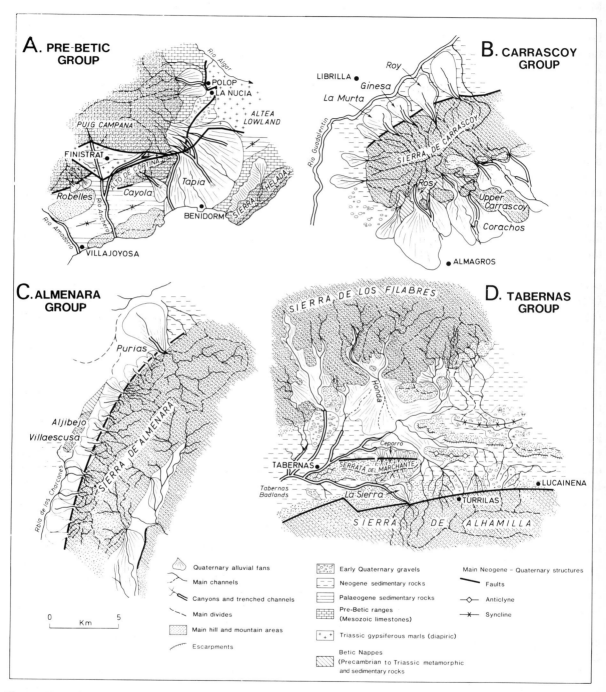

Figure 13.3. Geology/topographic relationships for the four main study areas (locations in Figure 13.1): (A) Pre-Betic group, (B) Carrascoy group, (C) Almenara group, (D) Tabernas group

Miocene to Plio-Quaternary sedimentary strata to the south. Maximum uplift is along the northern fault-bounded margin adjacent to the downfaulted Guadalentin depression (Bousquet, 1977). The alluvial fans occur in three groups (Harvey, 1986), along the faulted northern margin, mantling intermontane pediment surfaces, and mantling mountain-front pediments on the southern and western margins. The deep dissection of the Rio Guadalentin has not cut back to dissect the Carrascoy fans.

The Almenara fans (Figure 13.3C) issue from medium to high grade Betic metamorphic rocks of the Sierra de Almenara and occur in two groups. The main group are mountain-front fans along the faulted western margin of the mountains, and the second are intermontane fans in basins to the east of the main divide. The fans occur in a regional watershed zone between the Guadalentin and Almanzora drainages and away from dissection associated with incision of these major rivers. Fans do not occur at the northern and southern ends of the Sierra de Almenara where tectonically induced incision is present.

The Tabernas fans (Figure 13.3D) are situated in the undissected upper part of the Tabernas basin (see above and Figure 13.2) and form two smaller groups. Large fans (e.g. the Honda fan) issue from the Betic schists of the Sierra de los Filabres to the north. Although Neogene folding occurred within the basin there is little sign of faulting here and the fans blanket the basin margin. Much smaller fans occur in the southern part of the basin, a group issuing northwards from the folded lower Miocene conglomerates of the Serrata del Marchante. Locally this small mountain front is fault bounded, but no disturbance of the fans can be identified. Both Filabres and Marchante fans occur away from zones of tectonically-induced dissection. La Sierra fan, to the south of the Serrata del Marchante is currently undergoing dissection by the headward extension of the deep dissection of the Tabernas drainage.

All the fan groups illustrate a similar influence of tectonics on fan location. The fans occur where steep and/or substantial mountain catchments issue into lowland areas, and to that extent are influenced by the gross patterns of tectonic uplift of the mountain ranges. However, where tectonically-induced dissection has caused deep incision of the main rivers, few if any fans have accumulated. These river systems had sufficient stream power during the Quaternary to transport all sediment supplied, to incise their beds and to prevent alluvial fan accumulation.

TECTONIC DISTURBANCE

The regional tectonic patterns have influenced the location and preservation of the alluvial fans but tectonics seem to have had little *direct* influence on the *style* of fan formation or on fan morphology. Despite ample evidence for continuing Quaternary tectonic activity there is little direct evidence for tectonic disturbance of fan sequences. In most areas, even along faulted mountain fronts, there is little obvious tectonic expression in either fan morphology or sedimentary sequences. There are a few exceptions. These fall into three categories: areas of active faulting with vertical displacement, areas of strike-slip faulting with lateral displacement, and areas of diapiric uplift.

Many of the mountain-front faults displace early Quaternary gravels and, locally, mid–late Quaternary river terraces (Harvey and Wells, 1987; Dumas et al., 1978) but only in a few localities are faulted fan deposits evident. Bousquet (1977) refers to displacement of fan deposits in the Lorca area. On the Carrascoy mountain front, faulting is evident on two fans (Harvey, 1986), Ginesa (see Figure 13.4A), and la Murta (locations on Figure 13.3B). Despite displacement of the fan deposits there is no surface expression of the faults: subsequent erosion and calcrete formation have obliterated any trace of a fault scarp. However it is possible that the different morphometric properties of northern and southern Carrascoy fans may be partly attributable to tectonics. The northern Carrascoy fans have anomalously steep slopes, even when the steepness of the drainage basins is taken into account (Harvey, 1986). Displacement along strike-slip faults is evident in channel plan views, especially along the Carboneras strike-slip faults at La Serrata (Figure 13.4B). Diapiric activity has

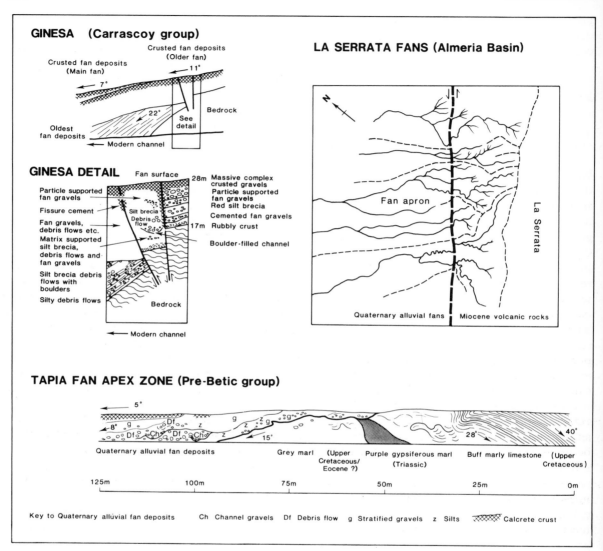

Figure 13.4. Examples of tectonic and diapiric disturbance of fan deposits

also disturbed fan deposits, locally along the Carrascoy fault, and in a more widespread manner at the apex of the Tapia fan in the Pre-Betic group. Exposed in sections along the Benidorm to Finistrat road (Figure 13.4C) are fan deposits disturbed by diapiric activity along a major fault line. At Nucia Park 1 km to the north are further diapirically disturbed fan deposits. However, the disturbance occurs only in the immediate vicinity of the rising gypsum diapir, or of the fault zone, and in midfan there is no trace of disturbance.

These tectonic and diapiric disturbances, while demonstrating continuing Quaternary tectonic activity, have only highly localized expressions in the alluvial fans. Tectonism is important primarily in creating the appropriate relief configuration for the formation of alluvial fans, rather than in influencing the style of fan development. Its influence is primarily on the distribution of alluvial fans, in a positive sense, where steep, substantial mountain catchments issue into lowland areas. Interestingly, in a negative sense, areas of greatest tectonic uplift, where associated with deep dissection by major rivers, are not areas conducive to alluvial fan accumulation.

Fan Sequences

THE FAN DEPOSITS

As in other alluvial fan environments (Blissenbach, 1954; Hooke, 1967; Bull, 1977), the fan deposits can be ascribed to debris flow, sheetflood, and streamflood depositional environments (Harvey, 1978, 1984a). Matrix-supported gravels are interpreted as debris-flow deposits. Clast-supported, structureless gravels may occur in sheets as sheetflood deposits, or in thicker units resembling those described by Wells and Harvey (1987) as transitional debris-flow, or hyperconcentrated flow deposits. Locally those showing upward fining as well as a packed texture may be sieve deposits (Bull, 1977). Silt sheets are interpreted as low energy distal sheetflood deposits and stratified, locally cross-bedded, locally imbricate gravels as streamflow deposits.

Depositional facies will reflect clast and matrix characteristics and water: sediment ratios (Wells and Harvey, 1987). Within any one fan there may be proximal-to-distal facies variations, including not only a down-fan diminution of clast size, but also a change in the relative importance of depositional environments. Debris-flows may be more common proximally, silt sheets distally. At Tapia fan (Pre-Betic area), for example, distal silt and gravel sheets appear to be the lateral equivalents of proximal debris-flows (Harvey, 1984a).

Source area characteristics, particularly bedrock geology, and drainage basin size and relief appear to be the major controls over variations in sedimentology between fans (Harvey, 1984a). On the basis of the relative proportions of debris flow, sheet, and channel deposits in the aggradation phases exposed in fan head trenches, 31 fans have been classified into three groups: those rich in debris-flows, intermediate, and fluvially-dominated fans (Harvey, 1984a). The debris-flow group can be differentiated from the others on the basis of drainage area and basin relief by the equation

$$R = 300A^{0.69}$$

where R is drainge basin relief (m) and A is drainage basin area (km^2). With only five anomalies two of which can be explained by source area geology (Harvey, 1984a), there is a clear indication that small, steep drainage basins produce debris-flow-rich fans. The influence of source-area geology can be seen in Table 13.1 where the fans have also been classified according to fan groups, i.e. Pre-Betic region, Carrascoy region, and the southern fans subdivided into those draining medium-to-high grade metamorphic rocks and those draining sedimentary and low grade metamorphic rocks. It is clear that the areas of higher grade metamorphic rocks rarely produce debris-flow deposits, but that the other areas, particularly the Pre-Betic and Carrascoy areas, commonly do. In these areas, the intermediate fans tend to be those issuing from larger drainage basins than do the debris-flow fans.

THE FAN DEVELOPMENT SEQUENCES

The overall fan sequences have been described already (Harvey, 1978, 1984b). Early, dominantly aggradational phases are followed, in many cases,

Table 13.1. Occurrence of fan depositional types (31 fans)

Fan Group	Depositional type		
	Debris-flow fans	Intermediate fans	Fluvial fans
Pre-Betic region fans	5	1	0
Carrascoy region fans	5	2	0
Southern fans sedimentary areas	3	5	0
Southern fans metamorphic areas	0	4	6

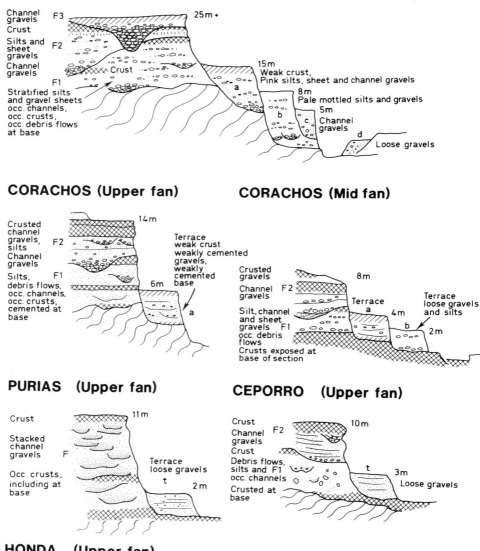

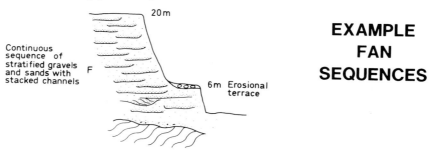

Figure 13.5. Example fan sequences representative of the Pre-Betic fans (Tapia, group a), the Carrascoy fans (Corachos, group b), the southern sedimentary area fans (Ceporro, group d) and the southern matamorphic area fans, Almenara area (Purias, group c), Tabernas area (Honda, group d). For locations of groups A–D see Figure 13.1

by surface stability and the formation of calcrete crusts, and are then succeeded by later, dominantly degradational phases. Figure 13.5 illustrates representative sequences from the major fan groups. In most cases the aggradational phases show a sequence of episodic aggradation punctuated by periods of stability, represented by buried calcrete crusts, and by periods of partial dissection. In addition, many sequences show vertical facies variations with a progressive diminution of fines-rich sediments. Debris-flows proximally, and silt-sheets distally, give way up-sequence to sheet and channel gravel units (e.g. see Figure 13.5, Tapia fan: contrast units F1, F2 with F3; and Ceporro fan: contrast unit F1 with F2). These two trends are characteristic of all the fan groups except the fluvially dominant fans from the southern metamorphic rock source areas. Of these the Almenara group (represented by Purias in Figure 13.5) show occasional buried crusts but no major within-sequence trenching, but fan sequences in the Tabernas metamorphic group (represented by Honda in Figure 13.5) simply show continuous sequences of stacked non-crusted fluvial gravels.

The later stages of fan development are dominated by dissection, by the cutting of fanhead and, in some cases, through-fan trenches. Dissection has been episodic, punctuated by major periods of aggradation within the overall degradational trend. This has resulted in a series of cut-and-fill terraces inset within fanhead trenches. The complexity of the terrace sequence varies from north to south with two-to-four terrace fills present in the Pre-Betic region fans, two in the Carrascoy fans, one in most fans further south, but only an erosional terrace in the Tabernas metamorphic area fans. In the Pre-Betic and Carrascoy fans the earliest terraces show weak crustal development at the surface but otherwise the terrace deposits are unconsolidated gravels, sands, and silts. On fans with only fanhead trenching the terrace deposits can often be traced downfan, beyond the intersection points, where they overlie the (usually) crusted, aggradation phase deposits.

During both the earlier aggradational and the later degradational phases, periods of aggradation, whether on proximal fan surfaces or within fanhead trenches, appear to be the result of increases in sediment supply in relation to the threshold of critical stream power (Bull, 1979). The episodic nature of erosion and deposition would imply an overall climatic or vegetation control of erosion rates in response to Quaternary climatic fluctuations. The overall trend from net aggradation to net dissection, together with the progressive decrease in the availability of fine sediments, might imply either a response to a progressive climatic trend throughout the Quaternary, or a tectonically related aging response as first identified in alluvial fans by Eckis (1928). The detailed variability in the sequences from fan group to fan group suggests a threshold-controlled (Schumm, 1977) degradational behaviour, with spatial variations in the factors controlling trenching thresholds (Harvey, 1984b).

RECONCILIATION WITH QUATERNARY CHRONOLOGY

The fans rest on, or are inset within, and therefore post-date Pliocene to early Quaternary pediment surfaces and gravel deposits. As with most other dry region terrestrial deposits they contain little or no conventionally datable material and therefore their chronology within the Quaternary Period is uncertain. The key to the chronology may be through soil development, the calcrete crusts, and the relationships of the fan deposits to other established Quaternary sequences. There is controversy in the literature over the origin of surface calcretes (Durand, 1963; Butzer, 1964), their relationship to carbonate soil development (Gile et al., 1966), and whether they can be used as indicators of surface age. Lattman (1973), working in Nevada, argued against their use as simple age indicators but Dumas (1969, 1977) used crusts to differentiate multiple stages within the Quaternary deposits of southeast Spain. Whether local variations in crust characteristics allow a high degree of sensitivity is open to question but, with confidence, three degrees of crust development can be recognized and ascribed to approximate age ranges. The youngest indurated crusts show only intergranular cementation, incomplete development of a petrocalcic horizon of thickness no more than 20–30 cm and very limited (less than a few millimetres), discontinuous, laminar calcrete layers. These are described here as weak crusts, and are assigned by Dumas (1969, 1977) a Würmian age. Mature crusts show complete cementation of a massive petrocalcic horizon of up to about 50 cm thickness, secondary brecciation and recementation, and discontinuous, laminar calcretes several millimetres in thickness. Crusts of this type have been assigned by Dumas (1969, 1977) Rissian and early–to–mid Pleistocene ages. Ancient crusts are confined to early Quaternary surfaces and show massive petrocalcic horizons of 1 m or more in thickness and laminar calcretes several centimetres in thickness.

Correlation of fan deposits with established Quaternary sequences is possible along the Mediterranean coast where there are numerous localities with established stratigraphic sequences of Quaternary raised beaches (Imperatori, 1957; Overjero and Zazo, 1971; Dumas, 1977; Zazo et al., 1981), many of which have been radiometrically dated to Tyrrhenian Stages I, II, and III (age ranges >200 to ca. 30 Ka) (Thurber and Stearns, 1965), or of regressional aeolianites (Butzer and Cuerda, 1962) generally ascribed to early Rissian or early Würmian ages. Local correlations with fan sequences can be seen at a number of localities. In the Benidorm and Alicante areas (locations, Figure 13.1), maturely crusted fan deposits predate Tyrrhenian II beach deposits and weakly crusted colluvial and alluvial fan deposits post-date the same beaches (Dumas, 1977). Near Aguilas (location Figure 13.1), is a complex sequence of interdigitated beach and alluvial fan deposits (Bardaji et al., 1986) with major pre-Würm fan episodes and minor Würm fans, and near Mojacar (location, Figure 13.2) weakly crusted fans post-date both Tyrrhenian (II?) raised beach deposits and overlying aeolianites (Harvey, 1987b).

Correlations of the fan sequences with established terrestrial sequences are rare. The correlation of pre-Würm crusted deposits, and Würm and Holocene terraces, between the Pre-Betic region fan sequence (Harvey, 1978, 1984b) and dated river terrace sequences near Alicante (Cuenca Paya and Walker, 1976) can be suggested. In the Feos Valley, Almeria, small, weakly crusted fans have been assigned, on the basis of soil chronosequence relationships, to the Würm phase (Harvey and Wells, 1987).

It is clear that both Würm and pre-Würm fan deposits are present which can normally be differentiated on the basis of calcrete crust development and, although the absolute chronology is uncertain, a general regional picture is emerging. In the Pre-Betic and Carrascoy areas and in some southern fans the aggradation phases and the upper fan surfaces appear to predate the Würm phase and can be ascribed to Riss age and earlier. The dissection phases appear to be Würm and Holocene, with perhaps the earliest within-dissection terrace fill, at least in the Pre-Betic fans, ascribable to the early Würm phase. The other fills possibly correlate in general terms with Vita Finzi's (1972) Würm and Holocene fills. On many southern fans net aggradation continued later and the fan surfaces appear to be of Würm age. In the Tabernas area the absence of surface calcrete crusts and the very limited soil development on the Honda group fans suggest that aggradation may have continued late in the Würm with only very recent, largely Holocene, dissection.

Two general conclusions may be made about the fan sequences. First, the episodic nature of both the aggradation and dissection phases, the general similarities between different fan groups, and the scale of individual aggradational units suggest that major changes in sediment supply were climatically induced rather than related to intrinsic thresholds (Schumm, 1977). Second, these shorter-term, climatically induced fluctuations in aggradation and dissection behaviour are superimposed onto a longer-term, progressive change in sediment availability from mid to late Quaternary time. This is expressed by both the longer-term progression from dominant aggradation to dominant dissection and by the progressive change from fines-rich to gravel-rich sediments.

Fan Morphology

FAN MORPHOMETRY

The progressive and episodic tendencies identified in the sedimentary sequences are expressed in the fan morphology. The crusted fan surfaces are deeply incised by fanhead trenches (Figure 13.6A and B) and in some cases by through trenches. Terraces are inset within these trenches. Fan surface, terrace, and modern channel profiles converge towards mid-fan intersection points (Figure 13.7) beyond which, as on many other fans (Wasson, 1974), distal aggradation occurs (Figure 13.6A and C). On some fans renewed incision by headcut development takes place at the intersection point (Figure 13.6D; see profiles for Cayola, Corachos and Ros, Figure 13.7), eventually leading to through, distal trenching (Harvey 1984a, 1984b, 1987a).

Fan surface, terrace, and modern channel profiles would have developed in response to the prevailing sediment transport processes at the time of formation. Slopes of the fan surfaces would have been adjusted to aggradation by debris-flow, sheetflood, or streamflow in wide, unconfined, braided channels ultimately controlled by the volume of sediment input, water–sediment ratio, clast: matrix ratio and clast calibre. Terrace and channel profiles were adjusted to sediment transport by streamflow in narrower channels within fanhead trenches. Higher slopes would be expected in response to debris-flow and sheetflow than to streamflow, and braided distal fan channels differ from narrow, fanhead trench channels by their much higher widths per unit discharge and therefore by steeper adjusted slopes (Harvey, 1984c). On fans for which both fan slope and channel slope data are available channel slopes are invariably less than fan surface slopes, with a relationship that may be expressed by

$$S = 0.429 G^{0.89}$$

($n = 38$, $r = 0.829$, $SE = 0.141$ log units)

where S is channel slope and G is fan surface slope (Harvey, 1987a). For all the fans for which complete profiles have been surveyed, and for which there is sufficient continuity of terrace surface for slope measurement, there is a progressive decrease in slope from fan to terrace to channel (Harvey, 1984b). This supports the interpretation of a long-term, progressively decreasing trend in sediment production indicated by the fan sequences (see above).

Fan morphometric properties in general have been shown to reflect source area characteristics (Blissenbach, 1952; Bull, 1962; Denny, 1965; Hooke, 1968). This can clearly be demonstrated on the Spanish fans. Drainage basin geology, size, and relief have been shown to influence the occurrence of debris-flow-rich fans. Fan surface slope is steeper per drainage area on debris-flow fans than on intermediate and fluvially dominant fans (Harvey, 1984a). Since a detailed analysis of the morphometry of drainage basin, fan, and channel characteristics has been presented elsewhere (Harvey, 1987a), only a brief summary need be given here (Figure 13.8). Fan slope, channel slope, and channel width correlate well with drainage area, the first two inversely, the last directly. These trends reflect the influence of larger discharges and the likely higher water: sediment ratios derived from larger drainage areas (Wells and Harvey, 1987). In most cases the correlations improve when fan groups are treated separately. Fan slopes and channel slopes tend to be higher per unit drainage area in the northern (Pre-Betic and Carrascoy) fan groups than elsewhere (Figure 13.8). Channel width per unit drainage area tends to be higher for the non-crusted fan channels of the southern metamorphic group (Figure 13.8). This agrees with Van Arsdale's (1982) observation of narrow channels on crusted fans. When drainage basin relief and/or mean drainage basin slope are also taken into account, most of the multiple correlation coefficients significantly improve and standard errors are reduced in comparison with those for the bivariate relationships (Table 13.2).

FAN DISSECTION

The modern regime of the fans is largely degradational but two styles are apparent. The first, as

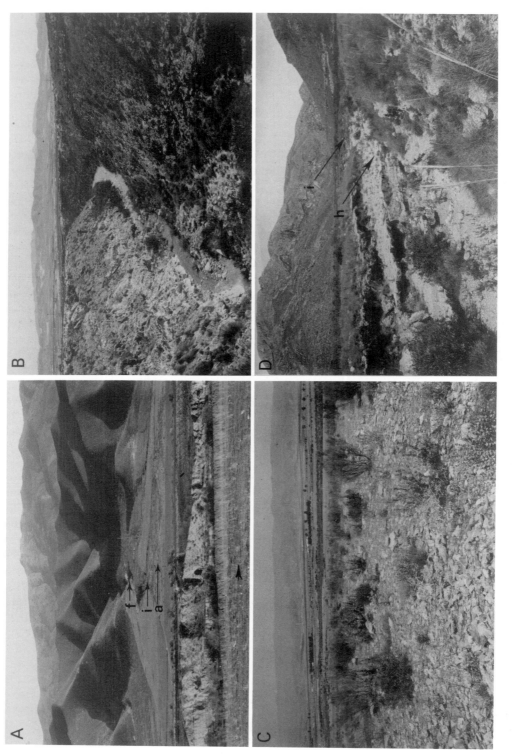

Figure 13.6. Illustrations of alluvial fan morphology, Almeria and Murcia. (A) A small distally aggrading alluvial fan; Ceporro fan, Almeria. Note: fanhead trench (f), intersection point (i) and distal aggradation (a). Despite incision of the main stream channel (foreground) there is no continuity between the fan and the main channel system; (B) Deep fanhead trench near Nijar, Almeria, through crusted fan deposits (f) into underlying Miocene bedrock; (C) Intersection-point and distal aggradational lobes: Ceporro fan, near Tabernas, Almeria; (D) Intersection-point (i) headcut (h), and distal fan trench on small fan east of Mazarron, Murcia

QUATERNARY FAN DEVELOPMENT

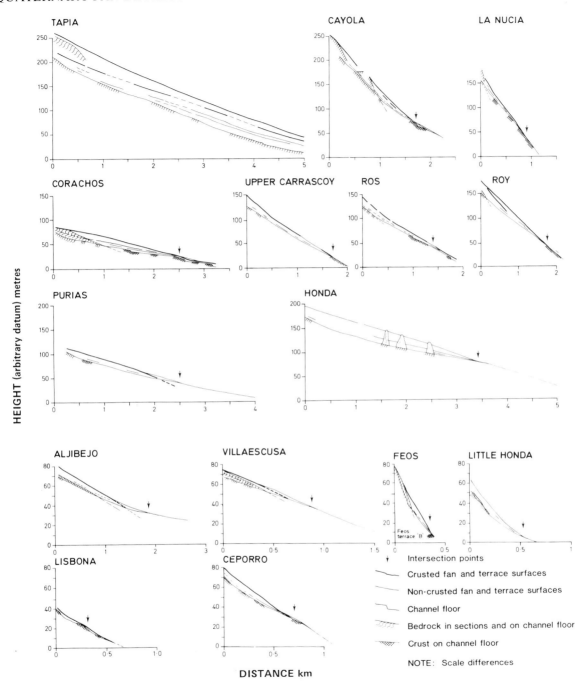

Figure 13.7. Selected fan profiles, showing progressive changes in fan morphology. Tapia, Cayola and La Nucia (Pre-Betic area); Corachos, Upper Carrascoy, Ros, Roy (Carrascoy area); Lisbona, Ceporro (Southern sedimentary area); Purias, Alijbejo, Villaescusa, Feos, Honda, Little Honda (Southern metamorphic area). Note scale differences

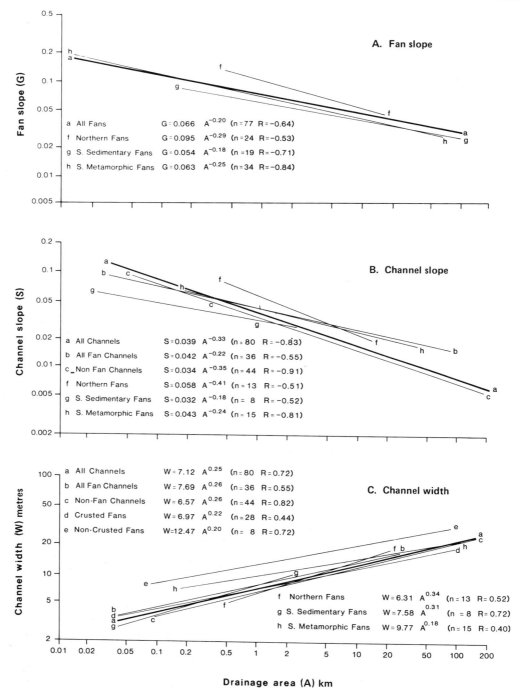

Figure 13.8. Morphometric relationships; summary regression relationships for (A) fan slope, (B) channel slope, and (C) channel width *vs* drainage area for selected fan groups. Regression lines for each group of fans are indicated. For details of multiple regression analyses, see Table 13.2. Full data are given in Harvey (1987a)

Table 13.2. Multiple correlation and regression analysis: fan morphometry for selected fan groups

Fan Groups	n	Correlation coefficients (standard errors of estimate, log units, in brackets)		Multiple regression equations*			
		Simple (vs A)	Multiple				
For Fan Slope (G)							
All fans	77	−0.638 (0.169)	0.801 (0.133)	$G = 0.0145$	$A^{-0.28}$	$R^{0.35}$	$B^{0.28}$
Northern Fans (Pre-Betic and Carrascoy)	24	−0.533 (0.162)	0.878 (0.096)	$G = 0.0047$	$A^{-0.46}$	$R^{0.64}$	$B^{0.55}$
Southern Fans (sedimentary areas)	19	−0.708 (0.171)	0.744 (0.167)	$G = 0.0044$	$A^{-0.38}$	$R^{0.46}$	
Southern Fans (metamorphic areas)	34	−0.839 (0.106)	0.894 (0.089)	$G = 0.0066$	$A^{-0.36}$		$B^{0.40}$
For Channel Slope (S)							
All Channels	80	−0.831 (0.202)	0.884 (0.172)	$S = 0.0012$	$A^{-0.49}$	$R^{0.61}$	
Northern Fan Channels	13	−0.511 (0.265)	0.776 (0.222)	$S = 0.0003$	$A^{-0.68}$	$R^{0.87}$	
Southern Channels (sedimentary areas)	21	−0.895 (0.169)	0.902 (0.171)	$S = 0.0082$	$A^{-0.42}$	$R^{0.10}$	$B^{-0.33}$
Southern Channels (metamorphic areas)	45	−0.883 (0.173)	0.908 (0.158)	$S = 0.185$	$A^{-0.18}$	$R^{-0.10}$	$B^{0.53}$
For Channel Width (W), m,							
All Channels	80	−0.721 (0.218)	0.740 (0.214)	$W = 0.111$	$A^{-0.03}$	$R^{0.55}$	$B^{-0.58}$
Northern Fan Channels	13	−0.504 (0.265)	0.519 (0.229)	$W = 0.072$	$A^{-0.03}$	$R^{0.67}$	$B^{-0.33}$
Southern Channels (sedimentary areas)	21	−0.835 (0.184)	0.906 (0.152)	$W = 0.017$	$A^{-0.13}$	$R^{0.76}$	$B^{-0.97}$
Southern Channels (metamorphic areas)	45	−0.676 (0.235)	0.681 (0.240)	$W = 0.204$	$A^{-0.03}$	$R^{0.48}$	$S^{-0.53}$

*A is drainage area (km^2); R is drainage basin relief (m); B is drainage basin mean slope.

on many other dry region alluvial fans, is of fanhead trenching but distal aggradation (Figure 13.6A, C). The second, which has received very little attention in the literature, is of fanhead trenching and intersection-point headcut development (Figure 13.6D), in some cases leading to through-fan distal trenching. This distinction has important implications, first for sediment transport and for connectivity through the fluvial system (Brunsden and Thornes, 1979), and second, if these contrasts apply also to the past sequences, in the implications for fan stratigraphy. Conventional, distally aggrading fans would produce stacked sequences of sediments distally, but those prone to distal trenching would produce more complex, inset stratigraphies distally as well as proximally (Harvey, 1987a). Contrast the inset stratigraphies of the Tapia and Corachos aggradation sequences, both fans prone to through trenching, with the simple stacked stratigraphies of Purias and Honda, both distally aggrading fans (Figure 13.5).

The intersection-point trenching mechanism appears to be associated with an increase in stream power at the intersection-point. This may occur as the result of an increase in slope as the modern channel emerges from the fanhead trench onto the steeper slope of the fan surface inherited from the aggradation phases. There appears to be a threshold between intersection-point trenching or aggradational tendencies controlled by: (1) volume of sediment throughput; (2) discrepancy between fan and channel slopes; and (3) cross-sectional geometry of the channel. A schematic model illustrating the relationships between sediment throughput, fan and channel slopes, and intersection-point behaviour is illustrated in Figure 13.9A (adapted from Harvey, 1987a). A profile of type A would be that of a conventional, 'American', distally aggrading fan, though in real-

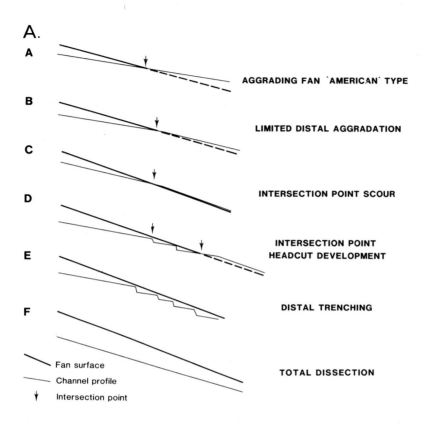
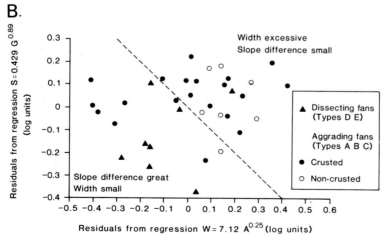

Figure 13.9. Schematic model for fan surface: (A) channel profile relationships for aggradation fans (Types A B C) and dissecting fans (Types D E F); (B) residuals from channel slope *vs* fan slope and channel width *vs* drainage area regressions for aggrading and dissecting types.
Adapted from Harvey (1987a)

Table 13.3. Occurrence of fan dissectional types (47 fans)

Fan group	Dissectional type (see Figure 13.9A for explanation)				
	Types A/B	Type C	Type D	Type E	Type F
Pre-Betic region fans	1	1	2	3	1
Carrascoy region fans	0	4	1	2	0
Southern fans sedimentary areas	7	2	0	0	3
Southern fans metamorphic areas	17	0	1	0	2

ity some of the classic American fans are more like type B (Harvey, 1987a). A profile of type B would be the result of limited distal aggradation, and one of type C would indicate a fan near the trenching threshold, with little aggradation but insufficient stream power for headcut development. A profile of type D would represent limited headcut development, and type E would result from continuous distal dissection. Type F has been added here to account for totally trenched fans, but in five of the six cases of this type observed in the field, trenching appears to be to base-level induced rather than mid-fan induced and therefore primarily related to local tectonics rather than to fan morphometry.

For the 47 fans on which fan channels have been studied there are clear spatial patterns in the distribution of dissectional types (Table 13.3). Intersection-point and distal trenching characterize the northern, Pre-Betic and Carrascoy fans, but are almost absent from the southern fans. Almost all the southern metamorphic fans are distally aggrading (Types A/B). Part of the explanation for these differences may lie in the tendency for the northern fans, all of which are crusted fans, to have steeper fan slopes and greater discrepancies between fan and channel slopes, and in the tendency for fans from southern metamorphic rock source areas to have wider channels, therefore channels less prone to incision. These trends are demonstrated in Figure 13.9B where residuals from channel slope: fan slope, and width: drainage area regressions are plotted by channel type. Seven of the eight distally trenching channels for which complete data are available plot to the lower left of the diagram in the zone of higher slope differences and smaller widths. Although there is overlap of plotting position, chi squared tests show these distributions to be significantly different to at least to 5% level (Harvey, 1987a). In summary, mid-fan trenching thresholds are most likely to be exceeded under conditions of relatively low sediment supply to fans where past aggradation produced steep crusted fan surfaces. The role of the crust appears to be to confine the channel during incision and thus to concentrate stream power at the intersection-point.

Discussion

The three sets of factors identified earlier, geologic/tectonic, climatic, and morphological, each operate over a range of temporal and spatial scales in influencing the geomorphology of the alluvial fans of southeast Spain. The effective temporal scales range from that of the Quaternary as a whole, through that of episodic fluctuations within the Quaternary, to that of short-term, threshold controlled behaviour. Relevant spatial scales range from the regional scale for locational factors or for widespread general fan behaviour, through the subregional scale of differences between fan groups, to local scales of variation between individual fans or within individual fans.

Geologic and tectonic factors are most important over the long-term and at regional and subregional scales. Regional patterns of tectonic activity, uplift, and dissection are important for fan location, and for the overall late Pliocene to Quaternary tectonic sequence. These, when combined with the climatic sequence may be important in the overall aggradation, dissection, and 'aging' behaviour of the fans (Eckis, 1928). Subregional and local scales may be important in influencing the overall timing of the aggradation/dissection sequence and, where relevant, the base-level induced dissection of some fans. The continued aggradation of the Tabernas fans may in part be due to their position in the upper part of the sedimentary basin, away from areas of rapid dissection. At the subregional and local scales, when combined with drainage basin size and relief, geology—especially rock type—is of major importance in determining the nature of sediment production, thereby influencing morphological development during fan aggradation and dissection and hence, in the short-term, influencing the sensitivity of trenching thresholds.

Climatic factors are important in three main ways. Over the Quaternary Period as a whole there appears to have been a long-term trend towards aridification from a relatively humid early Quaternary, with a fuller vegetation cover and more active soil formation (Butzer, 1964), through the more arid, treeless 'glacials', with high rates of sediment generation, in mid and late Quaternary times. This may be an important factor influencing the progressive diminution in the availability of fine sediments and the progressive trend from fan aggradation to fan dissection. The sediment sequences suggest a long-term sediment exhaustion trend reflecting the long-term erosion of the drainage basins.

Over shorter timescales the episodic climatic fluctuations within the middle and late Quaternary Period are associated with episodic fluctuations in aggradation/dissection behaviour, during both overall aggradational and overall dissectional phases of fan development. Major periods of sediment production and, therefore, fan aggradation appear to be associated with arid Pleistocene glacials, periods with reduced vegetation cover (Amor and Florschutz, 1964) and greater erosional effectiveness of storm events (Rohdenburg and Sabelberg, 1980). Major aggradation episodes clearly occurred during the Würm glacial phase and in several episodes prior to the last interglacial, presumably during the Riss and earlier glacial phases. Interglacials appear also to be associated with relatively dry climates, like that of the Holocene Epoch, but with a more continuous vegetation cover than during the glacial maxima, resulting in a reduced sediment supply, fan surface stability, and the maximum development of calcrete crusts and, possibly, in-channel trenching. The Holocene Epoch has been dominated by dissection of Würm deposits, and the limited aggradation during the Holocene appears to be a direct response to human-induced vegetation removal (Butzer, 1964; Vita Finzi, 1972).

Climatic factors also have a spatial expression. At present there is a marked north to south aridity gradient, related to the rain-shadow effect of the Betic mountains. The same would have been true of Quaternary climates. This may be a factor influencing the greater relative importance of aggradation over dissection in the southern fans. Under consistently greater aridity the spatial continuity of the fluvial system would be less (Wolman and Gerson, 1978; Brunsden and Thornes, 1979), and the relative effectiveness of major storms for the transport of sediment proportionally greater (Baker, 1977), leading perhaps to a more consistent aggradational tendency in the drier regions.

There are interesting contrasts between the Spanish fans and those of the more arid American southwest (Harvey, 1987a). The Spanish fans are more prone to dissection and show sequences of greater complexity than do the American fans. This contrast in part reflects the greater sensitivity of the Spanish fans to trenching thresholds, manifested by their greater fan slope to channel slope ratios and by their narrower channels (Harvey, 1987a), and may be related to two other climatically controlled factors; the stronger development of calcrete crusts on Spanish fans and the apparently greater effective range of Spanish Quaternary climatic fluctuations, in terms of temporal variations in sediment production.

The morphological factors are important at the regional scale, when associated with long-term tectonic and climatic sequences, in influencing the overall 'aging' nature of the fans. At more local and short-term scales, through their adjustment to sediment transport processes, fan and channel slopes and channel geometry provide the controls over trenching thresholds.

The three sets of factors combine to produce

the contrasts in fan morphology and development evident between the northern and southern parts of the study area. In the north the earlier cessation of major tectonic activity, the less arid climates, and the dominantly sedimentary bedrock combine to produce the more complex fans. These fans show a greater complexity of sedimentary sequence during aggradation, stronger crustal development, an earlier onset of fan dissection, greater complexity of terrace development, and a greater sensitivity to trenching thresholds. Those further south show simpler development and more closely resemble classic arid-area fans, especially those of the American Southwest, described in the literature. It is clear, then, that to understand alluvial fan morphology account must be taken of tectonic, climatic, and morphological factors over a wide range of temporal and spatial scales.

Acknowledgements

The author is grateful to the University of Liverpool research fund for grants towards the costs of the fieldwork and to the drawing office, particularly to Sandra Mather, and photographic section of the Department of Geography, University of Liverpool for producing the diagrams.

References

Alvarado, M. 1980. Espagne. In Dercourt, J. (Ed.), *Géologie de pays européens, Espagne, Grece, Italie, Portugal, Yougoslavie.* 26ᵉ Congres Geologique International, Paris. 1–54.

Amor, J. M. and Florschutz, F. 1964. Results of the preliminary palynological investigation of samples from a 50 m boring in Southern Spain. *Bol. Real Sociedad Espanola de Historia Natural (Geológica)*, **62**, 251–255.

Baker, V. R. 1977. Stream channel response to floods, with examples from central Texas. *Geological Society of America Bulletin*, **88**, 1057–1071.

Bardaji, T. *et al.* 1986. Geomorfologia y estratigrafia da las secuencias marinas y continentales cuaternarias de la cuenca de Cope (Murcia, Espana). In Lopez Bermudez, F. and Thornes, J. B. (Eds), *Estudios sobre geomorfologia del sur de Espagna, Murcia.* 11–16.

Birot, P. and Sole Sabaris, L. 1959. La morphologie du sudest de l'Espagne. *Revue de Geographie des Pyrennes et du sudest*, **30**, 119–184.

Blissenbach, E. 1952. Relation of surface angle distribution to particle size distribution on alluvial fans. *Journal of Sedimentary Petrology*, **22**, 25–28.

Blissenbach, E. 1954. Geology of alluvial fans in semi-arid regions. *Geological Society of America Bulletin*, **65**, 175–190.

Bourrouilh, R. and Gorsline, D. S. 1979. Pre-Triassic fit and alpine tectonics of continental blocks in the western Mediterranean. *Geological Society of America Bulletin*, **90**, 1074–1083.

Bousquet, J. C. 1977. Quaternary strike-slip faults in Southeastern Spain. *Tectonophysics*, **52**, 277–286.

Brunsden, D. and Thornes, J. B. 1979. Landscape sensitivity and change. *Institute of British Geographers Transactions, New Series*, **4**, 463–484.

Bull, W. B. 1962. Relations of alluvial fan size and slope to drainage basin size and lithology in western Fresno County, California. *U.S. Geological Survey Professional Paper*, **430B**, 51–53.

Bull, W. B. 1977. Alluvial fan environment. *Progress in Physical Geography*, **1**, 222–270.

Bull, W. B. 1979. Threshold of critical stream power. *Geological Society of America Bulletin*, **90**, 453–464.

Butzer, K. W. 1964. Climatic–geomorphologic interpretations of Pleistocene sediments in the Eurafrican sub tropics. In Howell, F. C. and Bouliere, F. (Eds), *African Ecology and Human Evolution*. Methuen, London. 1–25.

Butzer, K. W. and Cuerda, J. 1962. Coastal stratigraphy of southern Mallorca and its implications for the Pleistocene chronology of the Mediterranean sea. *Journal of Geology*, **70**, 398–416.

Cuenca Paya, C. and Walker, M. J. 1976. Pleistoceno final y Holoceno en la cuenca del Vinalopo (Alicante). *Estudios Geológicos*, **32**, 14–104.

Denny, C. S. 1965. Alluvial fans in the Death Valley region, California and Nevada. *U.S. Geological Survey Professional Paper*, **466**.

Dumas, M. B. 1969. Glacis et croutes calcaires dans le levant espanol. *Association de Géographes Français Bulletin*, **375**, 553–561.

Dumas, M. B. 1977. *Le levant Espagnol, la genese du relief*. Université Paris, Thesè de doctorat d'etat, Centre National de Recherche Scientifique. 520 pp.

Dumas, M. B. *et al.* 1978. Géomorphologie et néotectonique dans la région d'Almeria (Espagne du sudest). In *Relief et Néotectonique de pays Méditerranèens. RCP (CNRS, Paris) Publication*, **461**, 123–170.

Durand, J. H. 1963. Les croutes clacaires et gypseuses en Algerie, formation et age. *Société Géologique Français Bulletin Ser. 7*, **5**, 959–968.

Eckis, R. 1928. Alluvial fans of the Cucamonga district, southern California. *Journal of Geology*, **36**, 225–247.

Geiger, F. 1970. Die aridität in sudostspanien. *Stuttgarter Geographische Studien Band*, **77**, 173 pp.

Gile, L. H. *et al.* 1966. Morphological and genetic sequences of carbonate accumulation in desert soils. *Soil Science*, **101**, 347–360.

Harvey, A. M. 1978. Dissected alluvial fans in southeast Spain. *Catena*, **5**, 177–211.

Harvey, A. M. 1984a. Debris flows and fluvial deposits in Spanish Quaternary alluvial fans: implications for fan morphology. In Koster, E. H. and Steel, R. J. (Eds), *Sedimentology of Gravels and Conglomerates. Canadian Society of Petroleum Geologists Memoir*, **10**, 123–132.

Harvey, A. M. 1984b. Aggradation and dissection sequences on Spanish alluvial fans: influence on morphological development. *Catena*, **11**, 289–304.

Harvey, A. M. 1984c. Geomorphological response to an extreme flood: a case from southeast Spain. *Earth Surface Processes and Landforms*, **9**, 267–279.

Harvey, A. M. 1986. Alluvial fans of the Sierra de Carrascoy. In Sala, M. (Ed.), *Excursion Guide Book, IGU Commission on measurement, theory and application in geomorphology*. Barcelona. 135–139.

Harvey, A. M. 1987a. Alluvial fan dissection: relationships between morphology and sedimentology. In Frostik, L. and Reid, I. (Eds), *Desert Sediments Ancient and Modern*. Geological Society of London, Sp. Publ., **35**, 87–103.

Harvey, A. M. 1987b. Patterns of Quaternary landform development in the Almeria region, southeast Spain: a dry-region tectonically active landscape. *Die Erde.*, **18**, 193–215.

Harvey, A. M. and Wells, S. G. 1987. Response of Quaternary fluvial systems to different epeirogenic uplift: Aguas and Feos river sytems, southeast Spain. *Geology*, **15**, 689–693.

Hooke, R. leB. 1967. Processes on arid region alluvial fans. *Journal of Geology*, **75**, 438–460.

Hooke, R. leB. 1968. Steady state relationships on arid region alluvial fans in closed basins. *American Journal of Science*, **266**, 609–629.

IGME (Instituto Geologico y Minero de Espania) 1:50 000 Geological Maps and accompanying memoirs, including Sheets 848 (Altea), 872 (Alicante), 933 (Alcantarilla), 934 (Murcia), 954 (Totana), 955 (Fuente Alamo de Murcia), 976 (Mazarron), 997 (Aguilas), 1013 (Macael), 1014 (Vera), 1030 (Tabernas), 1031 (Sorbas), 1043 (Almeria), 1059 (El Cabo de Gata), 1060 (El Pozo de los Frailes), and 1:250 000 Geological Maps and accompanying memoirs including sheets 72 (Elche), 73 (Alicante), 78 (Baza), 79 (Murcia), 84–85 (Almeria–Garrucha).

Imperatori, L. 1957. Documentos para el studio de cuatenario alicantino. *Estudios Geologicos*, **13**, 141–151.

Lattman, L. H. 1973. Calcium carbonate cementation on alluvial fans in southern Nevada. *Geological Society of America Bulletin*, **84**, 3013–3028.

Moseley, F. *et al.* 1981. Alpine tectonics and diapiric structures in the Pre-Betic zone of southeast Spain. *Journal of Structural Geology*, **3**, 237–251.

Overjero, G. and Zazo, C. 1971. Nivelos marinos pleistocenos en Almeria (SE de Espana). *Quaternaria*, **15**, 145–149.

Postma, G. 1984. Mass-flow conglomerates in a submarine canyon: Abrioja fan-delta, Pliocene, southeast Spain. In Koster, E. H. and Steel, R. J. (Eds), *Sedimentology of Gravels and Conglomerates. Canadian Society of Petroleum Geologists Memoir*, **10**, 237–258.

Rios, J. M. 1978. The Mediterranean coast of Spain and the Alboran Sea, in Nairn, A. E. M., Kanes, W. H., and Stehli, F. G. (Eds), *The Ocean Basins and their Margins, Volume 4B, The Western Mediterranean*. Plenum, New York. 1–16.

Roep, Th. B. *et al* 1979. A prograding coastal sequence of wave-built structures of Messinian Age, Sorbas, Almeria, Spain. *Sedimentry geology*, **22**, 135–163.

Rohdenburg, H. and Sabelberg, U. 1980. Northwestern Sahara margins: Terrestrial stratigraphy of the Upper Quaternary and some palaeoclimatic implications. In Van Zinderen Bakker, E. M. Sr. and Coetzee, J. A. (Eds), *Palaeoecology of Africa and the Surrounding Islands*. 267–276.

Schumm, S. A. 1977. *The Fluvial System*. Wiley, New York.

Smith, A. G. and Woodcock, N. H. 1982. Tectonic synthesis of the Alpine-Mediterranean geodynamics. *American Geophysical Union and Geological Society of America, Geodynamics Series*, **7**, 15–38.

Thurber, D. L. and Stearns, C. E. 1965. Th230–U234 dates of the late Pleistocene fossils from the Mediterranean and Morrocan littorals. *Quaternaria*, **7**, 29–42.

Van Arsdale, R. 1982. Influence of calcrete on the geometry of arroyos near Buckeye, Arizona, *Geological Society of America Bulletin*, **93**, 20–26.

Vita Finzi, C. 1972. Supply of fluvial sediment to the Mediterranean during the last 20 000 years. In Stanley, D. J. (Ed.), *The Mediterranean Sea, a Natural Sedimentation Laboratory*. Stroudsberg, Dowden, Hutchinson and Ross. 43–46.

Volk, H. 1979. Quartäre Reliefentwicklung in Südostspanien. *Heidelberger Geographische Arbeiten Heft*, **58**, 143 pp.

Wasson, R. J. 1974. Intersection point deposition on alluvial fans: an Australian example. *Geografiska Annaler*, **54A**, 83–92.

Weijermars, R. *et al.* 1985. Uplift history of a Betic fold nappe inferred from Neogene–Quaternary sedimentation and tectonics (in the Sierra Alhamilla and Almeria, Sorbas and Tabernas basins of the Betic Cordillera, SE Spain). *Geologie en Mijnbouw*, **64**, 397–411.

Wells, S. G. and Harvey, A. M. 1987. Sedimentologic and geomorphic variations in storm generated alluvial fans, Howgills Fells, Northwest England, *Geological Society of America Bulletin*, **98**, 182–198.

Wolman, M. G. and Gerson, R. 1978. Relative scales of time and effectiveness of climate in watershed geomorphology. *Earth Surface Processes and Landforms*, **3**, 189–208.

Zazo, C. *et al.* 1981. Ensayo de sintesis sobre el Tirrheniense Penninsular Espanol. *Estudios Geologicos*, **37**, 257–262.

CHAPTER 14

Long-term Palaeochannel Evolution During Episodic Growth of an Exhumed Plio-Pleistocene Alluvial Fan, Oman

Judith Maizels
University of Aberdeen, Aberdeen

Abstract

This paper presents a reconstruction of evolution during the Pliocene and Quaternary periods of superimposed palaeochannel systems now exposed on two exhumed alluvial fan systems in interior Oman. A model of long-term hydrologic change is proposed, based on the evolutionary stages of channel development and associated changes in palaeoflow characteristics, on the apparent mechanisms of channel change, and on the likely autocyclic and allocyclic controls on long-term channel changes.

Analysis of the morphology, sedimentology, and weathering characteristics of palaeochannel deposits indicates that alluvial fan development has followed a distinct series of evolutionary stages. At least 12 successive generations of palaeochannel systems have been identified, representing major periods of fluvial activity of varying duration and character. The earlier generations (Pliocene?) on Fan I are characterized by broad, sinuous, single-thread channels comprising chert-rich, trough cross-bedded sands and gravels, associated with low stream discharges. These deposits are underlain by diagenetically altered ophiolite gravels, fine overbank sediments, and aeolian silts of the Barzaman Formation. The later generations on Fan I (early Quaternary?) are steeper, narrower and less sinuous, thin out rapidly downfan, and comprise coarse-grained, poorly imbricated gravel beds associated with high flows. These later generations are dominated by ophiolitic lithologies. The latest deposits in the sequence form extensive ophiolite gravel spreads on wadi terraces of Fan II, and comprise coarse-grained, tabular cross-bedded cobble beds associated with high magnitude flows.

Flow activity was characterized by overall long-term aggradation, interrupted by episodes of incision. Only the last stages of fan evolution were represented by large-scale fluvial entrenchment and aeolian deflation of the interfluve areas. Channel changes occurred both through meander migration along well-established routeways, and through avulsion to create extensive distributary systems. The long-term change from the earlier seasonal (?) flows in the fine-grained meandering channels to flash-flood regimes in the later distributary channels and braided terrace gravel spreads, appears to reflect a combination of local and regional controls. The long-term channel changes associated with fan development represent a response to progressive fanhead aggradation and headward extension of the drainage net resulting in river capture, as well as to a range of climatic, tectonic, and eustatic changes occurring during the Cenozoic Era in the Arabian peninsula.

Alluvial Fans: A Field Approach edited by A. H. Rachocki and M. Church
Copyright © 1990 John Wiley & Sons Ltd.

Introduction

The objective of this paper is to reconstruct the long-term evolution of superimposed palaeochannel systems exposed on an exhumed alluvial fan, in relation to regional tectonic, climatic, and hydrologic changes during the Plio-Pleistocene period. Alluvial fan aggradation has been widely recognized as representing a response to an increased balance of sediment yield over river discharge from the source catchment, a balance that varies with changes in the main allocyclic controls of geology and tectonics, climate, and hydrologic conditions. The particularly significant geological factors are the rates and degrees of tectonic uplift and tilting, relative elevation above base level, and structural controls on drainage development. Bedrock geology and mineralogy, the rates of rock weathering, and resistance to erosion are important geological controls on sediment availability and hence on fan development; while the magnitude and frequency of storm events, the seasonality of rainfall, and evapotranspiration rates are important since they affect patterns of soil and vegetation cover, and hence rates of surface runoff and sediment yield. The structure of the river network, the magnitude and frequency of runoff events, the seasonality and duration of stream flows, the sediment availability and patterns of downstream sediment transport are in turn likely to be reflected in the nature of the channel system draining the source area and feeding the alluvial fan environment. Changes in the tectonic, climatic, and hydrologic catchment parameters can be expected to effect downstream and temporal changes in the nature of the channel pattern itself, in the morphology and sedimentology of the channel, and in local and regional channel stability.

The channel system may also undergo internal changes, reflecting autocyclic controls on development, through such local mechanisms as meander migration, channel avulsion, variations in degree of braiding, and downfan and vertical changes in channel characteristics associated with progressive fan aggradation, rather than with regional changes in catchment conditions. Long-term channel evolution within alluvial fans has been investigated by a number of workers in a wide variety of palaeoenvironments, ranging from cold (e.g. Ryder, 1971; Williams, 1973; Wasson, 1977) to semiarid (e.g. Denny, 1965; Beaty, 1970, 1974; Harvey, 1984) environments.

However, these studies have been faced with three major problems. First, it has often proved difficult to distinguish the allocyclic and autocyclic controls on observed changes in patterns of alluvial sedimentation within a fan depositional sequence (e.g. Heward, 1978; Flores and Pillmore, 1987; Kraus and Middleton, 1987). Second, it has rarely proved possible to model the long-term evolution of the channel system, since palaeochannels preserved within an alluvial fan sequence normally are represented in only few exposures or disconnected borehole records. Third, in the absence of a widely exposed evolutionary sequence of palaeochannels, it has proved difficult to identify the mechanisms and rates of channel change during progressive fan formation.

This paper attempts first to identify the evolutionary stages of channel development during the Plio-Pleistocene period on a piedmont alluvial fanplain system in Oman, based on analysis of the numerous superimposed palaeochannel systems which have been revealed by widespread landscape exhumation; second, to identify the likely mechanisms of channel change during fan formation, and in particular both whether channel changes occurred continuously or episodically during fan formation, and whether fan formation was accomplished by continued aggradation or by alternating erosional and aggradational episodes; and finally, to identify the likely autocyclic and allocyclic controls on long-term channel change on the alluvial fanplain system.

Geomorphological Setting of the Alluvial Fans

The alluvial fans are located on the southern flanks of the Eastern Oman Mountains (Figure 14.1), between latitudes 21° and 23°N. The Eastern Oman Mountains comprise a Precambrian to Cambrian basement overlain by two extensive allochthonous nappe systems, namely, the lower

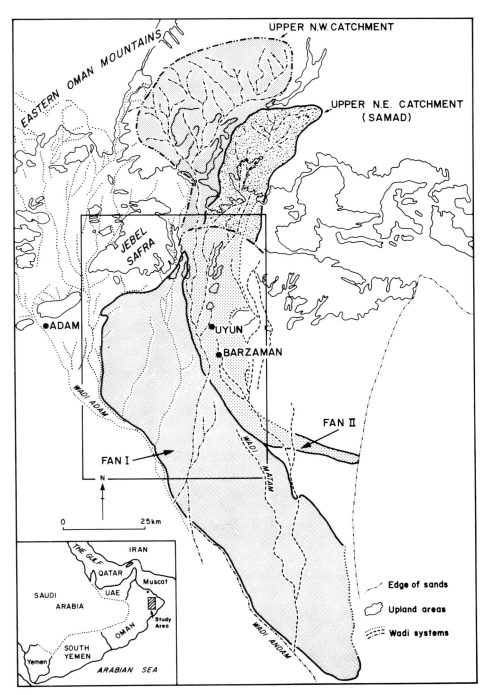

Figure 14.1. Location of major alluvial fans and wadi systems in the southern piedmont zone of the Eastern Oman Mountains. Area shown on Figure 14.3 is outlined

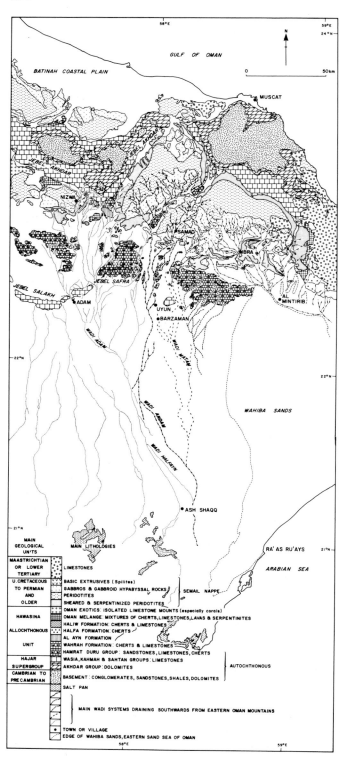

Figure 14.2. Bedrock geology of the Eastern Oman Mountains (after Glennie et al., 1974) in relationship to main wadi systems in the southern piedmont zone

Hawasina nappe and the upper Semail nappe (Figure 14.2; and see Glennie et al., 1974).

The lower Hawasina unit comprises largely a range of limestones and cherts, while the upper Semail nappe represents a major ophiolite massif, dominated by basic and ultrabasic lithologies including serpentine, gabbro, diabase, and basalt. The nappes were emplaced during the Late Campanian (Upper Cretaceous) from deep sea environments formerly lying to the northeast. Following emplacement, the nappe systems were mantled by early Tertiary shallow marine limestones, and subsequently uplifted and folded during the Oligocene and into the Miocene Period to form the mountain ridges and domes which now reach elevations of almost 3000 m. Large-scale erosion of these mountains during the Miocene–Pliocene resulted in the accumulation of continental clastic and playa-type sediments on the plains south of the mountains, to form the 'Fars' Group, implying a period of arid terrestrial sedimentation (Glennie et al., 1974). This early period of mountain erosion removed the Tertiary sedimentary cover to reveal the present pattern of geological outcrops in the fan source areas (Figure 14.2). The inner mountain zone of ophiolites is now bounded in the south by low, northwest–southeast trending ridges of thinly bedded Hawasina cherts and limestones, which in turn are flanked by more southerly outliers of autochthonous, massive limestones rising from beneath the nappe sequences.

Beyond the southern mountain edge alluvial fans and plains traversed by wadi systems (both relict and active) extend for over 200 km southwards towards the Indian Ocean (Figure 14.2), reaching thicknesses of over 280 m. The major drainage lines are cut transverse to the regional geological structure, forming a series of south-flowing wadi systems, the largest being (from west to east) the Wadis Adam, Andam, and Matam (Figures 14.1 and 14.2). To the northeast, the main, structurally controlled west–east drainage line of Wadi Batha bounds the northern and eastern margins of the Wahiba Sand desert. This study concentrates on the Wadi Andam drainage system which, unlike any other drainage system in the southern piedmont zone, extends far enough northwards through the Hawasina outcrops to drain the distinctive ophiolite core of the mountain region. The Wadi Andam system extends for over 250 km, draining a highly elongated catchment area of ca. 4500 km^2.

The Wadi Andam catchment currently experiences an arid climate, with significant rains generally occurring only once every three to six years, although estimates of rainfall occurrence are sparse and unreliable. Rainfall occurs as short, intense events associated with inland incursion of the southwest monsoon, or from penetration of depressions from the Gulf of Oman. With only a thin or absent soil cover, sparse scrub vegetation, and extensive and steep, barren mountain slopes, runoff is very rapid, and flash floods reaching discharges of up to 1000 m^3 s^{-1} have been recorded (see Maizels and Anderson, 1988). Discharges generally decrease downstream through evaporation and transmission losses, such that no surface runoff now reaches the sea.

Morphology of the Fan Systems

Two main fan systems have been identified from morphologic, stratigraphic, and lithologic criteria (Figure 14.1).

Fan System I is the oldest and most extensive, stretching southwards for over 120 km from the foot of the Oman Mountains to the desert margins and associated aeolianite outcrops in the south. PAWR (Public Authority for Water Resources) borehole records (Aubel, 1983; Jones, 1986) suggest that buried gravels extend to depths of up to 285 m in some places, although these gravels may also include the continental alluvial sediments of the Upper Fars Group. This fan system slopes gently towards the south and southeast with gradients decreasing from ca. 0.0030 west of Barzaman (Figure 14.1) to only ca. 0.0018 in the southeast, near the confluence of the Wadis Andam and Matam.

Fan System II forms a narrower, more recent system confined to the northerly, proximal parts of the present Wadi Andam piedmont zone (Figure

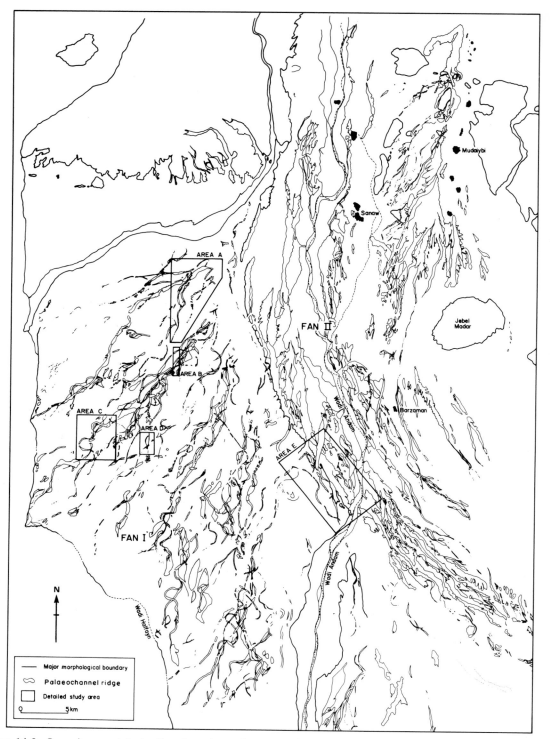

Figure 14.3. Superimposed 'raised' palaeochannel systems in the central parts of fans I and II. Location of detailed study areas of palaeochannel intersections indicated as areas A–E. Areas F–H lie further to the southeast. Map based on interpretation of 1:100 000 air photo mosaic

14.1). The deposits are relatively thin, averaging only *ca.* 8 m in thickness in the north (Jones, 1986), thinning to only 2 or 3 m downfan. Gradients are significantly higher than on Fan System I, decreasing from *ca.* 0.0046 in proximal zones to only *ca.* 0.0028 in the Barzaman area.

Morphology of the Palaeochannel Systems

The surface of Fan I is dominated by numerous linear gravel ridges, up to 5 m high and exhibiting markedly undulating crestlines (Figure 14.3). Many of these channels intersect or cross over one another such that successive channels form a series of superimposed or stacked channels (*cf.* Glennie, 1970). Single 'raised' channel deposits have also been described in parts of Arabia and North Africa by Miller (1937), King (1942), Knetsch (1954), Holm (1960), Butzer and Hansen (1968), Beydoun (1980), and in Texas and New Mexico by Reeves (1983), while similar multi-storey palaeochannel deposits have been identified in the Spanish Oligocene–Miocene sandstones of the Ebro Basin by Friend *et al.* (1981).

CRITERIA FOR ANALYSIS

The successive generations of palaeochannels are clearly exposed on the fan surface, and detailed mapping of each generation has allowed identification of the minimum number of palaeochannel generations, and hence the minimum number of periods of fluvial activity, during fan formation. Several major problems arise from this approach, however. First, identification of successive palaeochannel generations solely from their morphological relationships could mean that any gaps in the sequence are not accounted for. Not only will the number of periods of fluvial activity then be underestimated, but correlation between adjacent channel systems will be unreliable. Hence the number of palaeochannel generations estimated from morphological mapping must represent only the minimum number of periods of fluvial activity. Second, the relative duration and significance of each 'period of fluvial activity' remains unknown. One palaeochannel system may have been active for decades or centuries, while another may have been produced during only a few, or even a single rainy season. Third, morphologic relationships and measurements of palaeochannel characteristics have been established only from preserved and exposed fragments of the palaeochannel systems. Many of the older channel deposits, in particular, have been substantially eroded, reworked or buried, thereby reducing the accuracy of palaeochannel correlations both within and between successive generations.

Analysis of changes in channel planform morphology and pattern was based on measurement of a series of morphologic parameters derived from the palaeochannel systems mapped for each successive generation. These measures included the length of preserved channel fragments, the distance of exposure from the fan apex, the proximal and distal widths of the preserved channels, their sinuosity, wavelength, radius of curvature, and amplitude of the 'active zone' or meander belt (*cf.* Bridge and Leeder, 1979), and the degree of channel braiding or anastomosing. Measurements of channel widths may partly reflect the degree of post-formational dissection or truncation by wadi or aeolian erosion, thereby reducing the width of the preserved channel deposits. The extent of this modification appears, however, to have been relatively insignificant since numerous long and sinuous courses can still be so clearly identified. Finally, the number of active zones of a given generation was determined, in order to assess whether fluvial activity occurred within a single or multiple drainage system across the fan surface. Many of the morphological characteristics of the palaeochannels and particularly of palaeochannel intersections were checked in the field. The morphological field evidence provided additional information on topographic relations between successive palaeochannel generations.

MORPHOLOGY OF FAN I PALAEOCHANNELS

Twelve palaeochannel generations have been identified on the Fan I surface (Figures 14.4A,

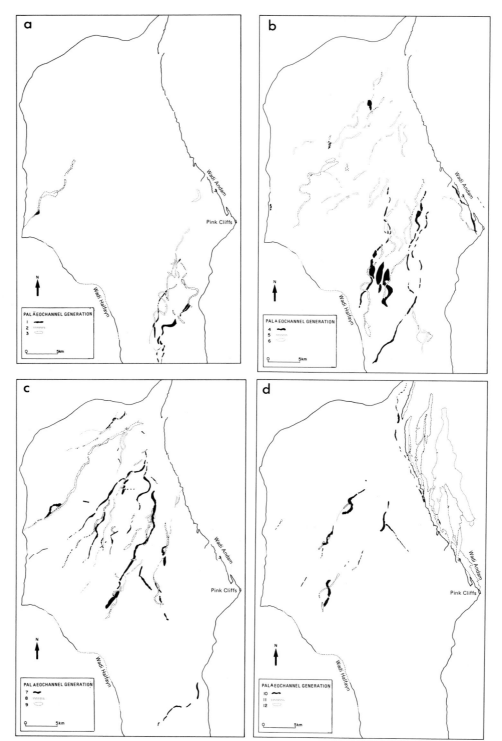

Figure 14.4. Successive generations of superimposed palaeochannel systems, Fan I. A. Palaeochannel generations 1–3; B. Palaeochannel generations 4–6; C. Palaeochannel generations 7–9; D. Palaeochannel generations 10–12

B, C, D). However, should each channel zone in fact represent a separate period (or subgeneration) of fluvial activity, a maximum of 37 such generations or periods of fluvial activity could be represented within the mapped channel sequence. The older generations of palaeochannels are generally the least extensively preserved, with generations 1, 2, and 3 (Figure 14.4A) being largely preserved only in the southern, distal fan area. Generations 4, 5, and 6 (Figure 14.4B), by contrast, extend for over 60 km from the fan apex in the north. Generations 7, 8, and 9 (Figure 14.4C) are also well preserved, but extend for only *ca.* 40 km south of the fanhead, while generations 10, 11, and 12 (Figure 14.4D) are of even more limited extent. Hence, the oldest channel systems appear to have been the most extensive, while the more recent have extended over a shorter downfan distance and contain fewer channel fragments. These changes in palaeochannel preservation and extent reflect either progressive weathering and erosion of the uppermost exposed deposits in the sequence, and/or more limited drainage during the latter stages of fan formation.

A distinctive feature of the palaeochannel courses is that the older channels tend to be highly sinuous, with maximum sinuosities exceeding values of 1.7, while the younger channels tend to exhibit low sinuosity courses. Generations 2, 3, and 5, in particular, have mean sinuosities exceeding 1.2, with maximum values of 1.73, while generations 7 to 12 all exhibit mean values lower than 1.1 (Table 14.1). This overall decrease in palaeochannel sinuosity during fan development is reflected also in an overall decrease in meander wavelength from over 4 km to about 3 km (Table 14.1), and in the width of the meander belt from 3 km to under 1.5 km.

The decrease in channel wavelength and amplitude may be associated with an increase in the number of active channel zones on the fan at a given period of formation. The older generations appear to have been represented by up to three or four apparently contemporaneous active channel zones on the fan, spaced *ca.* 4 to 5 km apart; the younger generations, by contrast, appear to exhibit up to six or seven such contemporaneous channel systems, spaced only 3.5 to 4 km apart.

Table 14.1. Morphological characteristics of successive palaeochannel generations

Palaeochannel generation (Fan I or II)	Total length of preserved channel (km)	Distal extent of channel systems (km)	Mean sinuosity	Mean meander wavelength (km)	Radius of curvature (km)	Width of meander belt (km)	Number of active channel zones	Maximum no. of channels in each active zone	Mean channel width (m) Proximal	Mean channel width (m) Distal
I 1	24.0	>60	1.17	2.8	1.42	1.3	2–3	2	–	136
I 2	36.6	>60	1.21	4.7	1.35	3.6	3	2	304*	105
I 3	28.8	>60	prox. 1.16 dist. 1.47	4.2	1.05	1.9	1	1	298*	102
I 4	73.0	>60	1.09	5.9	2.04	3.4	3	2	476*	241
I 5	79.0	>60	1.20	2.9	1.00	2.2	4	1–2	174*	164
I 6	89.0	>60	1.18	4.5	1.67	3.2	3–5	2–3	116	164
I 7	96.0	>60	1.09	4.4	1.29	2.3	6–8	2	173	140
I 8	80.4	40	1.07	3.3	1.58	1.5	5	1	313	122
I 9	69.3	40	1.04	2.7	1.65	1.3	6	2	148	75
I 10	29.1	40	1.07	3.2	1.49	1.2	2–3	1	30	34
II 11	14.9	40	1.07	3.2	–	1.4	2–3	1	–	23
II 11	56.1	40	1.07	–	–	>6.5	1?	3–4	881	866
II 12	48.3	30	1.09	–	–	>6.5	1?	>3	2185	1114

*Sample distances too small to be representative.

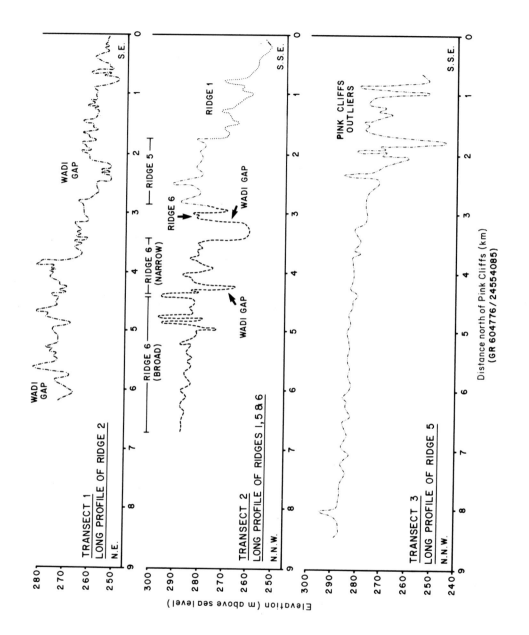

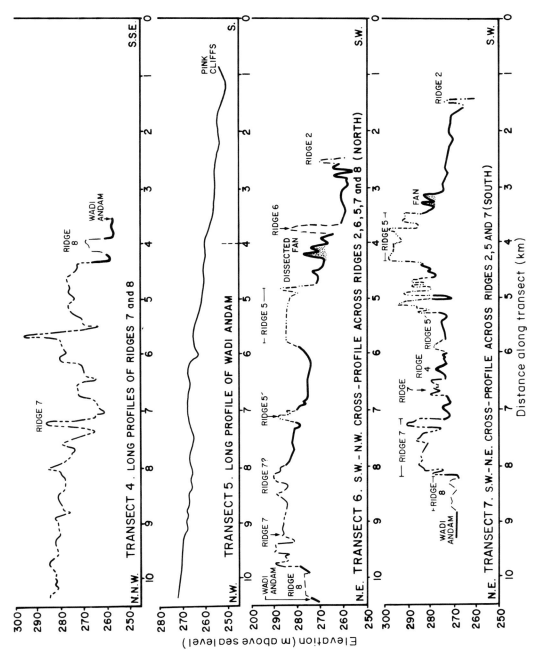

Figure 14.5. Longitudinal profiles of selected 'raised' palaeochannel ridges in area E, located at the Fan I/II boundary

These large distances apart indicate that separate channel systems existed, each comprising meandering to low sinuosity, single-thread channels; multiple or braided channels appear to have existed only locally (e.g. Figure 14.4B).

The number of active channel zones in the younger generations, (especially generations 6, 7, 8, and 9) increases substantially in the downfan direction. This downfan trend reflects the development of a distributary channel system, producing a series of distinct channel bifurcations associated with fanning out of drainage radially from the fan apex. In generation 7, for example (Figure 14.4C), two proximal palaeochannels progressively divide downfan to produce at least eight distal palaeochannel zones, some 40 km to the south.

A further distinctive feature of many of these palaeochannels is that they tend to become markedly narrower in the downfan direction (Table 14.1). The downstream narrowing in preserved channel widths is particularly evident in the later channel systems (generations 7, 8, and 9, Table 14.1), where proximal widths are 1.2 to 2.6 times greater than distal widths. Many of the channels which taper off and ultimately disappear downfan (Figure 14.4A) are those associated with the distinct distributary channel systems described as 'terminal fans' by Friend (1978). The oldest channels are generally the widest, with proximal widths exceeding 300 m, compared with those of the younger channels which generally average less than 175 m in width (except for generation 8, see Table 14.1). The youngest channels of generation 9 to 11, are well below 75 m in width. These narrow ridges contrast markedly with the broad gravel sheets of the later Fan II deposits (see generations 11, 12 in Table 14.1; and see below).

Field evidence indicates two additional significant features regarding the morphology of successive palaeochannel generations. One significant contrast between the older and younger generations is that the former exhibit markedly undulating crestlines, while the most recent deposits form smooth longitudinal profiles apparently representing the original depositional gradient (Figure 14.5). Secondly, the topographic relationship between successive palaeochannel generations at intersections are highly variable (Figure 14.6). Most commonly, the younger palaeochannel deposits are higher in the sequence, and overlie the older deposits. However, at a number of intersections the younger deposits lie at the same, or even lower, elevation than the older deposits (e.g. Areas F and G intersections, Figure 14.6F and G). These latter topographic relations occur particularly where successive channels intersect at relatively high angles, ranging between $ca.$ 80 and 90° (e.g. see Areas C and G, Figure 14.6). Hence, the relative height of a channel deposit need not be a direct reflection of its relative age, since periods of incision clearly occurred at certain stages during fan formation.

MORPHOLOGY OF FAN II PALAEOCHANNELS

The palaeochannel systems of Fan II are characterized in the east (Figure 14.3) by high relief (often over 10 m), low sinuosity channels that rarely intersect or are superimposed on one another; and in the west by extensive low relief, gravel spreads and sequences of three main terrace levels bounding the present course of Wadi Andam. The eastern channel courses form multiple channel systems extending from as far north as Samad (Figure 14.2) southwards towards Barzaman. From there the channels swing southeastwards, finally to disappear as they approach the boundary of the Wahiba Sand Sea, after a distance of $ca.$ 75 to 80 km from the most proximal deposits (Figure 14.1). Hence, the channel courses follow a broad southeast-trending arc, comprising low-sinuosity gravel spreads and ridges which narrow significantly and finally taper off and disappear downfan. Many channel courses narrow from over 40 m width to disappear within 6 km downstream (Figure 14.3).

The terraces and broad gravel spreads lying on the west of Fan II extend across the 6 to 7 km wide wadi system, with terrace widths averaging $ca.$ 700 m. Few individual channel courses can be identified on these terraces, although the higher deposits comprise several superimposed broad gravel spreads (see Area E, Figure 14.6).

Sedimentology of the Palaeochannel Deposits

FAN SYSTEM I

Facies Types and Sedimentary Structures

Four main facies types were identified in the Fan I channel deposits.

Facies 1. Trough cross-stratified sand and gravel The Fan I palaeochannel deposits are dominated by fine-grained, large-scale, trough cross-bedded units (Figure 14.7) up to 3 m thick, comprising individual sets 0.15 m to 1.6 m in thickness. The crossbeds generally consist of poorly-sorted basal gravels (location 18.6 on Figure 14.7) with clasts up to 200 mm in diameter, fining upwards into well-sorted fine gravels (18.7 on Figure 14.7). Clasts are well rounded, with Cailleux roundness indices exceeding 350.

Facies 2. Massive, clast-supported gravel This facies comprises thin units, up to 0.5 m thick, of poorly-structured, clast-supported cobble gravels, which can be traced longitudinally for tens of metres. The gravels contain clasts up to 200 mm in diameter, which are well rounded (Cailleux indices exceeding 350–400), but weakly imbricated. The gravels exhibit crude horizontal bedding, but no distinct internal sedimentary structures or grading.

Facies 3. Horizontally bedded sand Facies 3 comprises lenses of thin, horizontally bedded, medium sand, and is of only local, minor significance.

Facies 4. The 'Barzaman Formation' Facies 4 is the most extensive lower facies, underlying all sand and gravel facies types 1, 2, and 3, although they may also be locally interbedded with these deposits. This facies comprises a whitish-pink, massive, indurated, fine-grained material termed the 'Barzaman Formation' (Maizels, 1987, 1988). The Barzaman Formation is largely dolomitic in composition, with the younger horizons also being rich in clay minerals. The interbedded gravel pockets and channels exhibit a considerable degree of diagenesis; much of the Barzaman Formation exhibits 'ghosting', characterized by reddish-brown and orange mottling representing altered remnants of former ultrabasic clasts (and see Glennie et al., 1974).

Clast Size Characteristics

Maximum clast sizes (determined from the mean length of the intermediate axis of the ten largest clasts in each horizon at each site) exhibit a marked decrease over the 120 km of fluvial transport (Maizels, 1988). Maximum clast sizes in the proximal palaeochannel deposits reach ca. 200 mm in the older channel deposits and ca. 240 mm in the younger, and decrease to ca. 50 mm downstream, resulting in an overall mean rate of size decrease of 1.0 mm km^{-1}. However, there are significant differences in mean clast sizes between the older and younger channel deposits. Gravels associated with palaeochannel generations 1–5, for example, average 77 mm, while those of generations 7–9, at the same site, average 108 mm (Maizels, 1988). Hence, the older, more sinuous channels are associated with significantly finer bed materials.

Lithological Composition

The gravels exhibit a wide range of lithologies, reflecting the heterogeneous nature of the source materials. The major constituents of the clastic fraction are:
1. Ophiolites, which are dominated by diabases, gabbros, and troctolites, and also include serpentines, peridotites, diorites and basalts, derived from the Semail Nappe outcrops north of the Samad area (Figure 4.2);
2. Cherts, derived from the extensive Hawasina outcrops which flank the southern margins of the Oman mountains, which directly underlie most of the proximal channel deposits, providing an abundant local supply of resistant materials for fluvial transport; and
3. Limestones, derived from a variety of sources of different ages, ranging from the older up-domed massifs (Permian to Late Cretaceous),

284 J. MAIZELS

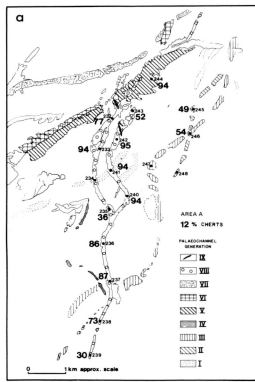
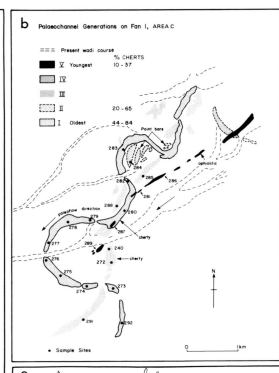
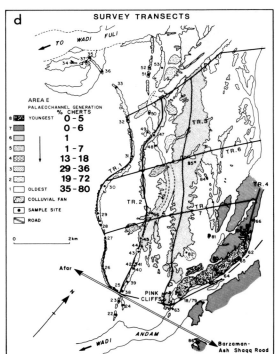
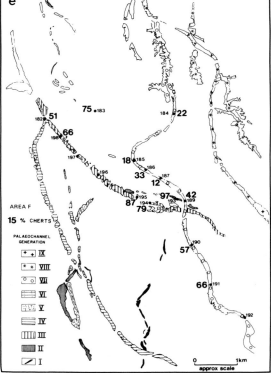

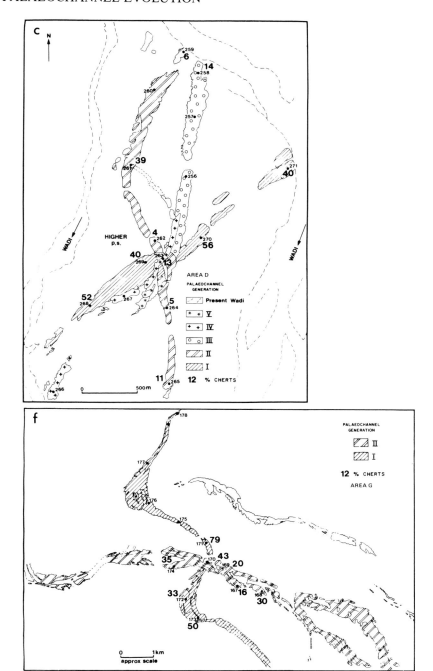

Figure 14.6. Morphologic and sedimentologic relations between successive palaeochannel generations at selected channel intersections. Palaeochannel generations indicated represent *local* generations. Small figures indicate sample sites. Large figures indicate percentage of chert clasts at individual sample sites. See Figure 14.3 for location of areas A–E. There is some variation in cartographic conventions between the maps

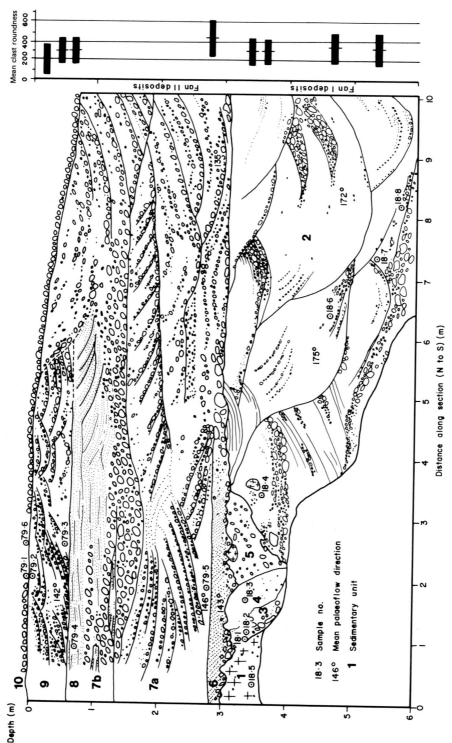

Figure 14.7. Stratigraphy and sedimentology of Fan I and Fan II deposits, site 18 near Pink Cliffs, Wadi Andam. A major unconformity separates the lower Fan I deposits from the upper Fan II deposits. The main sedimentary units are as follows: Fan I, unit 1. Barzaman Formation with partially weathered bleached basic clasts. 2. Trough cross-bedded sands and gravels, with cross-bed units 1.6 m thick; included cobbles <20 cm diameter, in friable, white calcite cement; numerous weathered basic rocks; R = 343 ± 144 and 335 ± 159. 3. Poorly-cemented weathered gravels, bleached cherts and ophiolites <8 cm diameter in sandy matrix; R = 309 ± 132. 4. Pseudobedded barzamanite, with a few scattered, bleached chert clasts <1 cm diameter. 5. Rubbly, structureless, poorly-cemented Barzaman Formation with partially weathered ultrabasic clasts <10 cm diameter, R = 308 ± 131. Unconformity. Fan II, unit. 6. Planar and cross-bedded sands, fining upwards from basal cobble layer; palaeocurrent vector 135°–146°. 7a. Large cross-bedded sand units, coarsening upwards R = 449 ± 186. 7b. Alternating sand and cobble beds. 8. Horizontally bedded fine sands. 9. Trough cross-bedded gravels with isolated sand lenses; unaltered pebbles cemented by opaque clayey-calcite; palaeocurrent vector 125°, R = 321 ± 133. 10. Surface lag of black, varnished, ophiolite-rich pebbles <16 cm diameter, R = 225 ± 161 (R—Cailleux roundness)

to the foothills of Hawasina turbidite limestone, interbedded with the cherts, and 'exotic' marbles (Figure 14.2; Hopson et al., 1981). A wide variety of other igneous and sedimentary rocks, with minor amounts of metamorphics, largely derived from the mountain source areas also occur within the gravels.

Analysis of the percentage of ophiolites, cherts, limestones, and other lithologies in channel deposits of different generations sampled at a series of channel intersections (e.g. Areas A–E in Figure 14.6) provides crucial clues to the mechanisms and patterns of palaeochannel evolution during fan development. The results of these analyses indicate that significant differences in lithological composition of the gravels occur between channel deposits of different relative age. The oldest, lowermost channel deposits, both in exposed sections and in superimposed palaeochannel sequences, exhibit the highest concentration of cherts. In Area C, for example, (Figure 14.6C), where a sequence of superimposed palaeochannels has been exposed, the mean percentages of cherts increase from 19% in the youngest deposits (generation 7), to 38% in the next oldest deposits (generation 5), to 65% in the oldest deposits (generation 2), representing a highly significant statistical trend (Figure 14.8; Maizels, 1987, 1988). Similar trends are found within all the distal channel intersection sequences (see Figures 14.6B–F). In Area F (south of Area D on Figure 14.3; see Figure 14.6E), the oldest channels exhibit chert percentages ranging between 66 and 97%, while the later generations exhibit values of *ca.* 50%. The most recent channel deposit (the north–south channel on the east of the site, Figure 14.6E), exhibits a marked increase in chert concentration downstream of the channel intersection: chert percentages rise from 12–33% upstream, to 42–66% downstream of the intersection. In Area D, the oldest channel deposits again exhibit chert percentages ranging between 40 and 65%, while the later deposits contain 5 to 39%, and the youngest contain only 13 to 14%.

In addition to significant changes in lithological composition with relative age of the deposit, significant changes also occur from west to east within gravel sequences of the same relative age. Hence while the gravels of the central fan are rich in cherts, those west of Wadi Adam are rich in limestones, with few cherts and no ophiolites present. Similarly, those in the far south and southeast, along Wadi Matam and beyond the Matam–Andam confluence zone, exhibit up to 80% limestone content.

Post-depositional Alteration

All the sand and gravel units of facies 1, 2 and 3 are cemented by clear sparry calcite, forming crystalline crusts around individual clasts and representing the main matrix material in the gravel (*cf.* Moseley, 1965; Vondra and Burggraf, 1978).

Gravels exposed on the channel surfaces exhibit evidence of long periods of subaerial weathering processes, including solution, dew etching and aeolian sand abrasion. Clasts exhibit evidence of desert varnish, case hardening, weathering rind development, of ventifacting and faceting, and of solution rilling and pitting, the latter particularly on limestone clasts (*cf.* Williams, 1970). The older gravels commonly form a layer or scattering only one particle thick (see McClure, 1978) over the Barzaman Formation, suggesting extensive removal of former channel sediments by decomposition, solution, mechanical breakdown, and deflation (*cf.* Denny, 1965; Williams, 1970).

FAN SYSTEM II

Facies Types and Sedimentary Structures

Two main facies types dominate the Fan II deposits and contrast markedly with those of the Fan I palaeochannels.

Facies 5. Crudely bedded, clast-supported pebble–cobble gravels Facies 5 comprises poorly-sorted, clast-supported gravel units up to 2 m thick, containing pebble to cobble sizes, but extending up to boulders 500 mm in diameter in proximal zones. The gravels exhibit crude horizontal bedding, with locally strong preferred fabric, but little in-

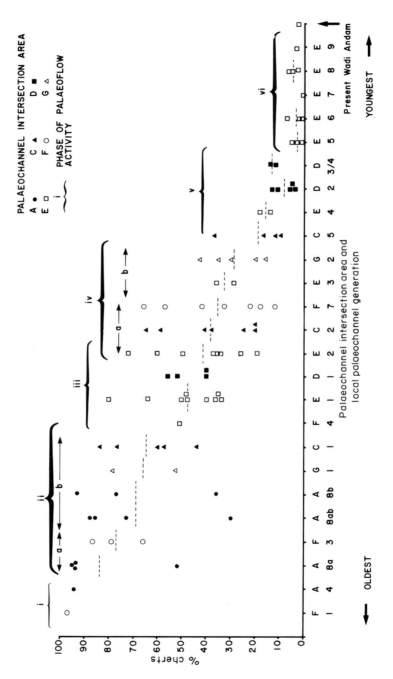

Figure 14.8. Increase in percentage of chert in palaeochannel deposits of successively older generations for areas A, C, D, E, F, and G. Major phases of fluvial activity based on significance tests of chert percentages between different deposits

ternal grading. Clasts are moderately well rounded, with roundness values exceeding 300.

Facies 6. Cross-stratified gravel Facies 6 includes both localized and extensive units of poorly-sorted, planar and trough cross-stratified gravels with total unit thicknesses exceeding 3 m. Each set ranges from 0.2 to 0.8 m thick, comprising basal clast-supported pebble gravels fining upwards into gravel and coarse sands (Figure 14.7). Individual units are laterally discontinuous and both upper and lower contacts are scoured. Clasts are moderately well rounded, with mean roundness values exceeding 350.

Clast Size Characteristics

Maximum intermediate particle diameters decrease at a maximum rate *ca.* 4 times greater than observed in the Fan I gravels, namely 4.1 mm km^{-1}. Mean clast sizes in the Fan II gravels are not significantly larger than those of Fan I at similar distances downfan. However, if there have been major changes in source area and in associated drainage nets through time (see below), the absolute distances of fluvial transport may be significantly larger for the Fan II deposits, such that they do form a substantially coarser deposit.

Lithological Composition

The clast fractions of the Fan II gravels are dominated by dioritic lithologies and other ophiolitic lithologies with only minor percentages of cherts, limestones, and other rock types. The younger deposits, forming the Wadi Andam terrace gravels, are wholly composed of ophiolites (see Figures 4.6D and 4.8).

Post-formational Alteration

The Fan II sands and gravels are extensively cemented by calcite but, unlike the Fan I cements, those of Fan II appear to exhibit an opaque, clayey appearance with high proportions of unaltered fine gravel particles in the matrix, and form a more friable and poorly-indurated deposit.

The surface gravels exhibit evidence of widespread disintegration marked by exfoliation, spalling, and *in situ* splitting, processes that appear to be active at the present time. These processes are acting to produce a much finer-grained surface deposit of small angular fragments and rock flakes. The clasts also exhibit desert varnish, faceting, and surface solution effects, according to lithology.

General Stratigraphic Relations

Fan II deposits overlie and are cut into the lower Fan I deposits along the Wadi Andam drainage system. The Fan II deposits are found at three main topographic levels along Wadi Andam, with the lower two levels cutting into both the Barzaman Formation and the channel gravels of the Fan II system. The contact between the two fan deposits forms a sharp, erosional boundary, often marked by a series of irregular, longitudinal flutes scoured into the underlying Barzaman Formation (see Figure 14.7). The boundary surface is generally coated with a thin microcrystalline layer of brown micrite (see Goudie, 1983). Boulders of older cemented gravels and the Barzaman Formation are occasionally incorporated into the basal Fan II gravels.

The Fan I deposits appear to overlie the continental alluvial and sabkha sediments of the Mio-Pliocene Upper Fars Group in the north, and Miocene limestones in the southwest (Aubel, 1983; Jones, 1986), indicating that the earliest possible date for commencement of gravel accumulation is early Pliocene. The latest date of formation is defined by the age of chert artefacts lying on the older, lower palaeochannel ridges of Fan I in Area E (see above). According to Edens (1988) the artefacts are likely to date from between 4 and 7 Ka, representing the latest period during which palaeochannel exhumation could have taken place.

Palaeohydrologic Interpretation

CHANGES IN CHANNEL PATTERN

Significant changes in channel morphology, pattern, and sedimentology have occurred during fan development. The earliest observed palaeochannels (generations 1–3, 5, 6) exhibit broad, sinuous, chert-rich courses that do not appear to narrow or taper off significantly downfan but extend continuously for over 60 to 100 km (e.g. Figures 14.3, 14.4; Area C, Figure 14.6B). By contrast, the later palaeochannels on Fan I (generations 7–11) exhibit low sinuosity, narrow, chert-depleted ridges that taper off downstream and disappear after only *ca.* 40 km of transport (e.g. Figure 14.4; Area C, Figure 14.6B). Significant downstream fining occurs in all channel deposits, but rates of decrease are highest in the youngest sediments. Finally, the palaeochannels of Fan II form broad but rapidly tapering channel courses in the east, and extensive, ophiolite-dominated gravel spreads and terraces in the west, within the Wadi Andam routeway.

The preserved sedimentological evidence suggests that the older, more sinuous courses are associated with finer-grained, large-scale, trough cross-bedded, ribbon conglomerates, reflecting the downstream migration of large fluvial dunes along the bed of a meandering channel. The younger, low sinuosity channels are associated with coarse-grained cobble deposits forming thin, tabular, planar cross-bedded units, representing longitudinal bars (e.g. Miall, 1978; Rust, 1978). The fine-grained overbank and floodplain deposits are likely to be largely represented by the compacted sediments of the Barzaman Formation. The Barzaman Formation appears to comprise a variety of compacted and indurated mudstones (*cf.* Bull, 1964), fluvial and aeolian silts (Stanger, personal communication) and sand, with cobble and pebble lenses, and indistinct subhorizontal beds characteristic of partially-vegetated low relief floodplains (*cf.* Yaalon and Dan, 1974; Vondra and Burggraf, 1978). Finally, the broad but thin gravel spreads across the Wadi Andam terraces accumulated as sheet gravels probably associated with flood flows filling the former wadi floor (*cf.* Ramos and Sopena, 1983).

CHANGES IN PALAEOHYDRAULIC CONDITIONS

Palaeodischarges associated with these different channel systems have been estimated for successive generations of palaeochannels on Fan I. Full details of the methods, sources of error, assumptions, and problems of interpretation are discussed by this writer elsewhere (Maizels, 1983, 1987, 1988). A brief summary of the approach is presented here.

Sources of Error

There are still many uncertainties and sources of error involved in the application of palaeohydraulic models to raised channel deposits. For example, estimates of former flow hydraulics, including flow depths, resistance to flow, velocity and discharge for different palaeochannels require estimates of the cross-sectional area of flow, former energy gradients, and maximum sizes of particles that the stream was competent to move. Accurate determination of these primary parameters may not prove possible for many of the palaeochannels. Overbank sediments have largely been removed, so that former flood flow widths can rarely be correctly determined. Former energy gradients can be estimated only very approximately from the published topographic maps. In addition, since many of the channels appear to have been dissected, diagenetically altered, and let down (both by deflation to form a pebble lag and by compaction associated with diagenesis) on to an underlying surface, the original channel gradients are unlikely to have been preserved. Maximum particle sizes, at least of the surface material, appear to decrease through time by progressive granular disintegration and hence measurements of maximum clast sizes may not always represent the largest materials that were originally transported and deposited in the channel.

Summary of Procedures for Palaeoflow Estimation

Two approaches have been adopted for determining former flow conditions. The first approach is based on initial estimates of the critical shear stress at which bed particles begin to move. This computation uses Shields' function and allows subsequent calculation of former flow depth using the Du Boys equation, and resistance to flow and flow velocity using Manning and Darcy-Weisbach type equations. This requires field determination of mean and maximum particle sizes to provide measures of particle sorting, bed roughness, former channel gradients, and former flow widths. The main assumptions of this critical shear stress approach are the former existence of steady uniform flow; the validity of the shear stress model itself; an estimated effect of packing and imbrication on Shields' coefficient; the absence of local bedform or coarse clast concentrations; and the availability of all particle sizes for transport by competent Newtonian fluid flows (see Costa, 1984). Most of these assumptions remain questionable. This approach has so far been applied to only two areas of raised channels: Area C on Fan I, and Area E which extends across the boundary between Fans I and II (see Figures 14.2, 14.6B, and D).

The second approach to the palaeohydrologic analysis of the raised channel deposits is based on the application of empirical discharge-form relations for meandering channels. These relationships require determination of meander wavelength, radius of meander curvature, and bankfull channel width, w, and hence major sources of error arise where former channel morphology is only partially or poorly preserved (see Ethridge and Schumm, 1978; Rotnicki, 1983). The equations adopted in this study are based on four empirical functions relating meander planform and channel morphological measures to discharge parameters for sand-bed rivers (see Table 14.2), derived from Carlston (1965), Schumm (1972), and Williams (1984). A second source of error therefore lies in the application of these derived equations to gravel-bed channels which are likely to have formed under significantly different environmental conditions. This empirical approach has been applied to each of the 12 main palaeochannel generations identified in Table 14.1.

Palaeoflow Conditions Associated with Fan I Channel Systems

Estimates of peak flow depths suggest that they averaged between about 1.2 and 4.5 m, although maximum depths may have exceeded 6 or 7 m in some channels during major flood events. Peak flow velocities averaged between about 3 and 7 m s^{-1} within the palaeochannel systems as a whole, with discharge ranging between about 160 and 1400 m^3 s^{-1}. In Area C of Fan I the two earlier palaeochannel generations exhibit significantly different channel form, sediment, and flow conditions from the youngest channel systems in the area (Figure 14.6B). The earlier channels are at least twice as wide as the later channels, significantly more sinuous, and comprise significantly finer clasts (Table 14.1), representing long-term rock breakdown and/or finer sediments in transport. Predicted flow depths and velocities were 30 to 85% greater in the younger, narrower, coarser-grained channels than in the broad, sinuous, fine-grained older channels. However, because of similarities in cross-sectional areas of flow, with the older channels exhibiting only marginally higher values, the peak discharges predicted for the different channels by the various models are of a similar order of magnitude. The older channels were associated with palaeodischarges reaching about 1150 m^3 s^{-1} (for Froude numbers of about 0.90), while the younger channels were associated with those of about 1330 m^3 s^{-1}. The palaeodischarges predicted by the two approaches, at least for the meandering channel generations exhibit a similar order of magnitude, although the empirical relationships of Q_6 to Q_9 tend to estimate slightly higher flows than predicted by the shear stress models (Table 14.2).

Hence, significant changes in palaeoflow conditions appear to have occurred during fan aggradation. Palaeochannels became increasingly narrow, less sinuous, and probably comprised

Table 14.2. Estimated palaeodischarge for successive palaeochannel generations, Fans I and II, Southern Piedmont, Eastern Oman Mountains

Palaeochannel generation (Fan I or II)		Palaeodischarge model (key given at base of table)									
		Q_1	Q_2	Q_3	Q_4	Q_5	Q_6	Q_7 proximal	Q_7 distal	Q_8 proximal	Q_8 distal	Q_9
I	1						2239		594		120	6270
I	2	1137*	898	873	1142*	470	4971		424		77	5848
I	3						4180		408		73	4134
I	4						7055		1249		320	10338
I	5	1176*	929	918	1181*	485	2363		757		165	3865
I	6						4649	483	757	92	165	7845
I	7	1350*	1066	1278*	1368*	553	4491	812	617	181	126	5493
I	8						2883	1755	516	500	100	7266
I	9						2117	663	274	139	43	7714
I	10						2750	83	98	9	11	6701
I } II	11	14390	11477	13292	14440	5752	2750	6737	59 6588	2933	6 2848	– –
II	12	30624	24424	28425	30717	12237	–	21943	9140	13862	4381	–

*Discharge estimates associated with Froude numbers 0.76–1.0. All values in m³ s⁻¹
Key to discharge models (Full details given in Maizels, 1983, 1987, 1988)
Q_1–Q_5 based on critical shear stress approach:
Q_1 based on Strickler (1923) resistance equation
Q_2 based on Limerinos (1970) resistance equation
Q_3 based on Jarrett (1984) empirical resistance equation
Q_4 based on Darcy–Weisbach–Hey (Hey, 1979) resistance equation
Q_5 based on Costa (1983) empirical resistance equation
Q_6–Q_9 based on empirical discharge–form relationships (see Maizels, 1983)
Q_6 based on Carlston (1965); Williams (1984)
Q_7 based on Williams (1984) discharge–width relationships
Q_8 based on Osterkamp and Hedman (Williams, 1984) daily discharge–bankfull width relation
Q_9 based on Osterkamp and Hedman (Williams, 1984) max. discharge–bankfull width relation

coarser sediments; flow depths and velocities during peak flow events increased; while maximum discharges remained of a similar order of magnitude. In addition, the older channel systems extended across the full downstream length of the fan, while progressively younger channel systems became of increasingly limited extent.

Palaeoflow Conditions Associated with Fan II Channel Systems

Fan II channel systems located at a similar distance from the mountain front as those examined on Fan I, exhibit significant contrasts with the latter. In particular, the great widths of the preserved Fan II palaeochannel deposits, together with the nature of the constituent sedimentary structures and facies types (see above), suggest that the deposits characterize broad, shallow braided channel systems (e.g. Miall, 1978; Rust, 1978). Estimates of peak flow depths of about 3.8 m suggest that width–depth ratios ranged between 200 and 400, almost an order of magnitude higher than in the Fan I channel systems. Gradients, particle sizes, resistance and velocity estimates are all similar in magnitude to those of the Fan I channels, but Froude numbers are much lower. The high palaeodischarges estimated by the different models largely reflect, therefore, the broad lateral extent of the former channel systems. Peak palaeodischarges averaged between about 13 000 and 31 000 m³ s⁻¹ according to most of the models (Table 14.2), particularly those with Froude numbers of over 0.7. The estimated peak flows therefore exceeded those of the Fan I channels by about one order of magnitude.

No detailed error analysis has yet been completed on these palaeohydraulic computations, but the palaeodischarge values were found to vary significantly according to the variability of the input parameters. For example, variability (measured by standard deviation values) of slope, width, particle size, and Shields function for the Fan II channels resulted in palaeodischarge differences of up to ±60%, ±48%, ±40%, and ±67%, respectively. Although there are few records that extend over more than a 5-year period, the available flood discharge data for recent floods in Wadi Andam suggest that peak flow conditions in the Fan I palaeochannels as a whole may have been fairly similar to those of the present day. Estimates of width, depth, and velocity during recent wadi flood events (Curtis, 1985; Maizels and Anderson, 1988) ranged between 5 and 405 m, 0.28 and 6.19 m, and 0.006 and 9.72 m s^{-1} respectively, while peak discharges reached *ca.* 1000 m^3 s^{-1}. These discharges are significantly lower than those estimated for the Fan II channel systems.

CHANGES IN SEASONALITY AND DURATION OF FLOWS

Long-term aggradation of the Fan I system appears to have been associated with relatively sinuous channels developed in comparatively fine-grained trough cross-bedded gravels, which maintain fairly constant widths for distances of up to 60 to 100 km downfan. The development of an equilibrium relationship between the meandering channel planform and associated dominant flows is likely to require a minimum period of time (Bridge, 1985), perhaps represented by minimum flow durations of at least two to three months per year (see Jackson, 1978; Arche, 1983). Such persistent and more uniform flows (e.g. Wells and Dorr, 1987) are also more likely to create, and sustain, an extensive channel system which exhibits little change in dimensions for many tens of kilometres (*cf.* Nichols, 1987). Hence, the earlier generations of sinuous palaeochannels, (generations 1–3, 5 in particular), are believed to have developed under at least seasonal flow (*cf.* Cotter, 1978) conditions, with bankfull flows averaging between *ca.* 400 and 800 m^3 s^{-1}, and total system flows of *ca.* 1000 to 2000 m^3 s^{-1}. A marked change from low sinuosity braided streams to high sinuosity meandering streams, recorded for the early Pleistocene of eastern Lake Turkana in Kenya by Vondra and Burggraf (1978) has also been attributed to increasing persistence of flow, possibly becoming perennial in the final stages of development.

The later stages of development of Fan I are represented by the more extensive growth of fan distributary channels. These channels generally extend only a limited distance downfan (*ca.* 40 km), and exhibit low sinuosity courses comprising coarse-grained deposits. The reduced sinuosity of the channels, and their ultimate downfan disappearance (*cf.* Friend, 1978), suggest that they may have been associated with more ephemeral flows (e.g. Steel and Aasheim, 1978) than the earlier Fan I channels (except perhaps generation 4?). Estimated bankfull discharges were higher (*cf.* Nami and Leeder, 1978), while discharges across the whole distributary network might have reached between *ca.* 2700–9830 m^3 s^{-1} in proximal zones and *ca.* 820–7000 m^3 s^{-1} in distal zones, before flows disappeared. Significant downstream decreases in flow would be promoted by the large number of bifurcations (e.g. Rachocki, 1981), by short-lived flow events, allowing rapid transmission losses into the dry, underlying gravels, and by high rates of evapotranspiration (*cf.* Stear, 1983; Nichols, 1987). These channel patterns, with downfan decreases in both sediment size and channel size, are characteristic of the semiarid 'terminal' distributary systems described by Nichols (1987) and by Parkash *et al.* (1983).

Water tables would probably have remained relatively high through the period of fan aggradation. Rapid cementation of the channel gravels through precipitation of calcite cement (e.g. see Stalder, 1975) probably occurred soon after streamflow, during periods of seepage and evaporation (e.g. Moseley, 1965). High water tables would also have allowed the large-scale alteration of the ophiolitic components of the gravels, particularly contributing to diagenesis of clastic lenses within the Barzaman Formation itself.

The eastern palaeochannels of the Fan II system also rapidly disappear southeastwards into the Wahiba Sands, again suggesting only intermittent or episodic flows. The subsequent concentration of flows into the Wadi Andam course was achieved through large-scale, sudden, and deep entrenchment, probably by large, rare flood events with peak flows ranging between *ca.* 13 000 and 31 000 m^3 s^{-1} (Maizels, 1987). Flood flows deposited extensive sheet gravels across the bed of the wadi (*cf.* Stear, 1983), and these were subsequently incised by later floods to form the main terrace sequence (*cf.* 'superfloods' of Schick, 1974). Large-scale entrenchment may have been accomplished relatively rapidly, since, as Baker (1977) argues, regions with seasonal aridity have particularly high potential for catastrophic response to runoff. This period of fan downcutting was probably associated with a relatively rapid fall in the watertable, and by extensive contemporaneous lowering of the fine-grained interfluve areas of the fans by deflation (Glennie, personal communication), to reveal the now exhumed palaeochannel landscape.

CHANGES IN CATCHMENT AREA

The distinctive vertical and spatial trends in the relative concentrations of ophiolite, chert, and limestone support the hypothesis that the source areas for the alluvial fan gravel changed both through time (Fan I–Fan II) and space (west–east). The lithology of the gravels appears to be dominated by the lithology of the source areas being drained, such that the western and eastern limestone-rich fan gravels must have been deposited by former courses of the Wadis Adam and Matam, respectively. The major central zone of both the Fan I and Fan II gravels, by contrast, has been fed throughout the period of fan formation by the Wadi Andam system, which drains southwards through a narrow gap in the Hawasina and outer limestone hills (Figure 14.2; *cf.* Price, 1974; Nichols, 1987; Sutton, 1987).

Hence the long-term trend of decreasing chert: ophiolite ratios during fan evolution must reflect a major long-term change in availability of these two lithologies within the catchment. Such a dramatic increase in ophiolite concentrations from Fan I to Fan II gravels can best be explained in terms of changes in the drainage network. The Wadi Andam network was probably initiated on the uplifted Tertiary surface and evolved through progressive headward extension across the chert and limestone outcrops northwards towards the ophiolites (*cf.* Mensching, 1974). Headwaters of the Wadi Andam system would therefore reach the ophiolite outcrops much later during the Tertiary–early Quaternary period than streams exploiting the more southerly cherts and limestones; in addition the northward extending headwaters of Wadi Andam could have captured the headwaters of the adjacent Wadi Adam system lying to the west (Figure 14.1; *cf.* Sutton, 1987). This pattern of river capture may have been facilitated by continued mountain uplift in the headwater zones, or increased climatic aridity with more rapid runoff and erosion. The Wadi Adam headwaters would already have developed into an extensive and deeply dissected network, and their capture by the Wadi Andam headwaters would provide a new and abundant supply of ophiolites to the southern Fan I gravels, and throughout the Fan II gravel sequence, may well have been associated not only with this sudden influx of large volumes of sediment, but also with significant additional volumes of runoff (*cf.* Price, 1974; Baker, 1978). Not only would the added catchment area have represented *ca.* 30% of the present total catchment (i.e. *ca.* 1230 km^2), it would also have contained the highest mountain ridges, steepest slopes, most humid climate, and hence produce the highest rates of runoff. The high runoffs generated by the extended drainage network might well account for the large increase of runoff estimated above for the Fan II palaeochannel systems.

Mechanics of Fan Formation and Channel Change

TEMPORAL CONTINUITY OF CHANNEL ACTIVITY

Significant differences in the lithological composition of different palaeochannel generations at many intersections indicate that significant time intervals are likely to have occurred between successive periods of channel activity. These time intervals would have allowed for further headward extension of the network, possibly associated with continued mountain uplift in the headwater zone, and/or would have permitted the ophiolitic channel gravels to have become significantly altered before burial by a later channel gravel. The time interval would also have been sufficient to allow the channel gravels to become cemented prior to burial or to later incision.

In some cases, however, no significant differences exist in lithological composition between palaeochannel deposits of successive generations. The absence of a significant difference in composition may indicate that a relatively short time interval elapsed between each period of channel activity, or that earlier deposits were extensively reworked.

The present evidence does suggest, however, that the fans have evolved through discrete episodes of fluvial activity (cf. Beaty, 1970). Estimates of the number of such episodes, based on calculations of the differences in lithological composition from a number of sample intersections, suggest that at least six separate major periods of fluvial activity occurred during the latter, now-exhumed stages of fan development (Figure 14.8).

ALTERNATIONS OF AGGRADATIONAL AND EROSIONAL EPISODES

The morphological relationships between different channel generations at channel intersections indicate that at a number of sites the younger channels are incised into the older channel deposits, and occur at similar or lower elevations. Hence, incision must have occurred within overall aggradation of the fan. Where incision appears to have occurred, increases in the concentration of lithologies which are more common in the older deposits are found downstream in the younger channel deposits. Since many of these incised intersections occur where the two palaeochannel systems intersect at a high angle (up to 90°), incision was probably more readily accomplished where the underlying cemented gravels presented the minimum length of bed and bank resistance. Incision may have been related either to local channel scour—perhaps following avulsion farther upstream (e.g. Hopkins, 1985)—followed by aggradation and sedimentation (Friend, 1978), or to more regional climatic or baselevel changes (see below).

MULTIPLICITY AND CONTINUITY OF ACTIVE CHANNEL ZONES

Mapping of palaeochannel systems of similar generation suggests that up to eight major channel routeways coexisted during periods of channel activity. These were associated with the downfan expansion of a distributary channel system. The most extensive distributary systems appear to have developed during palaeochannel generations 6, 7, and 8 (Figure 14.4B, C), although the contemporaneity of all these distributaries has not yet been confirmed. The proximal deposits of the early palaeochannel generations are only poorly preserved making it difficult to estimate downfan changes in channel characteristics and long-term stability (see Jackson, 1978). By contrast, the fewer active zones associated with the well-preserved younger channel systems do appear to represent a reduction in either the volume or the proportion of flows extending on to Fan I.

Comparison of the location of the active channel zones for each palaeochannel generation indicates that the major channel zones were normally reoccupied during successive periods of fluvial activity (see Figure 14.4). Many of the meandering channels appear to have been relatively stable, such that flows tended to remain along established routeways for long periods, rather than migrating across the whole fan surface (cf. Leeder, 1978; Puigdefabregas and Van Vliet, 1978; Bridge and Leeder, 1979, Bridge, 1985; but see Rachocki, 1981). Hence the exhumed palaeochannel routeways now represent well preserved, complete multistorey fossil meander belts (cf. Puigdefabregas and Van Vliet, 1978; Stear, 1983). These routeways probably formed shallow longitudinal depressions between broad, low relief interfluves or adjoining floodplains.

During the intervals between periods of fluvial activity, the fine-grained interfluve and floodplain sediments may have been partially deflated, while the coarser channel gravels became cemented and indurated. Subsequent fluvial activity would depend on the degree of interfluve deflation, aeolian sedimentation, and channel cementation, as well as on discharge and sediment yield characteristics of the flow. Subsequent streams would then follow the route of the former channel course or active channel zone; or through avulsion (*cf.* Gole and Chitale, 1966; Bristow, 1987; Leeder, 1978; Wells and Dorr, 1987) create a new channel away from the now topographically higher, cemented gravels of the previous channel bed; or create a new channel on top of indurated aeolian sands and silts. Periodic avulsion is likely to have occurred, particularly during seasonal floods (e.g. Bridge and Leeder, 1979). Bridge (1985) contends that rivers on aggrading alluvial plains tend to abandon their channel belts in favour of new ones, rather than migrating across the whole plain and that this scale of avulsion occurs at time intervals of 100 to 1000 years (*cf.* Heward, 1978).

Where successive channels have been confined to a single active channel zone for several channel generations, little interfluve deflation is likely to have occurred between the periods of fluvial activity, and rates of aggradation in both the floodplain and channel zones probably remained relatively high. By contrast, where successive channels occupied new routeways, interfluve deflation and channel bed induration are likely to have been high, perhaps reflecting more arid or longer time intervals between the periods of fluvial activity. Hence established channel zones could be linked by the development of new spillway channels across the fan surface, and/or new channel zones could be created, which could then be reoccupied during later periods of fluvial activity.

Only with the development of the Fan II channel system did major changes in the location of active fluvial activity occur. As the period of Fan I aggradation came to a close, increasing volumes of water were diverted through the Wadi Andam routeway and on to fan surfaces farther to the east (Figures 14.4 and 14.6F). Gradually, a single major zone of fluvial activity became concentrated in Wadi Andam.

MIGRATION OF CHANNEL COURSES

The mapped evidence suggests that both abrupt and gradual channel pattern changes occurred from one period of fluvial activity to the next. In many cases, successive palaeochannel generations at individual intersections indicate significant changes in channel morphology, sinuosity, orientation, and sedimentology. Such major contrasts in palaeochannel characteristics between successive generations at a given reach suggest that significant differences in local or regional palaeodischarge, fan gradients, and/or sediment yields must have occurred between each period of fluvial activity.

However, evidence also exists for gradual migration of channels within an active channel zone (*cf.* Puigdefabregas and Van Vliet, 1978). At a number of sites older and younger generation channel meanders appear to be linked by a series of point bar deposits (e.g. see Area C, Figure 14.6), occurring at similar topographic levels. At other sites successive generations of meanders at different elevations appear to have migrated either downstream or laterally across the active zone (e.g. see Figure 14.4B, C). Hence, successive generations of flow commonly appear to have reoccupied fan depressions following former channel courses (Bristow, 1987), but with flows often creating new channels at a higher topographic level.

Controls on Longterm Palaeohydrologic Change and Fan Evolution

AUTOCYCLIC CONTROLS

Long-term Fan Aggradation

The palaeochannel changes observed through successive periods of fluvial activity can partially be attributed to the effects of autocyclic controls (Figure 14.9). Long-term fan growth resulted in progressive steepening of the fan surfaces (*cf.* Fan I and Fan II mean gradients, see above), thereby allowing coarser material to be transported, and hence promoting the development of lower sinuosity channels in the increasingly poorly-sorted, non-cohesive bed and bank materials (*cf.* Vondra and Burggraf, 1978). Increased volumes of coarse sediment transport could result in rapid rates of sedimentation as downfan gradients de-

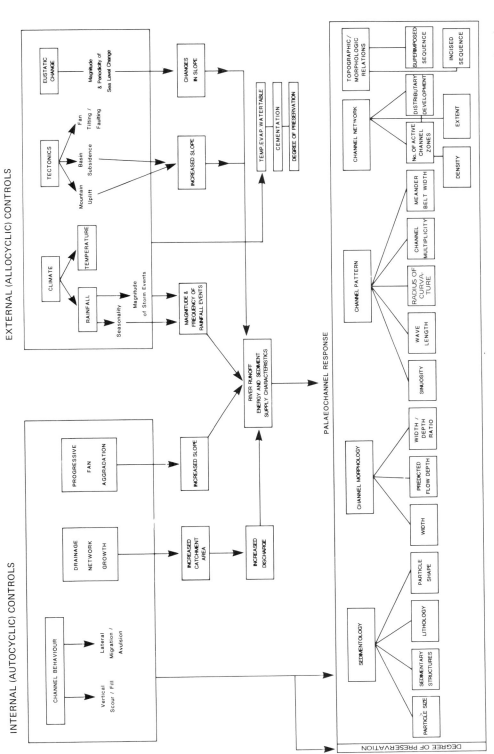

Figure 14.9. Simplified model of internal and external controls on channel changes within superimposed palaeochannel systems of Fans I and II, southern piedmont zone of Eastern Oman Mountains

crease, leading to rapid channel infilling, avulsion, and distributary development. Once, however, gradients reached critically steep slopes that exceeded the threshold for continued aggradation, excess stream energy would have been available for large-scale fan incision during larger flood events.

Hence many of the characteristics of palaeochannel change observed on the Fan I and II systems may be explained by long-term autocyclic response to fan growth itself. However, there are still a number of features which remain unexplained by this mechanism. In particular, fan growth alone does not explain the estimated increases in peak discharges and the increasingly limited distances of flow, during fan development; nor the marked changes in lithology between the older and younger palaeochannel courses; nor the extensive landscape lowering of the interfluve fan surfaces during large-scale fluvial incision of Fan II.

Changes of Catchment Area and Sediment Availability

Significant increases in the size of the catchment area during fan formation could account for both the large increases in estimated peak flows and the dramatic changes in lithological composition of the gravels (Figure 14.9). Progressive headward extension of the Wadi Andam source streams could have significantly increased the catchment area, both extending it northwards on to the ophiolites (Figure 14.2), and capturing the large headwater zone of the eastern Wadi Adam source streams. This capture would have more than doubled the total upland source area of the catchment (Figure 14.1), resulting in dramatic and highly significant increases in runoff, as well as influx of ophiolitic materials. Since larger catchments tend to be associated with lower fan gradients (e.g. Blissenbach, 1954; Bull, 1964; Melton, 1965; Rachocki, 1981; Harvey, 1984), incision could have resulted from channel adjustment to the larger volumes of runoff resulting from river capture (*cf*. Harvey, 1984; Bird, 1985; Nichols, 1987).

ALLOCYCLIC CONTROLS

No direct data are currently available on long-term changes in climate, regional tectonism, or eustatic fluctuations during the late Tertiary and Quaternary periods for this part of Oman. Furthermore, since no precise dates are yet available for the fan gravels, correlation with global or regional environmental changes cannot yet be attempted. Nevertheless, the nature of changes observed in palaeochannel conditions during long-term fan aggradation, may well be explained by significant climatic, tectonic, and eustatic controls. Some possible links between these allocyclic controls and the resulting field evidence are discussed here, but at this stage they represent hypotheses awaiting further exploration.

Climatic Change

Long-term increase in climatic aridity could partially account for the changes in flow regime and downfan extent of drainage which appear to have occurred during the period of fan development. Seasonal flows at least are likely to have been associated with the development of the early generation sinuous channels (see earlier discussion; *cf*. Moseley, 1965), while the associated influx of finer-grained sediments may have been a response to relatively well-vegetated catchment slopes. The high water table required for cementation and diagenesis of the gravels also suggests that humid climatic conditions existed during the earlier generations of palaeochannel development.

The later palaeochannel generations, by contrast, appear to reflect more episodic, 'flashy' flow regimes, associated with increasingly arid conditions (*cf*. Baker and Penteado-Orellana, 1978), and an associated change to infrequent, high magnitude flow events (*cf*. Baker, 1978). Such changes would account for the later palaeochannel deposits being significantly coarser, and with significantly higher flow magnitudes, than those of the earlier generations. In addition, the downfan narrowing and disappearance of palaeochannel courses, the development of extensive distributary systems, and the low sinuosity of the channels, are all characteristic of ephemeral drainage courses on semiarid fans (e.g. see Van de Graaf *et al.*, 1977; Friend, 1978).

Increasing climatic aridity can also be invoked to explain the large-scale lowering of the water table during fluvial incision of the Fan II deposits. Indeed, water table lowering due to diminished recharge may have contributed both to initial fluvial downcutting, and to the extensive contemporaneous lowering of the interfluve fan areas by wind erosion of successive horizons of exposed desiccated sediments, and by localized surface runoff. In proximal areas, a thickness of over 30 m of interchannel fines has been removed by long-term deflation and periodic runoff. According to rates of retreat during the Pleistocene Epoch of 0.1 to 2 m 1000 a^{-1} estimated by Yair and Gerson (1974) for a clastic escarpment in Sinai, removal of the Omani interfluve sediments could have required up to 300 000 years of arid or semiarid climatic conditions.

Tectonic Effects

There is little evidence at present of any significant tectonic movement in either the Wadi Andam source areas, or the fan areas in the south, since the end of the Oligocene–Miocene period of mountain uplift. However, it seems unlikely that this area has remained tectonically stable since the end of the Miocene, since foreland subsidence may well have occurred in association with the accumulation of the several hundred metres of clastic piedmont fan sediment (as reported elsewhere, by Melton, 1965; Williams, 1970; Heward, 1978; Kraus, 1984; Alexander and Leeder, 1987). The lower Wadi Batha valley, lying on the eastern edge of the Wahiba Sands, for example, has been subject to subsidence of over 1000 m since mountain emplacement (H. Weier, personal communication). However, although the effects of such tectonic movements cannot yet be fully assessed, it seems unlikely that they could account for all the observed palaeohydrologic changes (but *cf.* Friend, 1978; Varley, 1984).

Eustatic Changes

Global sea level change during the Pliocene and Pleistocene Epochs has followed an overall trend of high sea levels during the Pliocene followed by progressively falling sea levels with major fluctuations associated with glacial–interglacial episodes, during the Pleistocene. Hence the likelihood of progressively rising water tables during aggradation of Fan I is supported by the global high sea levels during much of the Pliocene.

The period(s) of fan degradation and land surface lowering is likely to have occurred during relatively arid climatic conditions with falling water tables, such as occurred during the glacial episodes characteristic of the continental tropics during the Pleistocene. Sea level was lower and desert winds are postulated to have been stronger at these times (Williams, 1973; Manabe and Hahn, 1977; Glennie, personal communication). The last period in which large-scale fluviatile incision and land surface deflation could have occurred would therefore have been during the last glaciation (*ca.* 25 000 to 10 000 BP) of the Pleistocene, but this possibility will remain uncertain until dates have been obtained for the fan deposits.

Summary and Conclusion

Analysis of the exhumed palaeochannel systems of the Omani piedmont fans has allowed identification of the main evolutionary stages of fan development in relation to the apparent mechanisms of channel change, and the possible hydrologic, climatic, and tectonic controls on these changes.

Alluvial fan development has followed a series of distinct evolutionary stages. The exhumed palaeochannel sequence indicates at least 12 successive generations of palaeochannels, associated with at least six major periods of fluvial activity. The earlier generations are characterized by broad, sinuous, fine-grained, chert-rich, single-thread channel systems that extend southwards across the full length of the fan. The later generations become increasingly narrow, less sinuous, coarser grained and chert-depleted; they exhibit steeper gradients and single-thread channels which rapidly narrow downfan and peter out. The latest stage of fan development is characterized by flows concentrated within the Wadi Andam system, producing a deeply incised routeway bounded by extensive, thin, coarse sheets of flood gravels, dominated by ophiolitic lithologies.

The evidence indicates that the changes actually occurred as a series of separate and discrete episodes of fluvial activity of varying duration and character. This activity was dominated during Fan I formation by long-term overall aggradation, although interrupted periodically by episodes of incision. Only the last stages of Fan II development were represented by periods of major fan entrenchment. Channel changes from one period to the next have occurred both through meander migration within a given active channel zone, and through major channel avulsion to create extensive distributary systems and transverse links between established and new channel zones. Hence, many flows remained along established routeways for long periods before avulsion created new channel courses across the fan surface. The degree of channel stability and frequency of channel avulsion depended on the rates of floodplain and interfluve deflation, and rates of cementation of the channel-bed gravels, as well as on the magnitude and frequency of high rainfall, stream runoff, and sediment yield events. These factors in turn would affect the rates of burial and preservation of channel and floodplain deposits, and hence the overall rate of fan aggradation.

The individual controls of palaeohydrologic and channel pattern changes during fan evolution cannot be isolated. It appears that the observed changes have resulted from a complex combination of both internally and externally generated environmental changes. Progressive fanhead aggradation produced long-term increases in channel gradients, possibly reflected in increased sediment size and decreased sinuosity. Headward extension of the drainage network, possibly associated with increased aridity and tectonic uplift, could have resulted in river capture and concomitant dramatic additional influxes of water and sediment to the fan systems. Such an increase in catchment area could act to increase sizes of material in transport, the magnitude of mean and peak flows, and may have been responsible for initiating large-scale fan entrenchment.

These changes internal to the hydrologic system must have occurred during periods of major climatic and eustatic changes, although the links between allocyclic controls and fan response remain to be fully tested. The early sinuous and continuous channels may have developed under seasonal flows, allowing sufficient time both for the observed meander morphology to develop, and for cementation of channel gravels to occur between flows. However, only a marked change in seasonality of rainfall, to produce a change from seasonal to ephemeral flows, could account for the progressive downfan disappearance of the later palaeochannel courses. Similarly, only a significant increase in aridity could account for the huge volumes of fine-grained sediments (30–40 km^3) that were deflated from the fans themselves, once fluvial aggradation had ceased. In addition, tectonic and eustatic changes are also likely to have been active throughout the period of fan formation, although the regional patterns of these changes remain uncertain.

A number of ideas and hypotheses raised in this paper remain to be fully explored and tested. In particular, the links between channel morphology and seasonality of flows need to be more rigorously tested. Second, a firmer chronologic framework needs to be established for each stage of fan development. Finally, the extent to which the different controlling parameters may have affected the channel systems at different stages of fan development need to be more precisely determined. A combined analysis of changes in each of the possible controlling variables through time, with the apparent effects on changes in channel morphology, pattern, sedimentology, and stability, would allow the construction of a more valuable model of long-term fan development during the Plio-Pleistocene period in the arid tropics.

Acknowledgements

The writer would like to express particular appreciation to Dr. K. Glennie for many invaluable discussions. Thanks are also extended to P. Considine, J. F. Jones, Dr. G. Stanger, and H. Weier, all of the Public Authority for Water Resources, Sultanate of Oman, and to B. Duff and Dr. M. Hughes Clarke of Petroleum Development Oman for their helpful and stimulating discussions. Thanks are also due to C. McBean for valuable field assistance and to J. Cutler and J. McIntosh for help with the ground surveys. The logistic support of the Sultan of Oman's Armed Forces and the Coastal Security Force is also gratefully acknowledged. The writer would like to thank all members of the Royal Geographical Society Eastern Sands Project for their enthusiasm and support during the progress of this research. Finally, the writer would like to acknowledge assistance with laboratory analyses from Prof. I. Parsons, Dr. G. Walkden, and Mrs M. Lamb.

References

Alexander, J. and Leeder, M. R. 1987. Active tectonic control on alluvial architecture. In Ethridge, F. G., Flores, R. M., and Harvey, M. D. (Eds), *Recent Developments in Fluvial Sedimentology. Society of Economic Paleontologists and Mineralogists Special Publication*, **39**, 245–252.

Arche, A. 1983. Coarse-grained meander lobe deposits in the Jarama River, Madrid, Spain. In Collinson, J. D. and Lewin, J. (Eds), *Modern and Ancient Fluvial Systems. International Association of Sedimentologists Special Publication*, **6**, 313–321.

Aubel, J. W. 1983. Bajada hydrological area. In *The Hydrology of the Oman Sultanate. A Preliminary Assessment. Public Authority for Water Resources Report*, **83–1**, 66–74.

Baker, V. R. 1977. Stream-channel response to floods, with examples from central Texas. *Geological Society of America Bulletin*, **88**, 1057–1071.

Baker, V. R. 1978. Adjustment of fluvial systems to climate and source terrain in tropical and subtropical environments. In Miall, A. D. (Ed.), *Fluvial Sedimentology. Canadian Society of Petroleum Geologists Memoir*, **5**, 211–230.

Baker, V. R. and Penteado-Orellana, M. M. 1978. Fluvial sedimentation conditioned by Quaternary climatic change in central Texas. *Journal of Sedimentary Petrology*, **48**, 433–451.

Beaty, C. B. 1970. Age and estimated rate of accumulation of an alluvial fan, White Mountains, California, U.S.A. *American Journal of Science*, **268**, 50–77.

Beaty, C. B. 1974. Debris flows, alluvial fans, and a revitalised catastrophism. *Zeitschrift für Geomorphologie N.F. Supplement Band*, **21**, 39–51.

Beydoun, Z. R. 1980. Some Holocene geomorphological and sedimentological observations from Oman and their palaeogeological implication. *Journal of Petroleum Geology*, **2**, 427–437.

Bird, J. F. 1985. Review of channel changes along creeks in the northern part of the Latrobe River basin, Gippsland, Victoria, Australia. *Zeitschrift für Geomorphologie N.F. Supplement Band*, **55**, 97–111.

Blissenbach, E. 1954. Geology of alluvial fans in semi-arid regions. *Geological Society of America Bulletin*, **67**, 175–189.

Bridge, J. S. 1985. Palaeochannel patterns inferred from alluvial deposits: a critical evaluation. *Journal of Sedimentary Petrology*, **55**, 579–589.

Bridge, J. S. and Leeder, M. R. 1979. A simulation model of alluvial stratigraphy. *Sedimentology*, **26**, 617–644.

Bristow, C. S. 1987. Brahmaputra River: channel migration and deposition. In Ethridge, F. G., Flores, R. M., and Harvey, M. D. (Eds), *Recent Developments in Fluvial Sedimentology. Society of Economic Paleontologists and Mineralogists Special Publication*, **39**, 63–74.

Bull, W. B. 1964. Alluvial fans and near-surface subsidence in Western Fresno County, California. *U.S. Geological Survey Professional Paper*, **437–A**.

Butzer, K. W. and Hansen, C. L. 1968. *Desert and River in Nubia*. University of Wisconsin Press, Madison.

Carlston, C. W. 1965. The relation of free meander geometry to stream discharge and its geomorphic implications. *American Journal of Science*, **263**, 864–885.

Costa, J. E. 1983. Paleohydraulic reconstruction of flash-flood peaks from boulder deposits in the Colorado Front Range. *Geological Society of America Bulletin*, **94**, 986–1004.

Costa, J. E. 1984. Fluvial paleoflood hydrology. *EOS*, **65**, 890 (abstract only).

Cotter, E. 1978. The evolution of fluvial style, with special reference to the central Appalachian Paleozoic. In Miall, A. D. (Ed.), *Fluvial Sedimentology. Canadian Society of Petroleum Geologists Memoir*, **5**, 361–383.

Curtis, W. F. 1985. Fluvial sediments in northern

Oman. *Public Authority for Water Resources Report*, 85-15.

Denny, C. S. 1965. Alluvial fans in the Death Valley Region, California and Nevada. *U.S. Geological Survey Professional Paper*, **466**.

Edens, C. 1988. Prehistory. *Journal of Oman Studies*.

Ethridge, R. G. and Schumm, S. A. 1978. Reconstructing paleochannel morphologic and flow characteristics: methodology, limitations and assessment. In Miall, A. D. (Ed.), *Fluvial Sedimentology. Canadian Society of Petroleum Geologists Memoir*, **5**, 703-721.

Flores, R. M. and Pillmore, C. L. 1987. Tectonic control on alluvial paleoarchitecture of the Cretaceous and Tertiary Raton basin, Colorado and New Mexico. In Ethridge, F. G., Flores, R. M., and Harvey, M. S. (Eds), *Recent Developments in Fluvial Sedimentology. Society of Economic Paleontologists and Mineralogists Special Publication*, **39**, 311-320.

Friend, P. F. 1978. Distinctive features of some ancient river systems. In Miall, A. D. (Ed.), *Fluvial Sedimentology. Canadian Society of Petroleum Geologists Memoir*, **5**, 531-542.

Friend, P. F. *et al.* 1981. Fluvial sedimentology in the Tertiary South Pyrenean and Ebro Basins, Spain. In Elliott, T. (Ed.), *Field Guides to Modern and Ancient Fluvial Systems in Britain and Spain*. International Association of Sedimentologists. 4.1-4.50.

Glennie, K. W. 1970. *Desert Sedimentary Environments*. Elsevier, Amsterdam.

Glennie, K. W. *et al.* 1974. The geology of the Oman Mountains. *Transactions of the Royal Dutch Geological and Mining Society*, **31**.

Gole, C. V. and Chitale, S. V. 1966. Inland delta building activity of Kosi River. *American Society of Civil Engineers Proceedings, Journal of the Hydraulics Division*, **92**, 111-126.

Goudie, A. S. 1983. Calcrete. In Goudie, A. S. and Pye, K. (Eds), *Chemical Sediments and Geomorphology*. Academic Press. New York. 93-131.

Harvey, A. M. 1984. Debris flows and fluvial deposits in Spanish Quaternary alluvial fans: implications for fan morphology. In Koster, E. H. and Steel, R. J. (Eds), *Sedimentology of Gravels and Conglomerates. Canadian Society of Petroleum Geologists Memoir*, **10**, 123-132.

Heward, A. P. 1978. Alluvial fan sequence and megasequence models: with examples from Westphalian D-Stephanian B coalfields, northern Spain. In Miall, A. D. (Ed.), *Fluvial Sedimentology. Canadian Society of Petroleum Geologists Memoir*, **5**, 669-702.

Holm, D. A. 1960. Desert geomorphology in the Arabian peninsula. *Science*, **132**, 1369-1379.

Hey, R. D. 1979. Flow resistance in gravel-bed rivers. *American Society of civil Engineers, Journal of Hydraulics Division*, **105**, 365-379.

Hopkins, J. C. 1985. Channel-fill deposits formed by aggradation in deeply scoured, superimposed distributaries of the Lower Kootenai Formation (Cretaceous). *Journal of Sedimentary Petrology*, **55**, 42-52.

Hopson, C. A. *et al.* 1981. Geologic section through the Samail ophiolite and associated rocks along a Muscat, Ibra transect, southeastern Oman Mountains. *Journal of Geophysical Research*, **86**, 2527-2544.

Jackson, R. G. II. 1978. Preliminary evaluation of lithofacies models for meandering alluvial streams. In Miall, A. D. (Ed.), *Fluvial Sedimentology. Canadian Society of Petroleum Geologists Memoir*, **5**, 543-576.

Jarrett, R. D. 1984. Hydraulics of high gradient streams. *Journal of Hydraulic Engineering*, **110**, 1519-1539.

Jones, J. R. 1986. Results of test drilling for water in northwestern Sharqiyah area, Sultanate of Oman, 1983-843. *Public Authority for Water Resources Report 85-21*.

King, L. C. 1942. *South African Scenery*. Oliver and Boyd, Edinburgh.

Knetsch, G. 1954. Allgemein-geologische Beobachtungen aus Ägypten (1950-53). *Neues Jahrbuch für Geologie und Päläontologie*, **99**, 287-297.

Kraus, M. J. 1984. Sedimentology and tectonic setting of Early Tertiary quartzite conglomerates, N.W. Wyoming. In Koster, E. H. and Steel, R. J. (Eds), *Sedimentology of Gravels and Conglomerates. Canadian Society of Petroleum Geologists Memoir*, **10**, 203-216.

Kraus, M. J. and Middleton, L. T. 1987. Contrasting architecture of two alluvial suites in different structural settings. In Ethridge, F. G., Flores, R. M., and Harvey, M. D. (Eds), *Recent Developments in Fluvial Sedimentology. Society of Economic Paleontologists and Mineralogists Special Publication*, **39**, 253-262.

Leeder, M. 1978. A quantitative stratigraphic model for alluvium, with special reference to channel deposit density and inter connectedness. In Miall, A. D. (Ed.), *Fluvial Sedimentology. Canadian Society of Petroleum Geologists Memoir*, **5**, 587-596.

Limerinos, J. T. 1970. Determination of the Manning coefficient from measured bed roughness in natural channels. *U.S. Geological Survey Water Supply Paper*, **1898B**.

McClure, H. A. 1978. Ar Rub'Al Khali. In Al-Sayari, S. S. and Zotl, J. G. (Eds), *Quaternary of Saudi Arabia Vol. 1*. Springer Verlag, Berlin.

Maizels, J. K. 1983. Palaeovelocity and palaeodischarge determination for coarse gravel deposits. In Gregory, K. J. (Ed.), *Background to Palaeohydrology*. Wiley, Chichester. 101-139.

Maizels, J. K. 1987. Plio-Pleistocene raised channel systems of the western Sharqiya (Wahiba), Oman. In Frostick, L. and Reid, I. (Eds), *Desert Sediments:*

Ancient and Modern. *Geological Society of London Special Publication*, **35**, 31–50.

Maizels, J. K. 1988. Paleochannels:Plio-Pleistocene raised channel systems of the Western Sharqiyah. *Journal of Oman Studies*.

Maizels, J. K. and Anderson, E. 1988. Surface water in the Sharqiyah: Flashfloods February/March 1986. *Journal of Oman Studies*.

Manabe, S. and Hahn, D. G. 1977. Simulation of the tropical climate of an Ice Age. *Journal of Geophysical Research*, **82**, 3889–3911.

Melton, M. A. 1965. The geomorphic and palaeoclimatic significance of alluvial deposits in Southern Arizona. *Journal of Geology*, **73**, 1–38.

Mensching, H. G. 1974. Landforms as a dynamic expression of climatic factors in the Sahara and Sahel—a critical discussion. *Zeitschrift für Geomorphologie N.F. Supplement Band*, **20**, 168–177.

Miall, A. D. 1978. Lithofacies types and vertical profile models in braided river deposits: a summary. In Miall, A. D. (Ed.), *Fluvial Sedimentology. Canadian Society of Petroleum Geologists Memoir*, **5**, 597–604.

Miller, R. P. 1937. Drainage lines in bas-relief. *Journal of Geology*, **45**, 432–438.

Moseley, F. 1965. Plateau calcrete, calcrete gravels, cemented dunes and related deposits of the Maalegh–Bomba region of Libya. *Zeitschrift für Geomorphologie N.F.*, **9**, 166–185.

Nami, M. and Leeder, M. R. 1978. Changing channel morphology and magnitude in the Scalby Formation (Middle Jurassic) of Yorkshire, England. In Miall, A. D. (Ed.), *Fluvial Sedimentology. Canadian Society of Petroleum Geologists Memoir*, **5**, 431–440.

Nichols, G. J. 1987. Structural controls on fluvial distributary systems—the Luna system, northeastern Spain. In Ethridge, F. G., Flores, R. M., and Harvey, M. D. (Eds), *Recent Developments in Fluvial Sedimentology. Society of Economic Paleontologists and Mineralogists Special Publication*, **39**, 269–277.

Parkash, B. et al. 1983. Lithofacies of the Markanda terminal fan, Kurukshetra district, Harayana, India. In Collinson, J. D. and Lewin, J. (Eds), *Modern and Ancient Fluvial Systems. International Association of Sedimentologists Special Publication*, **6**, 337–344.

Price, W. E. 1974. Simulation of alluvial fan deposition by random walk model. *Water Resources Research*, **10**, 263–274.

Puigdefabregas, C. and Van Vliet, A. 1978. Meandering stream deposits from the Tertiary of the southern Pyrenees. In Miall, A. D. (Ed.), *Fluvial Sedimentology. Canadian Society of Petroleum Geologists Memoir*, **5**, 469–485.

Rachocki, A. H. 1981. *Alluvial Fans*. Wiley, Chichester.

Ramos, A. and Sopena, A. 1983. Gravel bars in low-sinuosity streams (Permian and Triassic, central Spain). In Collinson, J. D. and Lewin, J. (Eds), *Modern and Ancient Fluvial Systems. International Association of Sedimentologists Special Publication*, **6**, 301–312.

Reeves, C. C. Jr. 1983. Pliocene channel calcrete and suspenparallel drainage in west Texas and New Mexico. In Wilson, R. C. L. (Ed.), *Residual Deposits: Surface Related Weathering Processes and Materials. Geological Society of London Special Publication*, **11**, 179–183.

Rotnicki, K. 1983. Modelling past discharges of meandering rivers. In Gregory, K. J. (Ed.), *Background to Palaeohydrology*. Wiley, Chichester. 321–354.

Rust, B. R. 1978. A classification of alluvial channel systems. In Miall, A. D. (Ed.), *Fluvial Sedimentology. Canadian Society of Petroleum Geologists Memoir*, **5**, 187–198.

Ryder, J. M. 1971. The stratigraphy and morphology of paraglacial alluvial fans in south-central British Columbia. *Canadian Journal of Earth Sciences*, **8**, 279–298.

Schick, A. P. 1974. Formation and obliteration of desert stream terraces—a conceptual analysis. *Zeitschrift für Geomorphologie N.F. Supplement Band*, **21**, 88–105.

Schumm, S. A. 1972. Fluvial palaeochannels. In Rigby, J. K. and Hamblin, W. K. (Eds), *Recognition of Ancient Sedimentary Environments. Society of Economic Paleontologists and Mineralogists Special Publication*, **16**, 98–107.

Stalder, P. J. 1975. Cementation of Pliocene–Quaternary fluviatile clastic sediments in and along the Oman Mountains. *Geologie en Mijnbouw*, **54**, 148–156.

Stear, W. M. 1983. Morphological characteristics of ephemeral stream channel and overbank splay sandstone bodies in the Permian Lower Beaufort Group, Karroo Basin, South Africa. In Collinson, J. D. and Lewin, J. (Eds), *Modern and Ancient Fluvial Systems. International Association of Sedimentologists Special Publication*, **6**, 405–420.

Steel, R. J. and Aasheim, S. M. 1978. Alluvial sand deposition in a rapidly subsiding basin (Devonian, Norway). In Miall, A. D. (Ed.), *Fluvial Sedimentology. Canadian Society of Petroleum Geologists Memoir*, **5**, 385–412.

Strickler, A. 1923. Beiträge zur Frage der Geschwindichkeits-Formel und der Ruhigkeitszament für Ströme, Kanäle und geschlossene Leitungen. *Mitteilungen des Amtes für Wasserwirtschaft (Bern)*, **1b**.

Sutton, S. E. 1987. Changing drainage patterns in a semiarid area and their effects on groundwater resources. *Earth Surface Processes and Landforms*, **12**, 567–570.

Van de Graaf, W. J. E. et al. 1977. Relict early Cainozoic drainages in arid Western Australia. *Zeitschrift für Geomorphologie N.F.*, **21**, 379–400.

Varley, K. J. 1984. Sedimentation and hydrocarbon distribution of the Lower Cretaceous Cadomin Formation, N.W. Alberta. In Koster, E. H. and Steel, R. J. (Eds), *Sedimentology of Gravels and Conglomerates. Canadian Society of Petroleum Geologists Memoir*, **5**, 175–187.

Vondra, C. F. and Burggraf, D. R. 1978. Fluvial facies of the Plio-Pleistocene Koobi Fora Formation, Karari Ridge, East Lake Turkana, Kenya. In Miall, A. D. (Ed.), *Fluvial Sedimentology. Canadian Society of Petroleum Geologists Memoir*, **5**, 511–529.

Wasson, R. J. 1977. Catchment processes and the evolution of alluvial fans in the lower Derwent valley, Tasmania. *Zeitschrift für Geomorphologie, N.F.*, **21**, 147–168.

Wells, N. A. and Dorr, J. A. 1987. A reconnaissance of sedimentation of the Kosi alluvial fan of India. In Ethridge, F. G., Flores, R. M., and Harvey, M. D. (Eds), *Recent Developments in Fluvial Sedimentology. Society of Economic Paleontologists and Mineralogists Special Publication*, **39**, 51–61.

Williams, G. E. 1970. Piedmont sedimentation and late Quaternary chronology in the Biskra region of the northern Sahara. *Zeitschrift für Geomorphologie Supplement Band*, **10**, 40–63.

Williams, G. E. 1973. Late Quaternary piedmont sedimentation, soil formation and paleoclimates in arid south Australia. *Zeitschrift für Geomorphologie N.F.*, **17**, 102–125.

Williams, G. P. 1984. Paleohydrologic equations for rivers. In Costa, J. E. and Fleisher, P. J. (Eds), *Developments and Applications of Geomorphology*. Springer Verlag, New York. 343–367.

Yaalon, D. H. and Dan, J. 1974. Accumulation and distribution of loess-derived deposits in the semi-desert and desert fringe area of Israel. *Zeitschrift für Geomorphologie N.F. Supplement Band*, **20**, 91–105.

Yair, A. and Gerson, R. 1974. Mode and rate of escarpment retreat in an extremely arid environment. *Zeitschrift für Geomorphologie N.F. Supplement Band*, **21**, 202–215.

CHAPTER 15

The Leba River Alluvial Fan and its Palaeogeomorphological Significance

Andrzej H. Rachocki
University of Gdańsk, Gdańsk

Abstract

The Leba River alluvial fan is one of the more controversial landforms in northern Poland. After eighty years of discussion it seems to be partly an alluvial fan, but issues from quite another valley. Both geomorphological and sedimentological evidence derived from the fan itself as well as from the adjacent areas, confirm that in the Late Glacial chronozone there never existed conditions for the accumulation of such a form. This statement forces rethinking and reinterpretation of the development of the whole, complex network of ice marginal spillways (pradolinas) in that area. Just as the Leba River alluvial fan is not a fan in the classic meaning of this word, the system of ice marginal spillways also has not been eroded in Late Glacial time. This system is far older and survived the last advance of the ice, as is confirmed by the presence of three depressions which extend tens of metres below the level of the recent floodplain. There is no explanation for their origin in terms of Late Glacial erosion by meltwater shaping the whole syste of ice marginal spillways.

The Leba River alluvial fan is a Late Glacial landform but it was deposited as a part of a valley outwash plain at the place where the valley joins the ice marginal spillway.

Introduction

Although alluvial fans are a common feature of the relief of northern Poland, being marked in dozens on every sheet of the geomorphological map, they have aroused little scientific attention because they are small, very young in geological sense, and the mechanism of their deposition always seemed too simple to be of real interest to geomorphologists. But among these thousands of really very simple forms there are at least a few whose setting makes them of significance from a palaeogeomorphological viewpoint. Without doubt, one of these is the so-called alluvial fan of Leba River.

The fan, consisting of only water laid deposits, has an approximate radius of 2 km and is situated at the place where the Leba River debouches into an ice marginal spillway. The problems of its origin and age have attracted the attention of geomorphologists for over eighty years.

In this paper the writer presents both some new facts concerning the fan topography and sedimentology, and a reinterpretation of known facts.

Alluvial Fans: A Field Approach edited by A. H. Rachocki and M. Church
Copyright © 1990 John Wiley & Sons Ltd.

Geographical Setting

The northernmost part of Poland (Figure 15.1) presents a characteristic landscape pattern of large plateau segmented by a system of wide, rather shallow valleys. Numerous authors associate the origin of these valleys with the Late Glacial chronozone, the masses of meltwater from the vanishing ice cap of the last glaciation—the Würm or Vistulian—being commonly accepted as the shaping medium. These forms are generally termed 'ice marginal spillways', by analogy to those very large drainage forms running across the North European Lowland (Figure 15.2) which accompanied the earlier stages of deglaciation. The German term *Urstromtal* (pl. Urstromtäler) and the Polish equivalent *pradolina* (pl. pradoliny) are mentioned even in English textbooks (Robinson, 1969). Being much shorter than 'ice marginal spillway', the Polish term will be used here.

As shown in Figure 15.1, only parts of the previous drainage network survive; others are submerged beneath the sea. The sole continuous form is the Reda–Leba pradolina, extending the total length of approximately 90 km and varying in width from less than 1 km at its eastern end to more than 5 km at its western end. Its recent floodplain is flat, covered with fields and meadows, and drained by two major rivers which flow in opposite directions, the Leba to the west and the Reda to the east.

In most prior papers erosional terraces have been mentioned, of absolute heights 60, 40, and 20 m a.s.l. The references originate from a paper by Sonntag (1914), who outlined the concept of Late Glacial development of the drainage network in this part of northern Poland. In the opinion of the writer, these terraces cannot be treated as erosional. Along the whole length of the pradolina of Reda–Leba there exist ia fact only a few very small 'shelves' of doubtful origin, which differ very much in height. It is impossible to correlate them into a system permitting reconstruction of a history of downcutting of the pradolina. One should say too that, against obvious evidence, these remnants intentionally have been termed erosional (Sonntag, 1914; Augustowski, 1965, 1972; Sylwestrzak, 1969, 1973).

The Leba River alluvial fan is situated at the mouth of the upper Leba River valley where the Lebt reaches the Reda–Leba pradolina, at the point where the watershed in the pradolina floor exists at present. The source of the Reda River lies in the central part of the fan in a shallow depression.

The depth of incision of the Leba River valley south of the alluvial fan is similar to that in the pradolina itself, in the range 60–70 m (Figure 15.8). The plateau, the top surface of which slopes gently toward the north, is composed of layered glacial lodgement tigl which varies in thickness locally from 0 to over 20 m, with interbeddings of glaciofluvial deposits (sands and gravels). The glacial and glaciofluvial deposits constitute a veritable mosaic both horizontally and vertically, and quite often different sediment sequences can be observed on opposite walls of the same gravel pit or larger excavation site.

History of Research on the Leba Alluvial Fan

Since it is practically impossible to separate research into the origin of the Reda–Leba pradolina from that concerned with the Leba River alluvial fan, both will be presented together in the following.

The Late Glacial age in the region as well as the erosional origin of the Reda–Leba pradolina were first discussed in 1899 by Keilhack, whose attention was attracted by the existence of the watershed on its floor. He explained it by reference to postglacial tectonic uplift. Wunderlich (1914) was the first to claim that the watershed is at the place where a 'very large' alluvial fan was accumulated. He estimated the thickness of fan deposits arbitrarily at 15 m, but did not mention to which part of the fan this value refers. Sonntag (1914) may be called the father of 'erosional terraces', being simultaneously the father of fundamental mistakes in the interpretation of geomorphic events of Late and Post Glacial his-

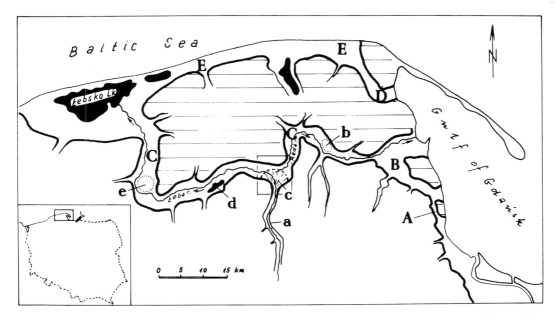

Figure 15.1. Segments of the ice marginal spillways in Northern Poland. A—Redlowo Depression, B—Kashoubian Meander, C—Pradolina of Reda–Leba, D—Pradolina of Plutnica River, E—Coastal pradolina. Details: a—Leba River valley, b—Orle depression, c—Leba River alluvial fan, d—Lubowidzkie Lake, E—Black Moore

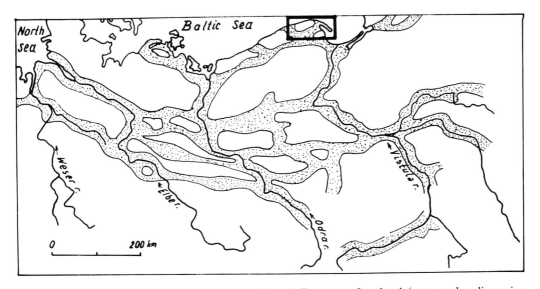

Figure 15.2. The ice marginal streamways of North European Lowland (area under discussion framed)

tory in the area under discussion. In his opinion the whole Reda–Leba pradolina is erosional, Late Glacial, and conveyed the meltwater towards the west, so all erosional terraces are sloped towards that direction. In the year 1931, Hartnack put forward the view that the flow across the pradolina of Reda–Leba was originally to the east rather than to the west. Zaborski (1933) gathered together existing theories but did not propose a new concept.

Such recounting of early ideas may seem pointless, but it has to be admitted that, although one might almost regard them nowadays as fables, they exerted a great and quite incomprehensible influence on all research that had been undertaken since the Second World War. In fact, with little or no modifications these ideas have been repeated in numerous theses, textbooks, and papers. Thus, Sonntag's notion of the Late Glacial erosional origin of the pradolina was entirely adopted by Augustowski (1965, 1972), whereas Sylwestrzak (1969, 1973) tried to connect all the different, and often opposed, concepts into one more or less coherent system.

All of the authors mentioned so far discussed the question of the general direction of palaeoflow at particular levels of Sonntag's terraces. All unquestionably accepted *a priori* their real existence as well as the erosional effect of meltwater in dissecting the plateau during the Late Glacial period (Figure 15.3). Only Hartnack (1931) noticed that the pradolina of Reda–Leba changes width too rapidly for a valley which has been eroded into relatively homogeneous material.

Two authors, Rosa (1963) and Marsz (1967), deny the notion of deepening the pradolina with breaks for lateral widening as manifested in the existence of Sonntag's erosional terraces, believing that the whole formation developed during one single, intensive, erosional episode in Late Glacial times. Marsz (1967) suggested that a convenient time could be the Bølling, or even perhaps its 200-year (sic!) climatic optimum.

Morphology and Deposits of the Leba River Alluvial Fan

The question of the contemporary watershed of the Leba River alluvial fan has arisen just as often as that of origin of the pradolina. The landform concerned in consistently marked on geomorphological maps as an alluvial fan (Figure 15.4), in an often somewhat sketchy manner which leaves little room for questioning whether this interpretation is really justified. Looking at Figure 15.4A one can see quite large fragments of terraces on both sides of the Leba valley at its mouth, and this may give the deceptive impression that the fan is built up of material originating from dissection of the terrace. It is only the very simplified map, or rather outline sketch, produced by Zaborski (Figure 15.4B) that represents the true situation namely that the western part of the fan has been destroyed by the Leba River.

In spite of the length of the time for which the Leba River alluvial fan has been under discussion, no geomorphological or other descriptive

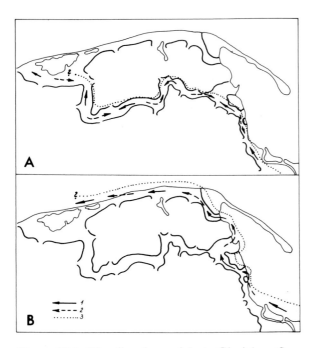

Figure 15.3. The directions of Late Glacial outflow. A—at the level of the 60 m terrace, B—at the level of the 20 m terrace. 1. By Sonntag and Augustowski; 2. By Sylwestrzak; 3. Inferred position of the ice margin

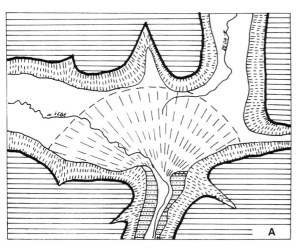

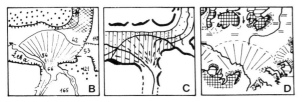

Figure 15.4. Leba River alluvial fan as presented on geomorphological maps. a and c—Augustowski, 1965; b—Zaborski, 1933; d—Sylwestrzak, 1969

work has ever been done on the form of the fan itself. The semicircular shape of the contour lines, especially on the smaller-scale survey maps, evidently led geomorphologists of at least three generations to the simple conclusion that it must be an alluvial fan. Slightly more detailed investigation of the fan and adjacent areas, however, discloses certain facts which were not known to previous authors.

The detailed contour map (Figure 15.5) shows that it is only in the central sector that the fan surface slopes uniformly towards the north, whereas the western part has been destroyed by fluvial processes associated with the Leba River and the eastern margin by the network of unnamed tributaries of the Reda River. The upper fragment of the western part has also suffered recently from erosion by the Leba River, since the floor of the Leba valley lies lower than the fan and is separated from it by a sharp escarpment which decreases in height from over 10 m at the apex of the fan to 0 m some two kilometres further down the fan. Another scarp with a height of about 2 m is formed on the fan surface between the radius lines D and E (cf. Figure 15.5). This also decreases in height to 0 m about 400 m down the fan.

The cross-sections of the fan surface (Figure 15.6) show the character of the surface fairly well, but it is important to remember that they are exaggerated by a factor 16 in a vertical direction. For a comparative purpose line A is redrawn (a in Figure 15.6) without exaggeration. In reality the fan surface is almost flat therefore, with a mean slope of $0°13'$, or 3.7 m km^{-1}. This means that it is utterly impossible to determine the position of the fan margin without additional sedimentological investigations. One should point out also another uncommon feature of fan surface morphology, the presence of depressions in the apical part of fan (cf. Figure 15.5). The deepest of them is about 4 m.

The Leba River alluvial fan is composed entirely of water-laid deposits, the most common of which are sorted sands of fine and medium size with a local admixture of gravels. This is revealed in a few exposures which are along the mentioned escarpment at the places where the river actively undercuts it. The exposures, especially those situated close of the fan apex, show vertical bipartition of material. In the upper part is a layer about 1.5 to 2 m thick which consists of medium sands with scattered gravels which do not create visible interbeddings. The lower part is built up of medium to coarse sands and fine gravels with clearly visible sedimentary structures.

To obtain more details on the process of fan deposition, the long axes of stones in the gravels were measured at six points. The measurements were made in shallow excavation pits and concern only the topmost layer of deposits where this is undisturbed by soil processes. A hundred readings were obtained at each site. The results of this palaeocurrent analysis show the compass roses (Figure 15.7) for the gravel axes to be unimodal in direction in the apical part, the main vector changing from northwest at sites 1 and 2, to northeast at site 3, and north at site 5. At the sites

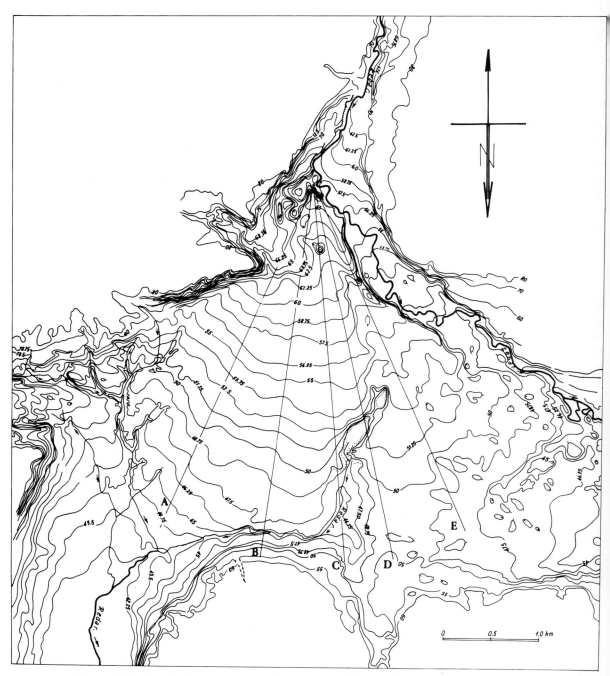

Figure 15.5. Detailed contour map of Leba River alluvial fan (contour interval 1.25 m)

THE LEBA RIVER ALLUVIAL FAN

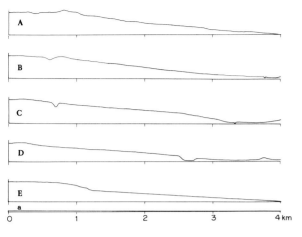

Figure 15.6. Cross-sections of the fan surface (for location see Figure 15.5)

4 and 6 distributions are polymodal, with no predominant direction of palaeoflow. Such a picture confirms the supposition of the writer that these sites are beyond the outer fan margin. They are situated on the floor of the pradolina which, in the past, has been influenced by quite different conditions of palaeoflow independent or only slightly dependent on those creating the alluvial fan.

Looking at the geomorphological map (Figure 15.8) we note the following additional features which might be helpful in explanation of the fan origin:

- Next to the mouth of the upper course of the Leba River valley is the mouth of another, unnamed valley (valley X) more than 4 km in length.
- Paired terraces at an absolute height of 67 m a.s.l. are located at the mouth of the Leba valley. These were also noted by Augustowski (1965) but he did not mention their height.
- The surface of the 67 m terrace is continuous with that of the alluvial fan.
- There are kames with heights of 62 and 66 m a.s.l. in a depression on the northern side of the pradolina.

What is the Leba River Alluvial Fan?

In order to answer this question it is necessary to return for a while to the relief of the pradolina itself, and especially that of its floodplain. As it has been stated in all references cited in the section on the history of research, the pradolina is a 'typical erosional valley' with steep slopes, a system of at least three different levels of erosional terraces, and a flat floor—also erosional—levelled by the shifting of the rivers meandering across it (Zaborski, 1933; Augustowski, 1965, 1972). Unfortunately this picture is not entirely accurate, for the floor also contains three highly interesting hollows (b, d, e in Figure 15.1). Some data on these hollows are presented in Table 15.1. It can be seen that the bottoms of two of these depressions lie beneath the recent level of the Baltic Sea. It is obvious that the existence of these hollows might be troublesome for the proponents of a Late Glacial, erosional origin of the pradolina.

If the pradolina of Reda–Leba had really been eroded during the Late Glacial chronozone, and in addition eroded with breaks when the floor had been levelled (as witnessed by the erosional terraces), what is the origin of these large and deep (in the scale of pradolina) depressions, and why were they not filled for instance with bedload? Similarly, why did the Leba River not flow through the Black Moore, as would be quite natural since this must have been the lowest point in the western part of the pradolina? Instead it appears to have passed to the west of it (Figure 15.9).

In order to elucidate the above points, borings were made in Black Moore depression. The mineral material of the sandy bottom of the depression which underlay the organic deposits (peats), was found to contain numerous fragments of fresh water shells. The lowermost layer of peat filling that depression was dated to 8130 ± 110 B.P. (Hel.-2170).

There seems to be only one logical explanation for this and other facts which concern all three hollows, that the depressions must be far older than Late Glacial and have been protected

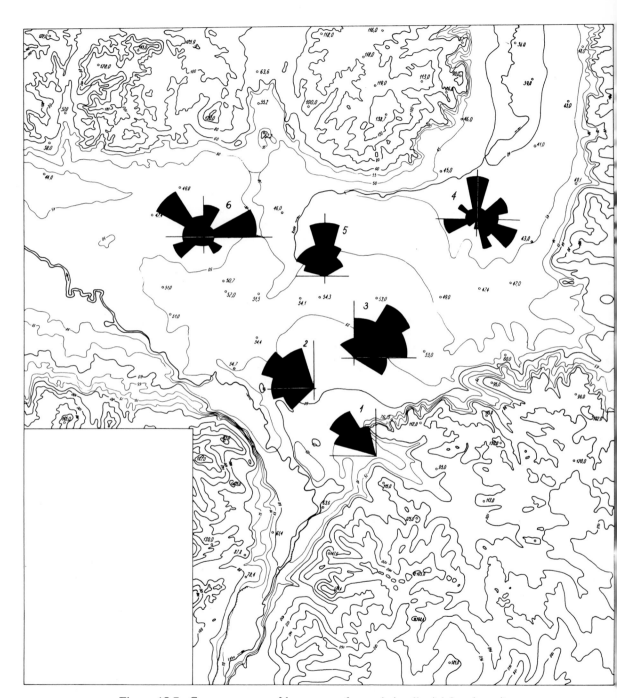

Figure 15.7. Compass roses of long axes of gravels in alluvial fan deposits

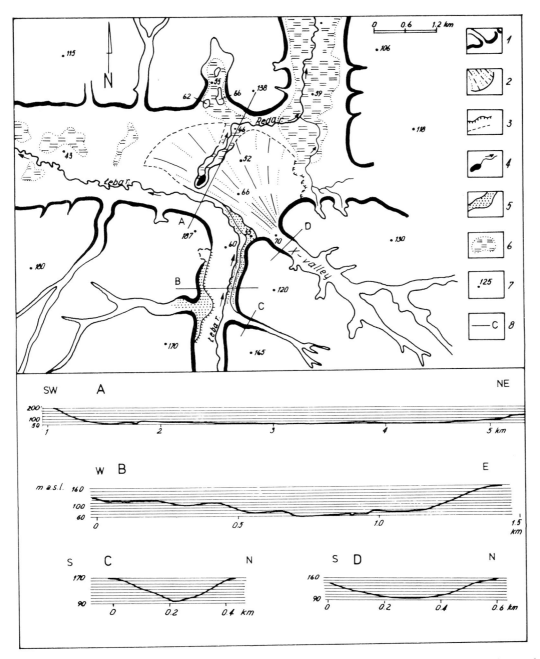

Figure 15.8. Geomorphological map of the Leba River alluvial fan. 1. The edges of pradolina and main valleys; 2. Alluvial fan; 3. Scarps of terraces (a); Inferred fan margin (b); 4. Ponds and rivers; 5. The terrace 67 m a.s.l.; 6. Peats: 7. Heights in m a.s.l.; 8. Cross-section lines

Table 15.1. Data of floor depressions in the Reda–Leba pradolina

Site	Area (km²)	Surface level (m a.s.l.)	Depth of depression (m)
Orle Lake depression	0.6	26.0	over 33*
Lubowidzkie Lake	1.2	24.0	15
Black Moore	3.3	8.0	13

*Bore hole did not reach the base of lacustrine deposits.

against infilling by blocks of dead ice. These blocks must have been of approximately the same size as the depressions in plan view. This also means that the blocks of the dead ice must have been grounded on the preexistent floor of the pradolina. This in turn forces us to accept that the proto-pradolina existed in the same area prior to the advance of the ice cap which marked the last episode of glaciation and must have been much deeper than the pradolina at present.

Regardless of the shape of the preexistent pradolina both it as well as the present pradolina of Reda–Leba must be considerably older than Late Glacial. Taken all together, the facts presented here indicate that the Reda–Leba pradolina must have existed before the last glaciation, or at least before its final episode of ice advancing, and must have survived that event due to protection from glacial ice or later dead ice. One can of course reject such an interpretation but it does not help us in explanation for the existence of hollows beneath the recent floor of the pradolina, which are filled with evidently postglacial organic deposits like peats in Black Moore or lacustrine chalk in the depression of the Orle Lake.

It is now possible to go on to propose an answer to the question of the origin of the Leba River alluvial fan. In view of the fact that the Reda–Leba pradolina is not a Late Glacial erosional form, the Leba River valley south of the alluvial fan is also far older than the Late Glacial age supposed by all previous authors. In fact the Leba River did not erode its own valley at the end of the last episode of glaciation but linked only older glacial or subglacial erosional formations (tunnel valleys).

The local 'tectonic uplifting' of part of the pradolina floor suggested by Keilhack (1899), partly repeated by Wunderlich (1914), and reiterated by Sylwestrzak (1969, 1973), may only be regarded as a misunderstanding. If such an event had taken place it would have been a classic example of antecedence, and in such a case would not have allowed any space for accumulation of an alluvial fan but would instead have promoted erosional dissecting of the obstacle which had arisen along the course of the river.

The form referred to as the Leba River alluvial fan cannot have been the fan of this river, as conditions appropriate for the accumulation of such a fan simply never existed. It could be possible only if the Leba River valley were younger

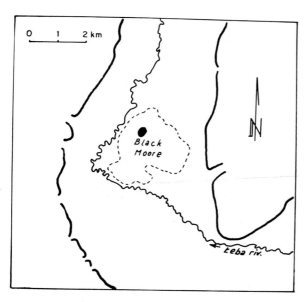

Figure 15.9. Course of the Leba River next to the Black Moore

THE LEBA RIVER ALLUVIAL FAN 315

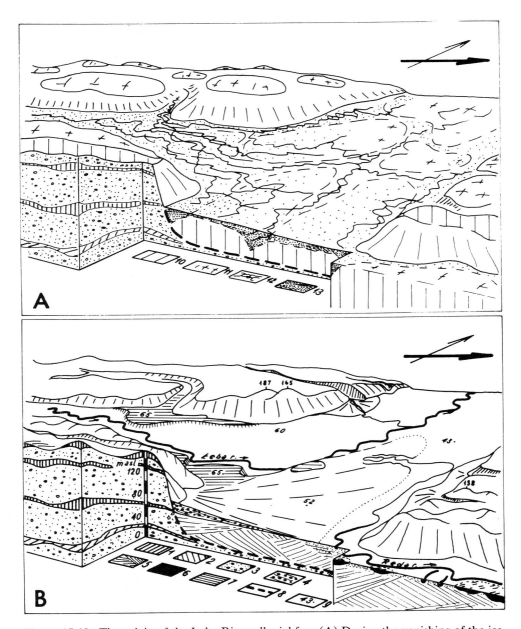

Figure 15.10. The origin of the Leba River alluvial fan. (A) During the vanishing of the ice cap; (B) Situation at present. 1. Glacial till of last glaciation; 2. Grey glacial till (Riss ?) glaciation; 3. Glaciofluvial deposits (sands and gravels); 4. Alluvial fan deposits; 5. Material deposited on or between the blocks of dead ice; 6. Organic deposits (peats); 7. The terrace 65 m a.s.l.; 8. Inferred floor of proto-pradolina; 9. Heights in m a.s.l.; 10. Dead ice (buried in deposits); 11. Dead ice on the surface; 12. Braided rivers; 13. Deposits of braided rivers

than the pradolina itself, but the valley is as old as the pradolina and in such a case there is no extraordinary source of material to be deposited in the form of an alluvial fan, nor a sufficient reduction of the slope of the Leba River to promote the deposition.

If one accepts the survival of the Reda–Leba pradolina from an earlier period one can probably accept the origin of the fan in the form in which it is presented in Figure 15.10. Part A in the drawing represents the situation at the time of the vanishing of the ice cap during Late Glacial time. On the surface of the plateaux, the floor of the proto-pradolina and the Leba River valley were a number of blocks of dead ice. The material supplied by the ice-free erodible surface, which exceeded the transport capacity of the ancient Leba River, was then deposited on the surfaces of these dead ice blocks or between them, giving rise to a crevasse-fill morphology. The best record of this kind of form on the floor of the whole Reda–Leba pradolina is that provided by Marsz (1967).

Climatic conditions during Late Glacial and early Holocene times permitted braiding of the rivers which were overburdened with bedload, of which there was a plentiful supply (Falkowski, 1972; Kozarski and Rotnicki, 1977). It is very probable that the upper Leba was a braided river too at that time. Such an assumption makes it possible to imagine that both the Leba River valley and the pradolina were at that time similar to valley outwash plains. Thus the accumulation of the material at the place where they join cannot be termed as an alluvial fan, at least it is not an alluvial fan in classical understanding of this term.

Whatever the pattern of the Leba might have been, the 67 m a.s.l. level was evidently maintained for long enough to form base level for side valleys of length 2 to 4 km (*cf.* Figure 15.8).

During the gradual melting of the dead ice, part of this material subsided and later became covered by a real alluvial fan. The only difference is that this fan has not its apex at the mouth of the Leba River valley but at that of the unnamed valley X (Figure 15.8). That explains the vertical bipartition of the material in the apical part of the fan. The palaeocurrent analysis, seems to confirm such a reconstruction of morphogenetic events. The present state of the relief is shown in part B of Figure 15.10.

The general warming of the climate will have stimulated further melting of the dead ice and the development of a plant cover, which upon becoming more dense will have retarded the activity of the slope processes. The resulting lowering in the rate of material supply will then have caused the river to revert to a meandering pattern. The dead ice buried within deposits must then have melted, involving changes in the slope of the river and occasioning downcutting in its channel. A lateral shift in the channel would account for the destruction of the western part of the fan, but the river evidently reached the level of the present pradolina floor before the remnants of the dead ice finally melted. This explains why the Leba River passed by the Black Moore depression.

The incision of the new floor into the Leba valley south of its confluence with the pradolina left side valleys at level 67 m a.s.l. which are thus in a hanging position.

In the light of the present interpretation the existence of erosional terraces in the Reda–Leba pradolina, which has been previously contested, indeed seems doubtful. If there are any terrace-like remnants there at all, they must have originated from local flow and deposition between the sides of the pradolina and the blocks of dead ice and will therefore be kame terraces and not erosional ones. The floor of the pradolina too, is not erosional but aggraded by accumulation of deposits similar to those known from recent valley outwash plains.

Some Further Implications

The statement that the Reda–Leba pradolina is not the result of Late Glacial meltwater erosion is of widespread palaeogeomorphological significance. To justify this claim it is necessary first to point out that just as the origin of this formation has customarily been explained by erosion, so starting from Sonntag (1914) the whole system of pradolinas in northern Poland as shown on Figure 15.3 has been treated as a fragment of a Late Glacial drainage system which developed simultaneously and consistently over the whole area. Acceptance of the notion of survival of the Reda–Leba pradolina allows us to understand how and why the other segments also survived for similar reasons. But it is only a beginning. One ought to remember that the North European Lowland is crossed by many pradolinas (Figure 15.2), and the explanation given for the origin of these, also in terms of Late Glacial erosion by combined meltwater from the north and river water from the south, are very often far from satisfactory or logical in many respects. This opens up new problems that are regrettably beyond the scope of the present paper.

The general conclusion to be reached is that it is frequently impossible or misleading to try to understand the origin of a landform in isolation from its palaeoenvironmental context. The Leba River alluvial fan is an excellent illustration of this principle.

References

Augustowski, B. 1965. Morphology and development of ice marginal streamways of the Cassubian Coast, *Zeszyty Geograficzne Wyzszej Szkoły Pedagogicznej w Gdańsku*, **4**, 9–92. In Polish with English summary.

Augustowski, B. 1972. Coastal Plains. In Galon, R. (Ed.), *Geomorphology of Poland*, Part 2, 111–128. In Polish.

Falkowski, E. 1972. Regularities in development of lowland rivers and changes in river bottoms in the Holocene, *Excursion Guide Book, Symposium of the INQUA Commission on Studies of the Holocene*, Part 2, 3–35.

Hartnack, W. 1931. Oberflächengestaltung der Ostpommerschen Grenzmark, *Der Nordosten*, **1**, 99–127.

Keilhack, K. 1899. Die Stillstandslagen des letzten Inlandeises und die hydrographische Entwicklung der Pommerschen Küstengebietes, *Jahrbuch des Königlishes Preussisches Geologisches Landesanstalt und Bergakademie zu Berlin*, **20**, 90–152.

Kozarski, S. and Rotnicki, K. 1977. Valley floors and changes of river channel pattern in the North Polish Plain during the Late-Würm and Holocene, *Questiones Geographice*, **4**, 51–93.

Marsz, A. 1967. An attempt to correlate geomorphic development of the Cassubian Pradolina with the Pradolina Reda–Leba, *Badania Fizjograficze nad Polska— Zachodnia»*, **19**, 55–91. In Polish with English summary.

Robinson, H. 1969. *Morphology and Landscape*, University Tutorial Press, London, 392 pp.

Rosa, B. 1963. O rozwoju morfologicznym wybrzeza Polski w świetle dawnych form brzegowych, *Studia Societas Scientiarum Torunensis*, Sectio C, **5**, 178.

Sonntag, P. 1914. Die Urstromtäler des Unteren Weichselgebietes, *Schriften des Naturforschungs Gesellschaft in Danzig*, **13**, 25–58.

Sylwestrzak, J. 1969. Meltwater outflow and the retreat of ice sheet from the eastern part of the Plain of Slupsk and Slovince Coast, *Zeszyty Geograficzne Wyzszej Szkoły Pedagogicznej w Gdańsku*, **9**, 9–18.

Sylwestrzak, J. 1973. The development of the valley network against the recession of the ice sheet in the northeastern part of Pommerania, *Gdańsk University Scientific Publications*, Special Issue, **14**, 204.

Wunderlich, E. 1914. Postglaziale Hebung in Westpreussen und Hinterpommern, *Zentralblatt für Mineralogie, Geologie und Paläontologie*, **12**, 398–401.

Zaborski, B. 1933. *Zarys morfologii pólnocnych Kaszub*, Wydawnictwa Instytutu Baltyckiego, Toruń, 56 pp. In Polish with French summary.

SECTION II

INTERPRETATION OF THE ENVIRONMENT OF ALLUVIAL FANS

PART B

**Alluvial Fans:
A Scene of Human Activity**

CHAPTER 16

Development of Alluvial Fans in the Foothills of the Darjeeling Himalayas and their Geomorphological and Pedological Characteristics

Subhash Ranjan Basu and Subir Sarkar
North Bengal University, Darjeeling

Abstract

The development of these fans has been closely controlled by local relief, climate, lithology, hydrology, and channel forms since their formation out of the periglacial debris and solifluction materials during the Pleistocene Epoch. The fan materials are coarse-grained, poorly-sorted, immature sediments. Maximum clast size and bed thickness decrease and roundness of coarse grains increases rapidly towards the toe of the alluvial fan deposits. Intermittent flash floods, stream action, stream floods, and mass movements are notable modes of deposition on alluvial fans. Stratification is moderately developed; boulder and pebble beds alternate with sandy, silty, and muddy beds rich in organic matter. The radial profiles of the fans consist of three well-defined straight line segments having approximately uniform slope. Excessive deforestation, increasing arable farming, and overgrazing have had profound effects on the existing soil cover. The removal of topsoil is so widespread that the present landuse is often being carried out on former subsurface horizons. The degree of pedogenesis seems to be maximum in the middle fan, moderate on the upper fan, and minimum on the lower fan. The unplanned use of the fans has been drastically disturbing the local ecological balance. The forces of nature have thus been prevented from maintaining any equilibrium.

Introduction

Rivers debouching from the Eastern Himalayas on to the wide-open plains of North Bengal have developed a number of alluvial fans. Between the 300 m contour in the north and the 75 m contour in the south, these alluvial fans coalesce to form a Piedmont zone. Textural diversity has led to the evolution of different landuse and sediment patterns on each side of the Tista, the main river of this region. The fans between the Rivers Balason and Rohini, bounded by the latitudes of 26°45′19″ to 26°50′23″N, and the longitudes of 88°14′35″ to 88°20′30″E were selected for detailed study (Figure 16.1). These are located on the right bank of the Tista, in the District of Darjeeling, West Bengal and cover approximately 40 km². The main purpose of this study is to describe systema-

Alluvial Fans: A Field Approach edited by A. H. Rachocki and M. Church
Copyright © 1990 John Wiley & Sons Ltd.

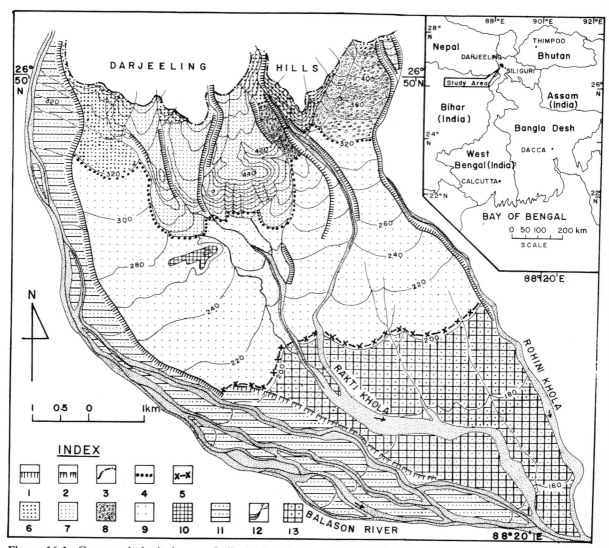

Figure 16.1. Geomorphological map of alluvial fans between Rohini Khola and Balason River. Key to legend: 1. scarp, 15–25 m; 2. scarp, 5–10 m; 3. limits of the Darjeeling Hills; 4. lower limit of the upper fan; 5. lower limit of the middle fan; 6. slope wash and mass movement materials; 7. landform due to mixed processes, including slope wash, mass movements and alluvial deposits; 8. boulder beds; 9. middle fan; 10. elevated tracts due to resistant rocks; 11. floodplain of Balason River; 12. rivers; 13. lower fan

tically the evolution and development of the alluvial fans together with their geomorphological, pedological, and cultural characteristics.

The methodology employed includes quantitative determination of slope, careful examination of soils, analysis of the geomorphological processes involved in sculpturing the land surfaces, together with the study of the nature of the existing landuse.

Evolution and Controls of Development of the Fans

EVOLUTION

The fact that thick boulder beds unconformably overlie the northward dipping Siwalik (Pliocene–Pleistocene) strata at the base of the Eastern Himalayas, indicates that the boulder formation originated sometime in the Pleistocene after the uplift, tilting and partial denudation of the Siwaliks. It is thus evident that during the Pleistocene, when the higher parts of the Darjeeling Himalayas were experiencing widespread glaciation, the Manebhanjan–Sukhiapokhri–Ghoom range (latitude 26°57′ to 27°00′N and longitude 88°00′ to 85°20′E) was subject to periglacial conditions. During this period, Balason, the main stream, together with its tributaries Rakti and Rohini, brought down a great volume of periglacial debris ands solifluction material which eventually were deposited as coalescing alluvial fans at their outlets (Kar, 1962, 1969; Godwin-Austin, 1968).

CONTROLS OF DEVELOPMENT

The geometry of the fans under study has been strongly controlled by the factors of local relief, climate, lithology, and the hydrologic characteristics of the streams:

1. Relief. The primary condition for the formation of the alluvial fans at the base of the Himalayas is a sudden change of slope (about 15° from the mountains to the foothills) where the emerging streams (Balason, Rakti, and Rohini) become unconfined and the hydraulic conditions require deposition of sediments being transported.

2. Climate. The climate is characterized by a generally hot–warm and humid summer with strong seasonal (June to October) distribution of rainfall. The highest mean annual precipitation has been recorded at Long View Tea Garden (4280.9 mm) and lowest at Simulbari Tea Garden (3690.3 mm), signifying that the apex zone of the fans receives more precipitation than the distal part, *cf.* Figure 16.4.

The flow of water shifts the sediments and the overall gradient gets reduced gradually from 10° to 3° and then to 1° only.

3. Lithology. The occurrence of boulders of granite and gneiss is quite common in the upper part of fans (Figure 16.2A) and such rocks have a strong control on the shape and size of the fans (Bull, 1964). Lithologically such boulders are comparable with those of the granite–gneissic areas of the higher, periglacially affected parts of the Darjeeling Himalayas.

4. Hydrology and channel forms. The main cause for the deposition of a large amount of debris in the alluvial fan region is the decrease in the velocity of flow of the stream as a result of the fall in the gradient and the subsequent spread-

Figure 16.2. Upper alluvial fans: (A) The fanhead of the Rakti–Rohini near Marchenbong is characterized by subangular gneissic boulders having an approximate diameter of 2 m. The provenance of such boulders is the higher Himalayas; (B) The fan deposits are coarse-grained, poorly-sorted, and immature sediments. Usually gravels, cobbles, and boulders predominate, with subordinate amounts of sand, silt, and some clay

ing of the stream water from the apex of the fans. On the fan surface the trunk stream feeds distributary channels and water discharge diminishes along its course. Moreover, as the fans consist of relatively permeable sediments additional discharge is lost because of downward water percolation. These discharge losses in downfan direction promote an increased sediment concentration within the water–sediment mixture during flow of any type; this leads to sediment deposition on fan surfaces.

Fan Materials

NATURE OF THE MATERIALS

In the area of study, the fan deposits are coarse-grained, poorly-sorted, and immature sediments. Usually gravels, cobbles, and boulders predominate with subordinate amounts of sand, silt, and some clay (Figure 16.2B). The coarsest and the thickest deposits occur near the fanheads. Maximum grain size and thicknesses of sediments decrease rapidly toward the base of the alluvial fan deposits. The roundness of coarse grains also increases with increasing distance from the apex of the fans.

MODES OF DEPOSITION

Field experiences show that there are four modes of deposition on alluvial fans in the area under study:

1. Flash floods. Flash flood deposits accumulate on the fans when a large amount of water charged with detrital sediments emerges from the mountains. Such a process is often accentuated when natural dams, formed due to excessive accumulation of large boulders across the rivers in the mountain, burst. The released water tends to spread out in sheets covering part of the alluvial fan deposits. Mostly such deposits are unsorted, coarse-grained sediments. As flash floods run for only a small distance and are of short duration, such sediments are mostly deposited at and near the fanheads. In 1899, 1950, and 1968 such flash floods occurred and modified the fanheads with the accumulation of coarse debris (Banerjee, 1980).
2. Streams. There exists a basic difference in flow characteristics as a drainage inversion occurs between flow in the source area and flow on the alluvial fan. Tributary streams in the source area continually enter and join the trunk channel along its course, and for this reason water discharge increases between the drainage divide at the mountain crest and the channel outlet at the mountain front. At this point the source area terminates and the fan essentially begins. At the fanheads the trunk streams, Balason, Rakti, and Rohini, have cut deeply. But once this channellized flow emerges onto the fan it is free to subdivide at some point down fan where it becomes unconfined. Thus, the tributary–trunk stream role in the mountain drainage basin becomes reversed; on the fan surface the trunk stream feeds distributary channels and water discharge diminishes along its course, favouring the formation of sandbars within the channels. Consequently the streams are braided in their lower sections (Figure 16.1).
3. Stream floods. The braided channels are quite incapable of containing the huge monsoon discharge (Table 16.1) in the reduced cross-sectional areas downfan and as a result the rivers often flood causing deposition in their respective floodplains, mostly in the middle and lower sections of these fans (Figure 16.3A). The sediments are generally fine-grained silts and sands. In exceptional cases, as in the year 1968, the upper fan area is also partly inundated by floodwater.

Table 16.1. Discharge of the rivers of the alluvial fans

Rivers	Place	Discharge ($m^3\ s^{-1}$)	
		Maximum	Minimum
Balason	Dudhia (260 m)	1,500	2.15
Rakti	W.Simulbari (205 m)	150	0.31
Rohini	E.Simulbari (208 m)	160	0.40

DEVELOPMENT AND CHARACTERISTICS OF HIMALAYAN ALLUVIAL FANS

Figure 16.3. Middle and lower alluvial fans: (A) The braided channel of the Rohini is quite incapable of containing the huge monsoon discharge in the reduced cross-sectional area and the river often floods, causing deposition in the floodplain, mostly in the middle and lower sections of the fans; (B) The lower fan area is without any proper vegetation cover and is utilized irregularly for the production of paddy and jute during rains; (C) With the gradual establishment of tea gardens on undulating terrain having suitable soil and adequate drainage, more and more forests were cut. At present, tea gardens cover about 40% of the total area under study

4. Debris-flows. Extensive masses of coarse sediments in a muddy matrix may move down the alluvial fans during the heavy downpour of the monsoon and may change the general slope of the fanheads. This area is included within the higher isoseismals of earthquakes, and severe shocks experienced in 1849, 1863, 1869, 1930, and 1934 were also invariably accompanied by several landslips (Banerjee, 1980). The fanhead near Bamanpokri Reserve Forest has been oversteepened up to the angle of 10° by such past debris-flows (Figure 16.1). The fanhead of the Rakti–Rohini near Marchenbong (Figure 16.1) is characterized by huge subangular gneissic boulders having an approximate diameter of 2 m (Figure 16.2A). The provenance of these boulders is the higher Himalayas.

One must keep in mind that the processes discussed above are essentially all contemporaneous. But there evidently is some difference in their areas of impact.

STRATIGRAPHY OF DEPOSITS

The stratigraphic arrangement of fan sediments reflects the characters of the sediments, climatic factors, and modes of deposition. In the upper part of the fans along the riverbanks sediments are found to be laid down in the beds more or less parallel to the surface of the fans, in this way repeating the surface angles of fans in the resulting deposits (Figure 16.5). Here the fans are characterized by occasional flooding and this has resulted in a vertical series of buried organic zones that exhibit considerable lateral continuity; sediments is successively deposited over each previous year's grass cover. Boulder and pebble beds alternate with sandy, silty, and muddy beds rich in organic matter (section A–A' in Figure 16.5). Such stratification is disturbed in places by occasional mass-movement and flash flood deposits which superimpose upon normal sequence huge boulder beds for example near Marchenbong and Bamonpokri Reserve Forest. Absence of stratigraphic sequence in the lower fan deposits is due to the continuous use of the land for agriculture.

The fans are built up by successive shifts in the loci of deposition in a lateral and downfan sense. Thus, nowhere is there likely to occur a vertical series of beds or layers that exhibit lateral continuity of deposition over the entire fan surface.

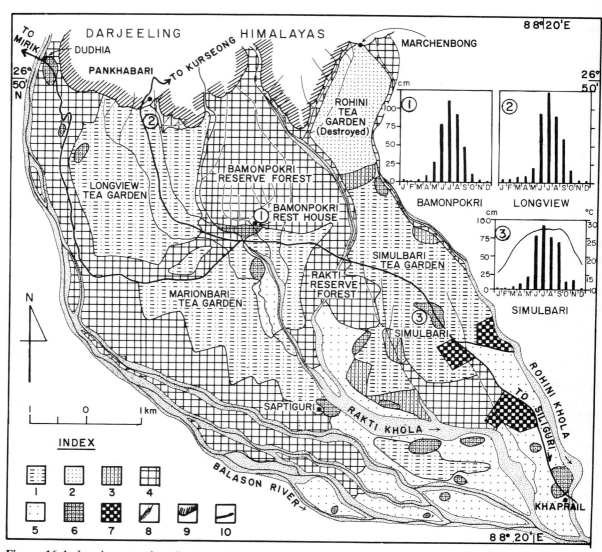

Figure 16.4. Landuse on the alluvial fans. Key to legend: 1. tea garden; 2. tea garden (deserted); 3. new plantation; 4. forests; 5. arable land; 6. settlements; 7. defence occupation; 8. rivers; 9. limits of the Darjeeling Hills; 10. road

The stratigraphic arrangement consists of a series of lenses or wedges of sediment because successive flow paths may cross and intertwine. Two successive floods along the same path may be of different magnitude, for example, and this will produce vertical alternation of coarse and fine sedimentary deposits at any particular distance from the source area.

Geometry of the Alluvial Fans

The radial longitudinal profiles of the fans under study consist of successive well-defined straight line segments. The surface represented by each of these segments makes a band of approximately uniform slope and runs concentrically away from the fan apex. By following Blissenbach, 1954 the

Table 16.2. Characteristics of alluvial fan segments

Segment	Area (km²)	Area (%)	Slope	Materials	Process	Major landuse
Upper	10.23	25.70	3.5°–10°	Coarse-grained gravels, cobbles and boulders	Flash floods, debris-flow, solifluction and stream action	Forests and tea gardens
Middle	18.26	45.87	1.5°–3.5°	Medium-grained sand–silt with occasional boulders	Stream action and stream floods	Forests, tea gardens, arable land
Lower	11.32	28.43	1°–1.5°	Fine-grained sand, silt and clay	Stream action and stream floods	Tea gardens, forests, arable land

fans under study may be distinguished into three broad zones (Figure 16.1). As each one of these has its own individual characteristics (Table 16.2), they have been treated separately in the following paragraphs.

UPPER FAN SEGMENT

This segment of the alluvial fans close to the apex falls within the contour limits of 320 to 460 m to the east and the west, but the lower limit goes down to 260 m in case of the central fan. The slope of the east and the west fans generally varies from 3.5° to 5°, while the slope has been accentuated on the central fan to about 10° by the accumulation of debris-flow materials. Seepage along this steep slope has given rise to parallel first order streams. The deposits of the upper fans are generally dominated by coarse gravels, cobbles, and boulders. Such areas are under constant modification due to intermittent stream action, flash flood, and mass movements. Due to stream action the fanheads are deeply entrenched. Huge flash flood debris and mass-movement materials are often seen scattered all over the surface. Well covered with vegetation except for the eastern segment, the upper fans have been able to retain their topographic characteristics since inception.

MIDDLE FAN SEGMENT

This zone between the fanhead and the lower margin of the fans, extends to the eastern and the western fans from 200 to 320 m. On the central fan the limit is between 200 and 260 m. The general slope of the middle fan segment varies from 1.5° to 3°. The size of the deposits, here, is mostly smaller. The presence of big boulders is quite uncommon except the riverbanks and soils are mostly sandy and silty. Being away from the apex, the middle fan segment is rarely affected by the hazard of flash floods and mass movements, but the stream action is quite competent in producing steep scarps (15 to 25 m) bordering the rivers. Stream floods often affect the lower part of this segment. During the catastrophic flood of 1968 most of the area of this region was inundated. Soil erosion is not appreciable due to the low gradient and the presence of vegetal cover in the form of tea gardens and forests.

LOWER FAN SEGMENT

The outermost area, south of the limiting contour of 200 m, grades into the zone of coalescence of all the fans. Coalescence is due to the frequent flooding of the existing braided streams and the mingling of their floodwater together with flood deposits. Here, the deposits are generally fine-grained silts and sands. The overall slope varies from 1.5° to less than 1°. Mostly, this area is without any proper vegetal cover and utilized irregularly for the production of paddy and jute during rains (Figure 16.3B). From November to May the lands remain totally unutilized for want of water and the loose sands and silts are often carried away by the strong winds. Rivers choked with sediments often change their courses during floods. Bank erosion producing scarps of 5 to 10 m high is not uncommon specially along the left bank of the River Balason.

Landuse

At the beginning of the 19th century all these fans were well covered with tropical forests of densely growing Sal (*Shorea robusta*), Segun (*Tectona grandis*), or sturdy Sheesham (*Dalberghia sisso*). The bases of these trees were also sheltered with tender creepers, herbs, and shrubs. But with the gradual establishment of an ordered government and with increase of prosperity, the population increased rapidly (to 327 persons per km^2 in 1981 in the District of Darjeeling) with the consequent increase of pressure on land for cultivation of food and other crops. Local inhabitants took possession of extensive areas including areas under forest. The early development of the Darjeeling District not having followed a plan drawn up in advance, the status quo had to be accepted when reservation started in 1866 and the consequent position today is that no forests exist on over 63% of this area. (Figure 16.4 and Table 16.3). Moreover, the poor farmers and agricultural labourers of this region are in constant need of fuel and livelihood. They encroach upon the forest, cut the valuable trees for fuel or sell wood illegally to owners of saw mills. Consequently, the proportion of forests even under official estimates has been steadily declining (*cf*. Figure 16.6).

With the gradual establishment of tea gardens after 1866 on the undulating terrain having suitable soil and adequate drainage, more and more forests were cut (Figure 16.3C). At present tea gardens cover about 40% of the total area under study. Moreover, tea gardens occupying about 5% of the region have been deserted by their owners as these were no longer economical.

Unemployment generated by the decline of the tea industry has further forced poor labourers to somehow maintain their livelihood by illegal felling of trees. Arable lands at present occupy about 11% of the territory and this is constantly increasing with the steady influx of settlers from the adjoining countries.

Since border difficulties with the Chinese in 1962, military installations have also been increased, occupying more and more of the forest land and uneconomical tea gardens. Thus, with the gradual reduction of the forests and with the introduction of a totally different landuse pattern, the natural environment is very much disturbed. The forest-covered slopes act as blankets for trapping rain water and maintaining the perennial character of the springs, as well as sustaining the flow of the non-glacial rivers even in the driest summer period. Since deforestation, many springs have dried up and many channels have contracted.

Excessive deforestation and increasing arable farming, as well as overgrazing, have had profound effects on the existing soil cover. The removal of topsoil is so widespread that the present landuse is often being carried out on former sub-surface horizons. The progressive breakdown of soil structure and tilth is also aiding the development of hard pans and subsoil compaction in suitable circumstances. Colluviation and slopewash are the two active processes operating on valley sides to produce soil having as its characteristic features solum mixing, truncation, and the presence of colluvium stone horizons. Such soils are quite different from those of the residual soils derived from weathering and soil formation on sedentary sites.

Table 16.3. Landuse on the alluvial fans

Types	Area (km^2)	Area (%)
Forests	14.74	37.03
Tea garden	16.10	40.43
Tea garden (deserted)	2.15	5.40
New plantation	0.81	2.04
Defence occupation	0.83	2.09
Settlements	1.05	2.64
Arable land	4.13	10.37
Total area	39.81	100.00

DEVELOPMENT AND CHARACTERISTICS OF HIMALAYAN ALLUVIAL FANS

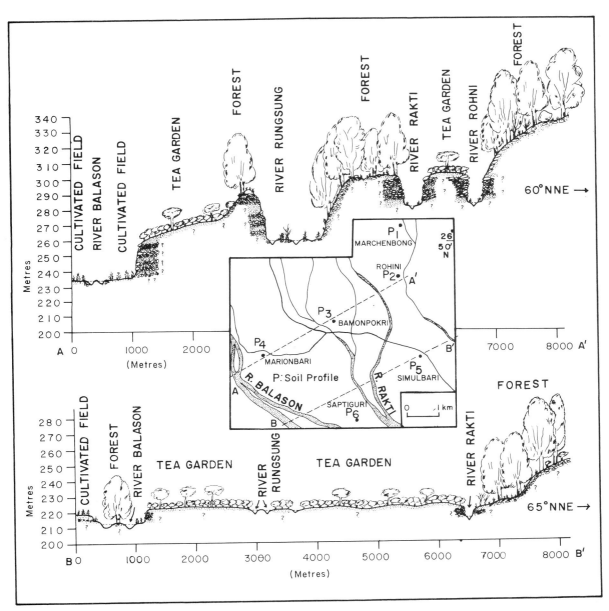

Figure 16.5. Cross-profiles showing stratigraphy and landuse on the alluvial fans

Pedogenesis

A number of soil samples (Figure 16.5) were collected from the field and investigated in the laboratory to determine the characteristics (Table 16.4) of the existing soils of the area under study.

PROFILE DEVELOPMENT

As the upper and the middle fan segments have good vegetal cover and moderate slopes receiving comparatively higher rainfall (Figure 16.4) than the lower fan segment, the process of leaching seems to be more active on the former than on the latter. Consequently, the finer particles of clay and silt move downward, leaving silica-dominated, coarse-textured surface soils. These are clearly evident in case of profiles 1 to 5 which show an increase in clay and silt content at a depth of 10 to 60 cm below the surface. This is a clear indication of B-horizon development within the same depth (Table 16.4). The distribution of sand, silt, and clay do not indicate such development on the lower fan (profile 6). Continuous alterations of surface soil by arable farming, increment of fresh deposits (mostly sands) via intermittent flooding, and overall deforestation hinder profile development and stand responsible for the non-existence there of a B-horizon.

Due to the presence of deciduous forests with considerable undergrowth, the accumulation of organic matter in the upper and middle fan segments is higher by 2 to 3% than in the lower fan segment where a few tall trees and bushes, found here and there, help very little in adding organic matter to the soil.

Higher rainfall on coarse-textured soils of the upper and middle fans helps considerably in the removal of soil-organic matter through leaching down to a depth of 20 to 60 cm, forming dark coloured B-horizons. As a result, mature to semi-mature soils develop with well-defined horizons. The continuous use of the deforested tracts for settlement and agriculture may be viewed as limiting factors for the accumulation of organic matter in the subsurface soils in the lower fan area, thereby hindering the formation of a well-defined B-horizon present in case of mature soils.

The different shades of brown within the range of hues 10YR and 7.5YR are the typical colours found in different horizons of almost all profiles except No. 6 (Table 16.4). The darker brown shades of the upper horizons of profiles 1 to 5 are due to the presence of organic matter. In profile 6, the colour of the surface horizon varies between pale brown and light yellow brown. Such colour is mainly due to the presence of accumulated mineral matter. A slight reddish tint in the lowest layer is due to the presence of iron translocated from the upper horizon. Moreover, the colour has also been affected by the deposits of finer alluvium brought down from the higher fans during rains.

SOIL-REACTION AND CLAY MINERALS

In the Balason–Rakti–Rohini alluvial fans the soils are slightly to moderately acidic in reaction. High rainfall on coarse-grained, acidic parent materials causes considerable leaching of bases from the surface to the subsurface at a depth of 60 cm. In the upper and middle fan area pH ranges from 5.8 to 6.1 on the surface to 6.0 to 6.5 at a depth of up to 60 cm, and 5.7 to 6.0 at a depth of up to 100 cm. In the lower pH varies from 6.2 on the surface to 6.0 at the depth of up to 100 cm. The slight increase in pH value at a depth of 20–60 cm is a clear indication of an enriched B-horizon in the upper and middle fans (Table 16.4).

From the low to moderate base-exchange capacity of the soils, it is evident that kaolinite is the predominant clay mineral in the lower fan area and this is due to the complex chemical weathering of feldspar and other feldspathic rocks. The base-exchange capacity of the soils of the upper and middle fans is to some extent higher due to the presence of a good amount of organic carbon in the profiles.

In the upper fan segment the percentage of the clay increases from 37.4 to 55.7 in the soils at the depth 10 to 30 cm from the surface whereas in the middle fan the same increases from 44.7 to 62.3 at the depth of 30 to 60 cm. This is due to the intensity of leaching, which is more widespread in the middle fan with much gentler slopes receiving

DEVELOPMENT AND CHARACTERISTICS OF HIMALAYAN ALLUVIAL FANS

Table 16.4. Soil characteristics of the alluvial fans*

Sample	Locality	Depth (cm)	Sand (%)	Silt (%)	Clay (%)	Hygroscopic moisture (%)	Organic matter (%)	pH	Water Holding Capacity (%)	Base exchange (meq)	Nitrogen (%)	Munsell colour notation
M_1	Marchenbong	0–10	35.2	17.9	39.9	2.91	3.23	6.1	51.3	29.3	0.29	2.5 YR 3/2
M_2	Altitude: 490 m	10–30	21.3	22.6	50.2	3.75	2.91	6.3	70.6	20.1	0.21	2.5 YR 3/2
M_3	Slope: 15–17° towards SSE	30–60	61.5	11.3	24.3	2.03	1.03	6.5	39.3	15.03	0.121	10 YR 4/2
M_4		60–90	73.9	8.9	15.7	1.75	0.51	5.9	30.2	11.7	0.072	2.5 YR 4/2
R_1	Rohini tea garden	0–10	31.7	18.7	43.1	3.10	3.17	5.8	50.7	30.1	0.292	10 YR 5/6
R_2	Altitude: 320 m	10–30	18.3	20.1	55.7	3.57	3.01	6.1	71.9	21.9	0.235	10 YR 6/6
R_3	Slope: 6° towards SSE	30–60	29.4	12.1	54.2	2.97	1.93	6.2	72.0	18.01	0.189	10 YR 4/2
R_4		60–100	69.4	6.3	22.1	2.03	1.01	5.7	49.2	12.81	0.102	10 YR 6/3
B_1	Bamonpokri R.H.	0–10	42.9	13.7	37.4	2.68	4.42	6.0	53.9	34.1	0.371	7.5 YR 4/4
B_2	Altitude: 250 m	10–30	35.4	16.7	42.1	2.73	3.01	6.4	64.9	27.6	0.263	5 YR 4/6
B_3	Slope: 15° towards S	30–60	39.6	13.9	40.7	1.91	2.63	6.0	60.4	19.5	0.324	2.5 YR 4/6
B_4		60–80	71.3	9.2	16.5	1.80	1.51	6.1	30.7	15.70	0.180	2.5 YR 4/3.5
Ma_1	Marionbari tea garden	0–10	28.3	20.3	44.7	3.51	3.73	5.6	63.8	28.7	0.315	7.5 YR 4/6
Ma_2	Altitude: 282 m	10–30	20.3	16.3	58.5	3.19	2.97	5.9	70.9	20.3	0.246	7.5 YR 5/6
Ma_3	Slope: 3° towards SSE	30–60	18.7	14.4	62.3	2.93	2.00	6.2	73.1	15.7	0.201	10 YR 5/4
Ma_4		60–80	47.3	11.1	37.3	3.10	1.31	6.0	50.3	10.8	0.146	5 YR 4/4
Ma_5		80–100	69.1	10.3	17.51	2.01	0.82	6.0	31.2	11.3	0.980	10 YR 8/3
S_1	Simulbari tea garden	0–10	21.7	20.9	50.3	3.21	4.37	5.9	61.7	30.1	0.368	7.5 YR 3/4
S_2	Altitude: 205 m	10–30	18.3	22.1	53.4	3.01	3.61	6.2	63.8	29.4	0.321	5 YR 4/4
S_3	Slope: 3° towards SSE	30–60	16.6	23.2	55.1	3.21	2.85	6.3	66.2	15.3	0.280	10 YR 4/4
S_4		60–90	31.7	15.1	48.3	2.73	1.10	6.0	59.3	12.7	0.173	10 YR 8/3
S_5		90–120	36.3	13.2	45.4	2.65	0.73	5.9	51.6	19.3	0.09	10 YR 8/3
Sa_1	Saptiguri	0–10	11.3	40.1	43.1	2.91	1.91	6.2	60.5	19.3	0.180	10 YR 6/3
Sa_2	Altitude: 190 m	10–30	20.1	35.3	39.3	2.63	1.72	6.4	56.1	13.7	0.165	2.5 Y 6/4
Sa_3	Slope: 1° towards SSE	30–60	25.3	31.4	40.1	2.53	0.91	6.2	58.3	9.7	0.081	2.5 Y 6/4
Sa_4		60–90	39.3	25.3	33.2	2.01	0.63	5.9	42.1	10.3	0.063	2.5 Y 6/2
Sa_5		90–120	43.4	20.7	32.9	2.12	0.23	6.0	36.9	6.5	0.031	2.5 Y 6/2

*Samples were collected during August 15–20, 1987 and were analysed during September to November, 1987, in the Pedology Laboratory, Department of Geography, North Bengal University.
†Based on material less than 2 mm in size only.

Figure 16.6. Apex of the alluvial fan near Marchenbong. In the background note the deforested Darjeeling Himalayas

a moderate amount of rainfall. In the middle fan, the B-horizon apparently lies between 30 and 60 cm below the surface, but in the upper fan area it probably lies between 10 and 30 cm below the surface. As the leaching of clay minerals to the subsurface is hampered due to secondary disturbances in the lower fan area, development of B-horizon is conspicuously absent here. Thus, the degree of pedogenesis seems to be the maximum in the middle fan, moderate in the upper fan, and the minimum in lower fan segment.

Concluding Remarks

The geometry of the Balason–Rakti–Rohini fans of the piedmont zone in the District of Darjeeling have been closely controlled by local relief, climate, lithology, hydrology, and channel forms since their formation out of the periglacial debris and the solifluction materials during the Pleistocene Epoch. The fan materials are coarse-grained, poorly-sorted, immature sediments. Usually gravels, cobbles, and boulders predominate with subordinate amounts of sand, silt, and some clay. Maximum size, thickness, and roundness of coarse grains decrease rapidly towards the base of the alluvial fan deposits. Intermittent flash floods, stream floods, debris-flows and braid-plain deposition are four notable modes of sediment deposition on the alluvial fans. These processes are essentially all contemporaneous, each one having its own area of impact. In the upper part of the fans stratification is moderately developed, boulder and pebble beds alternate with sandy, silty, and muddy beds rich in organic matter. But such stratification is not apparent in the lower reaches of the fans.

The longitudinal profile of the fans consists of three well-defined straight line segments (upper, middle, and lower) having approximately uniform slopes. Each one of these segments has its own topographic characteristics and landuse. The upper fans, spanning between 260–460 m, have a general slope varying from 3.5° to 5°. Due to the stream action the fanheads are deeply entrenched. Huge flash-flood, debris, and mass-movement materials are often seen scattered all over the surface. The middle fan extends from 200 to 320 m, having a general slope varying from 1.5° to 3°. The size of the deposits here is mostly smaller. Being away from the apex, the middle fan segment is rarely affected by the hazards of flash floods and mass movements, but stream action and floods often affect this region quite conspicuously. The outermost area, south of the limiting contour of 200 m, grades into the zone of coalescence of all the fans. Such coalescence is due to the frequent flooding of the existing braided streams and the mingling of their floodwater together with flood deposits. Here, the deposits are generally fine-grained silts and sands. The overall slope varies from 1.5° to less than 1°. Bank erosion and river migration are quite common in this region.

Excessive deforestation and increasing arable farming, as well as overgrazing, have profound effects on the existing soil cover of the fans. The removal of the topsoil is so widespread that the present landuse is often being carried out on former subsurface horizons.

Due to the presence of deciduous forests with considerable undergrowth, the accumulation of organic matter in the upper and middle fans is higher than in the lower fan. Higher rainfall on coarse-textured soils of the upper and middle fans helps considerably in the removal of clay particles and organic matter of soils through leaching down to a depth of 20 to 60 cm, forming dark coloured B-horizons. As a result, mature to semimature soils develop with well-defined horizons. The continuous use of the deforested tracts for settlement and agriculture may be viewed as limiting factors for the accumulation of clay particles and organic matter in the subsurface soils in the lower fan area thereby hindering in the formation of a well-defined B-horizon. The degree of pedogenesis seems to be maximum in the middle fan, moderate on the upper fan, and minimum in the lower fan segment.

The unplanned use of the fans has set gyrating the vicious cycle of deforestation, soil erosion, mass movements, and floods. Such a process begun by an ill-informed minority has been drastically disturbing the local ecological balance. The forces of nature have thus been prevented from maintaining any equilibrium. On the contrary, cultural impositions by man have induced the natural forces to remain ever active. As such the morphology of the fans under study is also being gradually modified, and continuous vigil is necessary to reveal the personality of such a dynamic landform.

References

Banerjee, A. K. 1980. *West Bengal District Gazetteer*. Darjeeling. Govt. of West Bengal. 39–47.

Blissenbach, E. 1954. Geology of alluvial fans in semi-arid regions. *Geological Society of America Bulletin*, **65**, 175–190.

Bull, W. B. 1964. Alluvial fans and near surface subsidence in Western Fresno County, California. *USGS Professional Paper*, **437A**, 1–71.

Godwin-Austin, H. H. 1968. Notes on geological features of the country near foot-hills in Western-Bhootan-Dooras. *Journal of the Asiatic Society of Bengal*, **37**.

Kar, N. R. 1962. Investigations on a piedmont drift deposit in the foot-hills of the Eastern Himalayas and its periglacial significance. *Biuletyn Peryglacjalny*, **11**, 215–216.

Kar, N. R. 1969. Studies on the geomorphic characteristics and development of slopes in the periglacial zones of Sikkim and Darjeeling Himalayas. *Biuletyn Peryglacjalny*, **18**, 43–67.

CHAPTER 17

Hazard Management on Fans, with Examples from British Columbia

Rolf Kellerhals
Heriot Bay

and

Michael Church
University of British Columbia, Vancouver

Abstract

Considerable development has occurred on alluvial fans in British Columbia because they often are the only gently sloping sites available. They also tend to appear deceptively safe because the two main hazards—channel avulsions and debris-flows—are highly intermittent and their traces disappear under dense vegetation after a few years. A sequence of events within the last decade has prompted several studies of fan hazards. Because of their destructive nature, clear identification of any potential for debris-flows to reach a fan is the key element of hazard assessment.

A hazard zoning system has been developed to deal with both fluvial avulsions and debris-flows. Mitigating measures implemented so far differ significantly from those commonly used in the European Alps or in Japan. Channelization and riprap-lined dykes are the most common measures because of the relatively low costs of land and earth moving. Warning devices, debris chutes, and debris interception barriers with debris-straining outlet structures have also been built. Upstream slope and channel stabilization is not usually feasible.

Introduction

The British Columbia land area of some 923 000 km^2 is made up primarily of mountains and dissected plateaus. While most human settlement is confined to large river valleys, to the coastal plain of the Strait of Georgia, and to the province's small share of the Northern Great Plains, important communication links such as railways, highways, pipelines, and transmission lines cross high mountain ranges. Many small towns with economic bases in forestry, mining, tourism, or communication are also located in mountainous terrain or along the steep, fjord coast.

As in many other mountainous areas, and particularly in lake or fjord-filled valleys, fans are often the only gently sloping surfaces below ridge top levels. To the untrained eye these fans may look like the most suitable sites for a wide variety of developments ranging from farming to housing, industrial facilities, or transportation routes.

Alluvial Fans: A Field Approach edited by A. H. Rachocki and M. Church
Copyright © 1990 John Wiley & Sons Ltd.

Needless to say, first impressions often deceive and fan surfaces are increasingly being recognized as potentially very hazardous. The indiscriminate development of fans has already led to considerable economic losses and loss of life.

Hazard problems associated with fans are not nearly as serious in British Columbia as in other mountainous areas with a long history of human settlement such as Japan (Erosion Control Engineering Society of Japan, 1985) or the European Alps (Eisbcher and Clague, 1984). Nonetheless, there are many situations in the province where alternate sites away from fans simply are not available, or where moving activities to an alternate site is not a reasonable option due to the level of existing developments. The degree of hazard on fans also varies widely from fan to fan and between different areas on any particular fan. Clearly, then, simple avoidance of the hazards associated with developments on fan surfaces is only one of several approaches that need to be considered. Procedures are required to classify, rank, and map the degree of hazard on fan surfaces and protective measures are needed for developments otherwise exposed to excessive hazards.

This paper addresses the general nature of hazards associated with developments on fans and describes field and office procedures developed in British Columbia for their identification and assessment. Finally, the two basic remedial options, avoidance through hazard zoning and construction of protective works, are discussed and illustrated with recent examples.

Both truly alluvial fans that have been built by fluvial sediment transport and deposition, and fans that are partially or entirely the result of debris-flow deposition are considered. Small, steep fans consisting primarily of debris-flow deposits are also known as debris cones.

Nature of Hazards

STREAM AVULSION

Fans are aggradational, fluvial features of the landscape and, as in the case of related features such as deltas, streams on fans are subject to avulsions; that is, sudden, often drastic shifts in channel position (Figure 17.1). Active alluvial fans are gradually being built up through the deposition of parts of the coarse sediment load of a stream. This deposition tends to take place in or near the stream channel with the result that the stream eventually flows at the top of a gentle ridge across the fan. During a major flood it tends to break out of its channel to find itself a new, more stable location. The exact route of the new stream channel is often quite unpredictable. It may be determined by a series of random events such as debris jams, or plugging of culverts and bridges. A very minor depression in the fan surface, such as a road ditch, can suddenly become a major stream course.

In contrast, 'sheetfloods'—laterally extensive, shallow flows—occur mainly on the distal parts of alluvial fans in deserts, where truly torrential downpours flow out onto surfaces with little or no vegetation and very attenuated relief. Even here, flows shift unpredictably.

Avulsions do not constitute 'flooding' in the classical sense of the word and the hazards associated with exposure to avulsions are not readily dealt with on the basis of floodplain mapping. Since fan surfaces are, by definition, unconfined, the normal backwater profile computational procedures of floodplain mapping are simply inapplicable, although this has not deterred many agencies from applying them. Many floodplain maps of fans have been published that give a mistaken impression of a 'normal' flooding hazard involving mainly inundation, when in fact the real hazard is related to potential avulsions. Avulsions normally are associated with high current velocities, erosion, and sediment deposition, including very coarse debris. In almost all instances avulsions are a much more severe hazard than normal 'flooding'.

As soon as the flood flows exceed the capacity of the stream channel on a fan the location, extent, and depth of flooding tend to become highly unpredictable. Since fan surfaces normally slope away from the stream in all directions and are often quite irregular on a microscale, with many old stream channel traces, the overflow is

HAZARD MANAGEMENT ON FANS

Figure 17.1. Modern alluvial fan of Cheakamus River, tributary to Squamish River near Squamish, British Columbia. Recent courses of the river are identified by immature and incomplete vegetation development (arrows). The stream moved suddenly to its present course by avulsion

likely to be concentrated into one or several channels. Initially this will be mainly water skimmed off the top of the existing channel, while the coarse bed material load will continue to move on down the channel. The loss of discharge immediately reduces its carrying capacity and this often results in rapid, complete plugging. The new outflow stream is likely to have a relatively steep slope, little coarse sediment, and therefore much erosive power. All of this tends to accelerate avulsions once they are initiated.

DEBRIS-FLOW

Steeper fans can also be exposed to various hazards associated with debris-flow processes. A debris-flow is a slurry of water and sediment, with possible addition of organic debris, avalanching down a steep slope or a steep, confined channel. These flows contain such a high percentage of solids that their mechanics differ greatly from normal, turbulent, open-channel flow of water (see Costa, 1984, for extensive descriptions). Since the hazards posed by debris-flows are different and generally far more serious than those associated with the above-described 'normal' alluvial fan processes, it is important to identify any potential for debris-flow to occur.

Debris-flow includes a considerable range of related flow phenomena, depending on the concentration and size distribution of the included solid material. Flows containing mainly cohesive fines (i.e. clay and silt) are normally referred to as mud flows. They are characterized by relatively slow motion and by their ability to maintain motion on remarkably low gradients of just a few per cent. The term 'debris-flow' normally refers to flowing mixtures containing a much wider range of grain sizes, extending to boulders (Figures 17.2A, B).

'Debris torrents', a term used mainly in British Columbia and in the adjacent Pacific Northwest of the United States, occupy the other end of the spectrum: they tend to contain very little solid material below sand size and to consist primarily of gravel to boulder sized material, with often a large percentage of organics ranging in size from mulch to logs and root boles of large trees (Figures 17.2B, C). Swanston and Swanson (1976) described debris torrents as follows:

> Debris torrents typically occur in steep, intermittent, first- and second-order channels. These events are triggered during extreme discharge events by slides from adjacent hill slopes which enter a channel and move directly downstream, or by a breakup and mobilization of debris accumulations in the channel. The initial slurry of water and associated debris commonly entrains large quantities of additional inorganic and living and dead organic material from the streambed and banks. Some torrents are triggered by debris avalanches of less than 100 m^3, but ultimately involve 10 000 m^3 of debris entrained along the track of the torrent. As the torrent moves downstream, hundreds of metres of channel may be scoured to bedrock. When a torrent loses momentum, there is deposition of a tangled mass of large organic debris in a matrix of sediment and fine organic material covering areas up to several hectares.

Several events involving more than 50 000 m^3 have been observed in British Columbia, and the largest documented event—in a wilderness area—is of the order one million cubic metres (Clague et al., 1985).

Most British Columbia mountains are characterized by resistant bedrock, thin or absent veneers of unconsolidated materials, dense forest cover, and steep slopes, all of which lead to a predominance of true debris torrents made up of very coarse material requiring steep, well-confined channels to maintain motion (Figure 17.2B). In heavily forested areas, the proportion of organic material can approach or exceed 50% by volume (cf. Figure 17.2C). Areas characterized by deep veneers of unstable, weathered, or unconsolidated materials, sparse or absent vegetation cover, and generally unstable slopes occur primarily in northern British Columbia and in the Rocky Mountains. They tend to produce debris-flows containing finer material and capable of travelling on lower gradients (Figure 17.2). True mud flows, consisting entirely of flowing cohesive material, are not normally associated with alluvial fans because source areas of fans tend to yield coarser material.

In a discussion of the practical aspects of dealing with debris-flows, Hungr et al. (1984) described the debris-flow process observed in British Columbia as follows:

> Many debris-flow events occur in two or more surges, spaced over several hours. Individual surges have short durations measured only in minutes, and are commonly associated with abundant water flooding. A typical surge through the lower reaches of a mountain creek begins by the rapid passage of a steep bouldery front, followed by the main body of the flow. This consists of unsorted coarse particles ranging from gravel to boulders and logs, floating in a slurry of liquified sand and finer material. Both the proportion of fines and the water content increase in the later stages of the surge, forming a liquid 'after

Figure 17.2. Aspects of debris-flows on fans: (A) Muddy, bouldery flow, Rocky Mountains near Golden, British Columbia; (B) Section cut through debris flow in Chilliwack Valley, near Vancouver, British Columbia, showing mud, boulders and wood debris, and lack of bedding; (C) Front of debris torrent in the Queen Charlotte Islands, showing very high wood debris content

flow', which gradually merges into normal flood flow. Upon reaching flatter gradients or a less confining channel, the surge tends to decelerate and deposition can occur. The liquid after flow may break through the coarse deposits and continue further down the cone.

They indicated that, upon the evidence of the deposits, there appears to be a continuous spectrum of phenomena ranging from debris-flows to fluvial bedload. True debris-flow surges are distinguished from 'mass bedload movements' by the substantial rise of the debris wave above the floodwater level at peak discharge.

Several different hazards occur on fans subject to debris-flow events. The debris-flow front itself tends to contain the very coarsest fractions of the slurry, often involving boulders of several metres in diameter. While travelling at speeds up to 12 m s^{-1} (VanDine, 1985), such bore-like surges of up to 5 m in height can be exceedingly destructive. Since debris surges tend to lose momentum quickly as soon as they escape confinement and can spread out and drain, the danger from direct impact of a surge front diminishes rapidly with distance from the fan apex. However, the relatively rapid arrest of coarse-grained surge fronts high on the fan leads to a second hazard. The arrested surge front often blocks the existing stream channel and the blockage can be far more massive than is achievable during the course of a normal flood event. Apparently well-incised channels can be instantly blocked. Such blockages tend to divert the after-flow of the debris surge and later streamflow to other parts of the fan, often into areas that appeared safe and well beyond the reach of the stream.

Along the coast and on lakeshores, many fans are largely subaqueous deposits. Then the entire terrestrial surface may lie within the fanhead region subject to direct debris-flow impact.

General Hazard Identification

British Columbia has had formal guidelines for the design and review of developments on floodplains since the mid 1970s. These guidelines address such problems as minimum freeboard above flood levels and obstruction of the floodway, but it soon became evident that they do not provide adequate guidance in the case of alluvial fans. In response to this concern, the provincial Ministry of Environment commissioned the writing of a simple manual to help field staff with fan-related problems (Thurber Consultants, 1983a). The manual outlines a systematic procedure for problem diagnosis, emphasizing proper identification and classification of fans.

Clearly the initial and most critical step in dealing with the hazards of alluvial fan development is the proper identification of fans. While this may appear trivial, many investigations have taken off in the wrong direction right at this juncture. Most landuse managers are not trained in geomorphology and when confronted with questions concerning development near streams tend to think in terms of flooding and floodplain mapping. Both are misleading concepts in the case of fans. Fan identification and delineation are deceptively easy looking tasks in sparsely vegetated desert or high alpine terrain—the classical environments for geomorphological studies of alluvial fans—but many significant fans are either densely forested or highly developed, and this often disguises critical features.

The second step involves the classification of fan surfaces into active and inactive areas. Almost all of British Columbia was ice covered only some 12 000 years ago and the period of deglaciation was associated with high sedimentation rates and the deposition of some large fans (Ryder, 1971). Since deglaciation, there have been further significant climatic changes. As a result of this climatic instability, there are now many fan surfaces in British Columbia that may appear fresh, but are no longer aggrading. In practice this means that once a fan has been delineated, its geometry must be determined in considerable detail to see whether there is a relatively uniform, single, conical surface or whether there might be two or more surfaces, as is quite common. In the latter case, only the lowest, actively aggrading surface would likely be subject to hazard from stream avulsions. The entire stream channel or parts of it might also be sufficiently well incised to make avulsions impossible. This would suggest that much or all of the fan is dissected, and therefore inactive and safe.

Channel incision can deceive, however, and it needs to be assessed cautiously. As pointed out above, debris-flows can clog very large channels instantly, but even channels not subject to debris-flows can clog remarkably quickly. The hydrologic regime of the source area might well result in irregular, multiyear periods of channel incision on the fan, followed by rapid channel aggradation and clogging under occasional, relatively rare flood conditions, or in response to an upstream change in landuse or to a forest fire.

Once the presence of an active or potentially active fan surface has been established and delineated, the third step of the investigation should address the nature of the active fan deposition process. All active fans are, by definition, subject to channel avulsions, but the potential for debris-flow magnifies the danger of avulsions and introduces other, more serious hazards.

The British Columbia manual lumps all active fan surfaces together, although the special danger from debris-flows is noted. In the writers' opinion, there is such a great difference with respect to both zoning and types of protective measures between purely alluvial fans and fans potentially affected by debris-flow, that detailed investigations aimed at identifying any potential debris-flow hazard normally are in order.

Debris-flow Indicators

Debris-flow potential may be identified through (a) office investigations and interviews; (b) field study of the upstream channel; or (c) careful examination of the fan deposits. A combination of all three approaches is often needed.

A review of available maps is the obvious starting point since debris-flows typically require steep channel sections of at least 15°, but more commonly 25° (VanDine, 1985) for initiation and, since they tend to start losing momentum and depositing material at about 12°, the apex slopes of fans affected by debris-flows generally exceed 8°. There is a good correlation between the steepness of drainage basins and the steepness of the alluvial fans, so this gradient criterion may be used in a preliminary discrimination in the office of potentially debris-flow prone basins using topographic maps (Jackson and Kostaschuk, 1987). This also constrains the maximum size of the drainage basin, in any particular landscape, that is apt to be debris-flow prone. If the channel and fan slopes fall within the indicated ranges, the obvious next step is to search for records of past events.

Since the phenomenon of debris-flows has only recently been recognized by the engineering community in British Columbia and is still not widely known by the public, past debris-flows are generally reported as either slides or floods. Reports of 'slides' and records of 'washed out' bridges and culverts across steep streams refer, in many instances, to past debris-flows.

Even though air photos are often available only as far back as the 1940s, and rarely to the 1920s (in Canada), they still provide the most reliable record in most instances because they provide unequivocal visual evidence. Because of the massive scouring, most debris-flow tracks remain easily identifiable on photos for several years. If the tree species that colonize scoured tracks differ from surrounding tree cover, the track may remain identifiable for 50 to 100 years (Smith et al., 1986). In the Coast Mountains and on Vancouver Island, small, steep channels in coniferous forest lined by strips of alder (*Alnus*) or willow (*Salix*) clearly indicate past debris-flow activity or—what is equally dangerous—snow avalanche tracks, since alder and willow are the local colonizing species on exposed mineral soils.

Figure 17.3 shows air photos of the coalescing fans of Harvey Creek and Alberta Creek, on the east side of Howe Sound some 15 km north of Vancouver, British Columbia. The 1939 photo (Figure 17.3B) shows clear evidence of a debris-flow in Alberta Creek that occurred after 1932 (Figure 17.3A). By 1954 (Figure 17.3C), only a detailed ground search for clues would have revealed the past debris-flow activity, and by 1966 substantial development had occurred on the site (Figure 17.3D). Figure 17.4 illustrates the result of the next major debris-flow in Alberta Creek on February 11, 1983. Needless to say, the old air photos were not reviewed prior to the development of this fan.

The desposition zone of the Alberta Creek debris-flows lies mostly below the water of Howe Sound. Where the deposition zone is exposed it may be characterized by stands of dead trees or by the above colonizing species (Figure 17.5). The typical lobate form of debris-flow deposits is not easily identified on air photos, but stands out clearly on large-scale maps of debris cones constructed by ground survey.

Field inspection must address both the channel and the deposition zone. Debris-flows leave many tell-tale signs along the channel that may remain identifiable for centuries and differ significantly from evidence left by normal 'water floods'. Careful inspection of the channel is therefore essential. Some of the signs to look for are:

- boulder levees (irregular lines or mounds of boulders) along relatively flat or less confined channel reaches (Figure 17.6A);
- logs or boulders deposited on either side of the channel or across the channel at elevations that cannot be reached by normal floods (Figure 17.6A);

Figure 17.3. The coalescing fans of Alberta and Harvey creeks: (A) Photo A4441: 75, taken in 1932 (National Air Photo Library), before any recent debris torrent. Alberta Creek cannot be detected under the trees; (B) Photo BC143: 80, taken in 1939 (British Columbia, Ministry of Environment, Surveys and Mapping Branch), showing a recent debris torrent running all the way to the sea; (C) Photo BC1682: 56, taken in 1954, showing the new crown closure of the forest vegetation, making it difficult to detect the former debris-flow; (D) Photo BC5175: 106, taken in 1966, showing development of the fan surface for suburban settlement and major transport routes with no protection against potential debris-flows

Figure 17.4. View of the track of the debris-flow of February 11, 1983, where it enters Howe Sound. It is evident that the roads and railway embankment interfered with the flow and affected its route

- debris jams of wood and boulders with some logs splintered, shattered, or broken (Figure 17.6A);
- boulders rolled against trees on either side of the channel (Figure 17.6B);
- pieces of wood buried under boulders (Figure 17.6B);
- scoured bedrock above the highest conceivable flood levels;
- scars on trees along the channel above maximum flood levels;
- differences in vegetation as discussed under 'office investigations'. If this is found, a tree corer can be used to determine ages of trees, hence the minimum age of past debris-flows.

The appearance of the deposition zone varies widely depending on the size of the stream, type and size of sediment and debris being supplied by the basin, and upon the relative importance of normal alluvial sedimentation versus debris-flow deposition. On active alluvial fans the occasional debris-flow deposit, which does not normally extend far down the fan, may be reworked by the stream and will then end up as a true alluvial deposit farther down the fan. In other words, failure to find debris deposits does not indicate security from debris-flow events.

Debris-flow deposits are characterized by boulder levees, irregular and often indistinct ridges or lines of boulders along either side of the runout zone. Debris-flow fronts often come to rest very abruptly, particularly if the debris is coarse, leaving a prominent debris lobe that may be several metres high. During transport, the coarsest materials in a debris-flow tend to ride to the top and to the front of the moving mass and this is likely to be preserved in the deposits (Miles and Kellerhals, 1981). Cuts into debris lobes tend to show boulders on top, supported by a matrix of finer materials with no bedding.

Buried logs, boulders, rolled against trees, bruised trees, and boulders too large to be carried by fluvial processes are other signs of debris-flow activity. If exposures are available in stream banks, road cuts, or other excavations, debris-flow deposits are distinguishable from alluvial deposits by their lack of stratification and imbrication, wide range of sizes, and, occasionally, inclu-

Figure 17.5. Vertical air photo of debris-flow fans in the Queen Charlotte Islands. (1) Lobate debris-flow deposit outlined by distinctive successional vegetation, less than 30 years old; (2) Recent debris-flow incompletely revegetated; (3) Debris tongue that stopped near the top of the fan. The surrounding hillside was logged about 40 years ago and has the same species in the successional forest as occur on the debris-flows. Nonetheless, the deposits remain distinctive because of the different successional stage

sion of organic debris (but it often has decayed and disappeared from ancient deposits). Debris-flow transport also produces less abrasion and, therefore, more angular material than stream transport.

It is naturally pointless to look for boulders if boulder sized material is not available for entrainment upstream, but the converse also is true: if boulders are available in the stream channel, they will almost certainly be moved. The mechanics of debris-flow motion exhibits little dependence upon grain size, so tractive force-based estimates of the maximum grain size that can be moved by a large flood are irrelevant.

Hazard Avoidance through Classification and Zoning

By definition, all active fan surfaces are subject to one or both of avulsion and debris-flow, but the frequency and severity of events vary by many orders of magnitude. Unlike the case of floodplains, where water levels and associated return periods provide an objective and widely accepted classification of the degree of exposure, there are no broadly accepted standards for classifying hazards on fans.

AVULSIONS

Although avulsions almost always occur during major floods, the probability laws governing avulsions differ greatly from those of floods. Dawdy (1979: *cf.* discussion in French, 1987) attempted to analyse the probabilistic aspect of flooding on alluvial fans but, in order to make the problem tractable, he had to adopt several restrictive and questionable assumptions. In particular, the assumption that 'the degree of flood hazard is approximately equal for all points that are radially equidistant from the fan apex' may have some

Figure 17.6. (A) Upstream view over the left bank boulder levee of a recent debris-flow: location near the fan apex; (B) A boulder at rest against a tree in the runout zone of a debris-flow, in the mid-fan area. Note also buried wood debris

merit on large fans with ill-defined, unstable stream channels and a dominance of sheet flooding in semiarid regions. However, on most British Columbia fans it is not realistic because of transient topographic constraints.

Contrary to most flood-generating processes, the process of channel avulsion can have a relatively long memory. When a stream finds itself a new path across a fan, it tends to follow a topographically low route and will therefore be initially quite stable in its new course. However, with time, aggradation will proceed along that route. Eventually it will become a topographic high and a new avulsion will become progressively more likely. These alternating periods of stability and instability may last from days to centuries. The consequence of the former situation is obvious in most hazard assessments, but the latter is not. On a fan with a history of infrequent avulsions, the stability of the existing channel location must be assessed carefully. At the same time one should look for steeper and more direct alternate routes across the fan. This is of interest both for design of remedial measures and for detailed hazard assessment on the fan. Any sites along alternate, more direct channel routes are particularly vulnerable. Figure 17.7 illustrates a fan on which the inevitability of eventual avulsion is avoided only through annual removal of gravel from the active channel along the entire length of the fan: the measure is necessary to protect the highway bridge at the toe.

The Thurber Consultants (1983a) report, 'Floodplain management on alluvial fans', recommends the classification of fan surfaces into 'very active', 'moderately active', and 'slightly active' subareas based on activity indicators such as the following:

- high sediment supply from the upstream catchment;
- degree to which the fan has displaced the main river to which the fan stream drains (an indicator of sediment supply rate);
- debris-flow potential;

Figure 17.7. Alluvial fan of Williscroft Creek at Kluane Lake, Yukon Territory. The channel on the right hand side of the fan must be cleared by bulldozer after major storms to maintain the stream in the course to the bridge which carries the Alaska Highway over it

- steep land and stream gradients;
- potential for flash floods;
- large number of abandoned or secondary channels;
- a high ratio of active channel area to total fan area;
- a high watertable on the fan surface as an indicator of a relatively high and therefore unstable channel position.

This classification obviously could be refined considerably further after detailed mapping of potential avulsion routes. A similar classification was developed in the detailed regional study discussed in the next section (see Table 17.2).

DEBRIS-FLOW

Debris-flow occurrence has an even more uncertain and complex probabilistic structure than the occurrence of fluvial avulsions because there are two types of debris-flow channels. Some, mostly bedrock controlled, channels are occasionally cleaned of accumulated debris by a debris-flow. In the period immediately following such an event the probability of further debris-flows may be very low due to lack of readily available debris in the channel.

Debris-flow channels incised into deep, unconsolidated materials can, however, act differently. They tend to be destabilized by the occurrence of a debris-flow and can then remain active for decades because there is an unlimited supply of debris right in the channel bed and banks. A long period of low flows eventually allows them to regain stability, usually by vegetation succession. As a consequence, debris-flows in this type of channel tend to occur in groups that may be separated by decades or centuries of deceptive tranquility.

Bovis and Dagg (1987) have made a detailed study of debris supply mechanisms to the Howe Sound creeks near Vancouver which illustrates the contrast in behaviour between the two types described, and also shows that the influence of geology on the calibre of debris introduced to the channel may be important. Very coarse, blocky debris cannot be moved by normal fluvial processes, and so it builds up over long periods to form a dangerous charge for debris-flows.

Besides having long-term memory, the stochastic process of debris-flow occurrence is strongly affected by landuse changes or forest fires in the upstream basin. O'Loughlin (1972) found, on the basis of data from the Coast Mountains in southwestern British Columbia including the Howe Sound sector, that clearcut logging and, particularly, logging road construction increased the occurrence of landslides. More recently, Rood (1984) found a 34 times increase in the frequency of landslides due to clearcut logging and the associated road building in an area of steep and generally unstable terrain in the Queen Charlotte Islands of British Columbia. Many of the landslides entered stream channels and triggered debris-flows.

Some channels are fed debris by an individual, very large failure in bedrock or unconsolidated material which may remain active for centuries or even millenia, then finally stabilize. Hence, very long-term episodic behaviour may be superimposed upon the patterns described above.

For fans exposed primarily to debris-flow hazards the approach adopted by Thurber Consultants (1983b; see also Hungr et al., 1987) is of interest. This second Thurber study was initiated after a debris-flow swept away a wooden bridge across M Creek, north of Vancouver, on a rainy autumn night. Nine lives were lost when several automobiles drove into the swollen stream and were swept into Howe Sound. M Creek is located along the same stretch of road as Alberta Creek, illustrated in Figures 17.3 and 17.4. The disaster focussed concern on the fact that there are many similar fans along Howe Sound, all crossed by a major highway and a railway line. Some are also developed extensively for housing. The Thurber Consultants study (1983b) was commissioned by the British Columbia provincial government in response to this concern. The objectives were to review and classify the flooding and debris-flow hazard along a 29 km stretch of the highway involving 26 stream crossings.

A two-step approach was adopted to debris-flow hazard classification. The first step involved a detailed examination of the channel and catchment upstream of each fan, combined with a review of all historical records and the identification and dating of debris-flow deposits on the fan. The outcome of this assessment was the assignment of each stream to one of the four debris-flow occurrence classes shown in Table 17.1 and an estimate of the 'design' debris-flow volume for each stream. This volume is defined as 'a reasonable upper limit of the quantity of material involved in future large debris torrents'. It is intended primarily for planning and bridge design purposes. A single design event may include several surges. The basis for this design volume is primarily an estimate of the total quantity of debris that is currently available for entrainment in the stream channel or, if a channel has recently been scoured by a debris-flow, it is the quantity of material that might be available in the foreseeable future. Only the normal debris-generating processes are considered: a large slope failure continuing down the channel as a debris-flow could result in greater volumes (Hungr et al., 1987). The debris volume estimates do not involve frequency considerations, but they are being applied to bridge and dyke designs for which the 200-year flood is the normal design criterion in British Columbia.

The second step in Thurber Consultants' hazard classification involved hazard mapping of the fan surface according to a classification like that of Table 17.2. The three T-type classifications codify debris-flow hazards while the three F-type classifications refer to hazards related to normal processes on alluvial fans; i.e. avulsions and flooding. All but two of the 26 study fans were hazard zoned according to debris-flow clas-

Table 17.1. Probability of occurrence of debris flows*

Category	Description	Category	Description
4	*Very high probability of occurrence:* indicates that debris flows of less than the design magnitude can occur frequently with high runoff conditions, and the design event should be expected within the next 10 years. It is applied to streams that have a recent history (<30 years) of more than one event involving greater than 500 m^3 or have physical characteristics that are comparable to such streams.		physical characteristics that fall well within the regionally observed threshold where debris flows are possible, although not in the range of category 4. Such streams generally have no recent history of debris flows, but may have experienced events of uncertain origin.
3	*High probability of occurrence:* indicates that debris flows of less than the design magnitude will occur less frequently than under category 4 but the design event should still be expected within the short term (<10 years). It is applied to streams that have a recent history of a single debris flow event and to streams that have no history of events but physical characteristics comparable to those of category 3 streams.	1	*Low probability of occurrence:* indicates a low potential for the design debris flow. It is applied to those streams whose physical characteristics place them at or close to the threshold where debris flows are possible. Although a significant debris flow is possible during the life of structures such as residences or bridges, it would require an unusually high (and thus infrequent) runoff condition.
2	*Moderately high probability of occurrence:* indicates that the design debris flow should be assumed to occur during the life of structures such as residences or permanent bridges. It is applied to those streams that have	0	*No risk:* indicates that there is virtually no potential for large debris flows, although small and local events may occur, and flows of varying magnitudes might develop in upper reaches and tributaries. It is applied to channel reaches whose physical characteristics fall well below the threshold where debris flows are possible.

*Modified from Thurber Consultants (1983b).

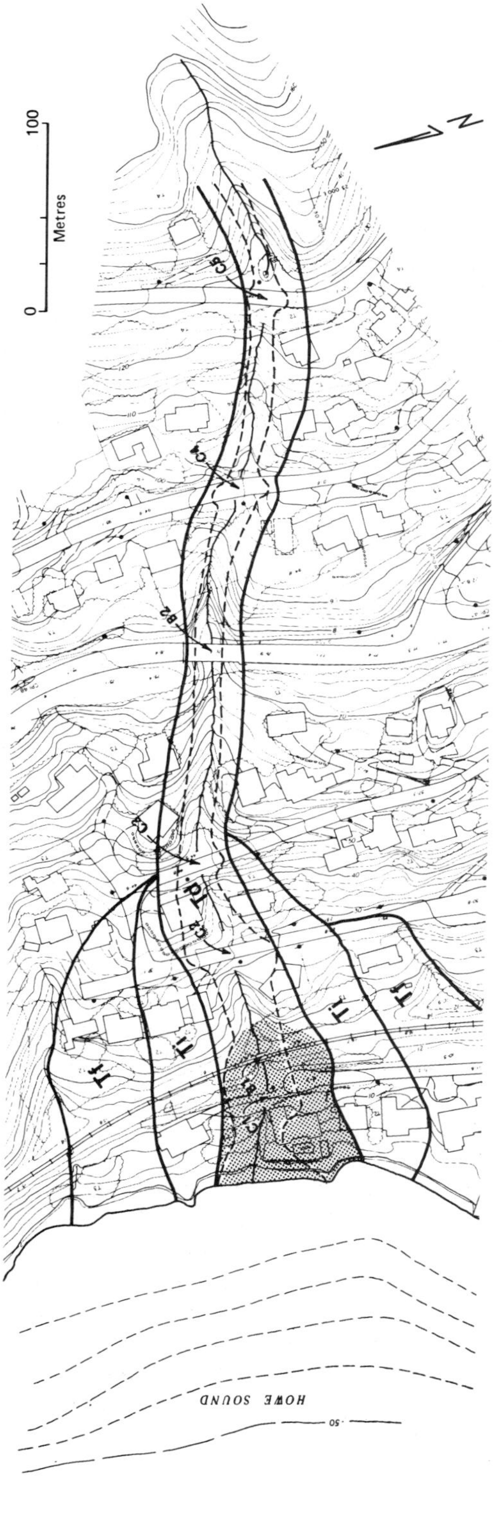

Figure 17.8. Hazard zones on the fan of Alberta Creek, as determined by Thurber Consultants (1983b).

Table 17.2. Fan hazard zone classification*

Category	Description
Td	*Direct impact zone of debris flows:* zone through which the debris surge may travel. The risk of impact damage is therefore high. Material transported through this zone could include boulders up to several metres in diameter and logs over 30 metres long.
Ti	*Indirect impact zone of debris flows:* zone through which later debris surges may be diverted and/or through which after-flow may travel. The risk of impact damage is lower. Material could include large rock and log debris, but is more likely to contain boulders of less than 1 m to fine-grained material and organic mulch.
Tf	*Flood zone due to debris flows:* zone that is exposed to flooding as a result of blockage of the main channel by debris-flow deposits. The risk of impact damage is low. Fine-grained material and mulch could be contained in the flood water.
	Area of potential deposition of debris: areas within which debris-flow materials could be deposited.
	Outline of area directly affected by known previous events: refers to historical events rather than to ones known only from morphology or stratigraphy, and of uncertain date.
Fh	*High flood hazard zone:* zone that has a high probability for flooding. In this zone, avulsions are possible.
Fm	*Moderate flood hazard zone:* zone that has a moderate to high probability of flooding. Avulsions could occur but are unlikely.
Fm	*Low flood hazard zone:* zone that has a moderate to low probability of flooding, but avulsions are unlikely.

*Modified from Thurber Consultants (1983b).

ses. The two largest fans were judged to be free from debris-flow hazard but exposed to avulsions and flooding, and were accordingly zoned into F-classes. By implication, the maximum size drainage basin which appears to be susceptible to debris-flow in the Howe Sound study area has an area of about 10 km². There probably are many fans that should be zoned into both T and F classes, but this did not appear to be necessary in this particular area. Figure 17.8 illustrates the hazard zoning for Alberta Creek, the stream illustrated in Figures 17.3 and 17.4.

Since completion of the zoning study, there has been only one debris-flow event—of 5000 m^3 in Sclufield Creek—on the 26 streams studied. The design volume listed in Thurber Consultants (1983b) for that fan is 6000 m^3. A map of the deposition is shown in Hungr et al. (1987). Approximately 10% of the actual deposition zone was found to lie outside the designated Ti and Td zones.

Protective Works

AVULSION HAZARD

In situations where it is not feasible to avoid the hazards associated with developments on fans, a wide array of protective devices have been used. Channel avulsions can be prevented by dyking the channel, by artificially incising it, or by increasing the transport capacity across the fan with measures such as channel straightening, channel lining, or channel enlargement. In British Columbia, where both land values and earth moving costs are relatively low, riprap dykes, channel enlargement, channel incision, and channel straightening, or a combination of these measures are the normal solutions. No smoothly lined, chute-like channels of the type so frequently seen in the Alps and in Japan have yet been built to deal with fluvial avulsion hazards.

Thurber Consultants (1983a) described five fans that were selected to exemplify the range of problems typically met in British Columbia. Besides hazard avoidance zoning, detailed mapping to identify potential avulsion routes, and the active measures mentioned above, the study also recommends large lot sizes combined with construction on slightly elevated pads to ensure that the fans do not become densely cluttered with buildings and that buildings will not be directly in the path of future avulsions.

In an increasing number of situations, dyking on partially developed fans has lead to rapid aggradation within the dykes. In such situations, debris interception basins are considered. However, there exist no straightforward criteria for determining the desirable basin capacity. Bed-load sediment transport equations are generally not valid in the steep upstream channels. The one purpose-designed equation extant (Smart, 1984) is based upon laboratory experiments with a plane bed. Irregular, step-pool mountain channels pass bed material at much lower rates. Indeed, even coarse debris transport rates may be supply limited. In this circumstance, regional storm magnitude-sediment yield correlations may provide a best basis for design, but such data are almost nonexistent.

DEBRIS-FLOW HAZARD

In situations where debris-flows are the main hazard, protective measures implemented in British Columbia range from simple tripwire warning devices to debris chutes, large diversion dykes, and debris storage basins. The long staircases of check dams and extensive slope drainage works that are so prevalent in Japan and in the Alps have no equivalent in British Columbia for reasons that are primarily economic but may be partially cultural. The centuries-old Asian and European traditions of landscape gardening appear to have been left behind by the emigrants to North America. Of course, sound forest land management is recognized as important to minimize erosional mobilization of debris in the first place. Hungr et al. (1987) provide a comprehensive, detailed description of all types of debris-flow defences implemented in British Columbia so far.

Debris chutes are effective in situations where debris deposition on the lower parts of the fan is acceptable. Figure 17.9A shows the debris chute of Alberta Creek (see Figures 17.3 and 17.4). Chutes are often a good solution in the case of bridge crossings on undeveloped fans since they confine the debris-flow to a fixed channel and help prevent bridge openings from becoming clogged (Figure 17.9B).

Large diversion dykes, as illustrated in Figure

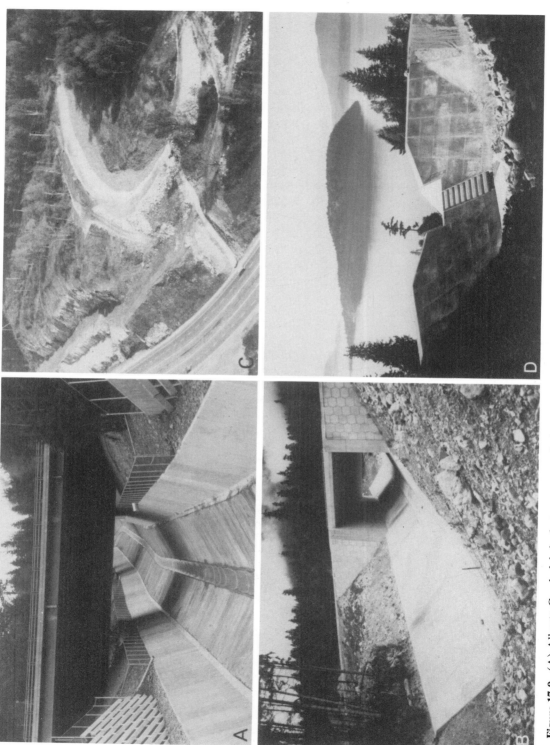

Figure 17.9. (A) Alberta Creek debris chute constructed after the February, 1983, debris-flow (Figure 17.4); (B) Bridge opening with debris chute; (C) Debris flow diversion dyke. A debris storage area of 12 500 m³ capacity is created by a 5 m high dyke. The dark material within the dyke is the deposit of a debris-flow. Photograph courtesy of Thurber Consultants, Ltd. and Dr. Oldrich Hungr; (D) Debris interception barrier on Charles Creek. See Hungr et al. (1987) for a detailed description of the design. Photograph courtesy of Mr. Michael Younie

17.9C, are often the preferred solution in British Columbia if space permits. They do, however, require careful maintenance. Any significant debris deposits behind the dykes must be removed if eventual overtopping is to be avoided. Nasmith and Mercer (1979) describe the design of a system of diversion dykes to protect the town of Port Alice. This industrial 'instant town' is located on a large debris-flow fan along one of Vancouver Island's west coast fjords. The debris-flow problem became apparent only some eight years after the town was built. Another case, on a much smaller scale, is described by Martin *et al.* (1984): in this instance, dykes and an interception basin were used to protect a prison built partly on a debris-flow fan.

Debris interception basins at or above the fan apex are generally the most expensive solution and have the most stringent maintenance requirements, but there is little interference with existing fan developments. Basin capacity may in this case be estimated on the basis of debris-flow volume criteria discussed above. Three of the 26 streams studied by Thurber Consultants (1983b) were eventually judged to pose sufficient hazards to existing residential developments and to the highway and railway crossings to warrant the construction of interception basins. Figure 17.9D illustrates the basin on Charles Creek. It is formed by a rockfill dam designed as a water retaining structure. The basin can hold the design debris volume and the overflow chute can accommodate a second design flow volume. The detailed design considerations are described by Hungr *et al.* (1987).

Conclusions

Fans can be deceptively hazardous sites. In British Columbia, much development has proceeded without a clear appreciation of these hazards and this has led to loss of life and much property damage, but it has also stimulated interest and encouraged systematic approaches to alleviating existing problems and to avoiding new ones.

In classifying hazards on fans, it is critically important to recognize any debris-flow potential since the problems associated with debris-flows differ greatly from the problems posed by normal alluvial fan processes. Fully objective fan hazard classifications that would correspond to floodplain delineation at a specific return period are not feasible because of the complex and ill-defined probabilistic structure of both 'avulsion occurrence' and 'debris-flow occurrence'. Even in the European Alps, records are generally too short and past climate and landuse too uncertain to define these processes. However, somewhat subjective classification schemes that incorporate observations and experience do offer practical alternatives to rigorous statistical analysis.

Protective measures implemented in British Columbia differ somewhat from those commonly employed in the European Alps and in Japan because of the relatively low land values, lower earth moving costs, and lower population densities typical of North America. Diversion dykes, debris chutes, and interception basins placed within the occupation zone on the alluvial fan are the major means selected. Upstream channel improvements are not attempted, and drainage basin land management is restricted to careful forest landuse.

References

Bovis, M. J. and Dagg, B. R. 1987. Mechanism of debris supply to steep channels along Howe Sound, southwest British Columbia. In Beschta, R. L., Blinn, T., Grant, G. E., Ice, G. G., and Swanson, F. J. (Eds), *Erosion and Sedimentation in the Pacific Rim. International Association of Hydrological Science Publication*, **165**, 191–200.

Clague, J. J., Evans, S. G., and Blown, I. G. 1985. A debris flow triggered by the breaching of a moraine-dammed lake, Klattasine Creek, British Columbia. *Canadian Journal of Earth Sciences*, **22**, 1492–1502.

Costa, J. E. 1984. Physical geomorphology of debris flows. In Costa, J. E. and Fleisher, P. J. (Eds), *Developments and Applications of Geomorphology*. Springer-Verlag, Berlin. 268–317.

Dawdy, D. R. 1979. Flood frequency estimates on alluvial fans. *American Society of Civil Engineers, Proceedings, Journal of the Hydraulics Division*, **105**(HY11), 1407–1413.

Eisbacher, G. H. and Clague, J. J. 1984. Destructive mass movements in high mountains: hazard and management. *Geological Survey of Canada Paper*, **84–16**, 230 pp.

Erosion Control Society of Japan 1985. *Erosion Control Works in Japan*. Japan Sabo Association, Tokyo. 67 pp.

French, R. H. 1987. *Hydraulic Processes on Alluvial Fans*. Amsterdam, Elsevier, 244 pp.

Hungr, O., Morgan, G. C., and Kellerhals, R. 1984. Quantitative analysis of debris torrent hazard for design of remedial measures. *Canadian Geotechnical Journal*, **21**, 663–677.

Hungr, O., Morgan, G. C., VanDine, D. F., and Lister, D. R. 1987. Debris flow defences in British Columbia. *Geological Society of America, Reviews in Engineering Geology*, **7**, 201–222.

Jackson, L. E., jr., Kotaschuk, R. A., and MacDonald, G. M. 1987. Identification of debris flow hazard on alluvial fans in the Canadian Rocky Mountains. *Geological Society of America, Reviews in Engineering Geology*, **7**, 115–124.

Martin, D. C., Piteau, D. R., Pearce, R. A., and Hawley, P. M. 1984. Remedial measures for debris flows at the Agassiz Mountain Institution, British Columbia. *Canadian Geotechnical Journal*, **21**, 505–517.

Miles, M. J. and Kellerhals, R. 1981. Some engineering aspects of debris torrents. *Canadian Society for Civil Engineering, 5th Canadian Hydrotechnical Conference, Fredericton, New Brunswick. Proceedings*, 395–420.

Nasmith, H. W. and Mercer, A. G. 1979. Design of dykes to protect against debris flows at Port Alice, British Columbia. *Canadian Geotechnical Journal*, **16**, 748–757.

O'Loughlin, C. L. 1972. A preliminary study of landslides in the Coast Mountains of southwestern British Columbia. In Slaymaker, O. and McPherson, H. J. (Eds), *Mountain Geomorphology*. Tantalus Research, Vancouver. *British Columbia Geographical Series*, **14**, 101–111.

Rood, K. M. 1984. An aerial photograph inventory of the frequency and yield of mass wasting on the Queen Charlotte Islands, British Columbia. *British Columbia Ministry of Forests, Land Management Report*, **34**, 55 pp.

Ryder, J. M. 1971. The stratigraphy and morphology of paraglacial alluvial fans in south-central British Columbia. *Canadian Journal of Earth Sciences*, **8**, 279–298.

Smart, G. M. 1984. Sediment transport formula for steep channels. *Journal of Hydraulic Engineering*, **110**, 267–276.

Smith, R. B., Commandeur, P. R., and Ryan, M. W. 1986. Soil, vegetation and forest growth on landslides and surrounding logged and old-growth areas on the Queen Charlotte Islands. *British Columbia Ministry of Forests, Land Management Report*, **41**, 95 pp.

Swanston, D. N. and Swanson, F. J. 1976. Timber harvesting, mass erosion and steepland forest geomorphology in the Pacific Northwest. In Coates, D. R. (Ed.), *Geomorphology and Engineering*. Dowden, Hutchinson and Ross, Stroudsburg, Pennsylvania. 199–221.

Thurber Consultants 1983a. *Floodplain Management on Alluvial Fans*. Report to British Columbia Ministry of Environment, Water Management Branch. 39 pp + appendices.

Thurber Consultants 1983b. *Debris Torrent and Flooding Hazards, Highway 99, Howe Sound*. Report to British Columbia Ministry of Transportation and Highways. 25 pp + appendices.

VanDine, D. F. 1985. Debris flows and debris torrents in the southern Canadian Cordillera. *Canadian Geotechnical Journal*, **22**, 44–68.

CHAPTER 18

Artificial Recharge of Aquifers in Alluvial Fans in Mexico

Napoleon Otero San Vicente
Torreon, Coahuila

Abstract

Rapid population growth and development in Mexico are straining water supplies for domestic, industrial, and agricultural purposes in arid and semiarid regions as groundwater is pumped in great quantities. The purposes of this paper are to review briefly the methods of artificial recharge of groundwater and to propose the first experimental massive artificial recharge project in Mexico's history.

The two main methods of recharge are well injection and surface spreading. The choice of method or combination depends upon the characteristics of the particular site. The author has investigated 42 small dams throughout Mexico that have strong seepage problems and so are accidentally recharging the groundwater.

The Laguna Seca zone, near Torreon, Coahuila, was chosen for an experimental recharge project. The site consists of an extensive alluvial fan/plain complex bordering the Sierra de Jimulco. Climate maps indicate that the probable range of precipitation in the drainage basin is 250 to 350 mm yr^{-1}. Using a value of 300 mm yr^{-1} it was estimated that, by constructing a series of small earthen dams, infiltration ponds, and trenches, it would be possible to recharge the aquifers by approximately 20×10^6 m^3 yr^{-1}. The estimated cost of this project is $US2.5–2.7 million.

Introduction

Mexico has approximately 81 million people, many of whom are poorly fed at the present time. The rate of population increase is approximately 2.5 to 3% per year, one of the highest in the world. About 65% of the population is under 25 years of age. The accelerating growth of Mexico's population, combined with an improving standard of living throughout Mexico, is greatly increasing the demand for water resources. This demand certainly will continue to grow. Even in the new cities, the development of additional water supplies is an urgent task. Where water for irrigation or even domestic uses comes from wells, there has been not only a serious decline in the watertable, but also significant problems of ground subsidence and decline in water quality.

At the same time, the search for new and more abundant water supplies is becoming more and more complex. In order to obtain sufficient water supplies for drinking and irrigation in the near future in arid and semiarid zones or for the cities, new ideas and techniques must be devised to complement the old ones.

The principal purposes of this paper are to

Alluvial Fans: A Field Approach edited by A. H. Rachocki and M. Church
Copyright © 1990 John Wiley & Sons Ltd.

review the methods of artificial recharge of groundwater applicable in arid and semiarid regions, and then to apply these methods to Laguna Seca, Coahuila, Mexico (Figure 18.2), which is herein proposed as the site for the first massive recharge experiment in the history of Mexico.

The site reported in this study is on a complex of alluvial fans, playas, and plain. At the present time in Mexico there exists a multitude of small dams which were constructed on similar landforms for the sole purpose of the surface storage of runoff for land irrigation, livestock watering, and domestic uses. However, because of unanticipated geomorphological and hydrogeological considerations, water which the dam was designed to store rapidly infiltrated into the subsurface, leaving the reservoirs dry. The subsurface behaviour of this water is unknown at present. By analysis of a similar situation it will be demonstrated that it is possible to recharge the aquifers at Laguna Seca with more than 20×10^6 m^3 water per year.

Artificial Recharge in Arid and Semiarid Regions

Artificial recharge of aquifers has been done since the beginning of the twentieth century. Surface runoff in arid and semiarid climates is often lost to evaporation or outflow to the sea (Böckh, 1977). Where the runoff is erratic, surface retention by dams seldom is justified. Under such circumstances, artificial groundwater recharge can be the ideal means of water conservation. Even in cases where dam construction is feasible, measures of artificial recharge and underground storage could be more economical because of the decreased water loss from evapotranspiration.

The arid and semiarid regions of the world are highly dependent on groundwater supplies. Alluvial fan deposits are part of the groundwater reservoirs in basins filled with alluvium, and much of the recharge of the groundwater reservoirs is through the permeable fan deposits that fringe the basins (Figure 18.1). Developed examples are found in the Los Angeles basin in southern California and the San Joaquin Valley of central California. Tucson, Arizona is one of the largest cities in the world totally dependent on underground water supplies. Much of the water used in Tucson is pumped from an alluvial fan deposit of late Cenozoic age.

Wherever heavy pumping of groundwater resources is undertaken for water supply, serious decline in the groundwater table has followed. Other problems follow. In Mexico, ground subsidence at rates of up to tens of centimetres a year

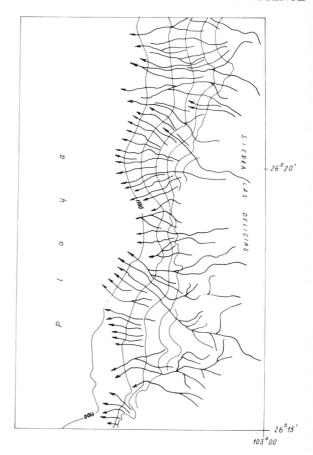

Figure 18.1. Alluvial fans at the mountain front, Valle de Acatita, Francisco I, Madero, Coahuila, Mexico. The radiating, disappearing streams indicate the characteristic situation whereby runoff from the adjacent mountains percolates through the fans to recharge groundwater in the adjacent basin. From Map No. G.13 B.76, 'Sierra de los Remedios' (Direccion Nacional de Estudios del Territorio Nacional, Secretaria de Programacion y Presupuestos)

has been experienced. For example, at Celaya, in Guanajuato state, subsidence of more than a metre in only five years destroyed part of the 16th century Iglesia de San Francisco. Serious deterioration in the quality of the pumped water may also be experienced. At Tlahualilo, in Durango, arsenic contamination of groundwater supplies has produced chronic health problems. Artificial recharge appears to be the ultimate solution to all of these problems.

For water to collect underground in exploitable quantities, there must be an intake area or 'recharge area', that is, a unit of land surface into which precipitation or runoff can infiltrate to charge the storage space in the soil and rock. The subsurface material must have sufficient porosity to contain an adequate volume of water, and it must be sufficiently permeable to allow water to flow to a well hole to replace water pumped out, and at a rate approximately equal to the rate of extraction (*cf.* Todd, 1959). Given a source of recharge, the occurrence of good groundwater supplies depends, then, upon geology. More particularly, the volume and texture of aquifers and their geomorphological history control the amounts of water that can be stored and released to wells and streams in amounts and at rates that are beneficial for human purposes. Stratigraphic relationships between aquifers and aquicludes also affect the pattern of movement and entrapment of groundwater. Most alluvial fans meet the permeability and porosity conditions.

Although Mexico has a long history of civilization, the first systematic investigation of artificial recharge of aquifers was begun only in 1977 when the writer, through the Instituto Nacional de Investigaciones Agricolas de Mexico in the Unidad de Manejo y Uso del Suelo; Agua-Planta, initiated a programme of studies of techniques of artificial massive recharge of aquifers for irrigation in different parts of Mexico (Figure 18.2). Prior to these studies, from 1971 to 1977, the writer worked for the Direccion de Aprovechamientos Hidraulicos of the Secretaria de Agricultura y Recursos Hidraulicos conducting hydrogeological studies and inspecting small dams throughout Mexico. These small dams, which were constructed to store excess runoff, have very strong seepage problems. Indeed, some of the reservoirs will not hold water at all.

Two examples will be mentioned here to illustrate the rapid infiltration that occurs on alluvium, because it is this very property that makes artificial recharge possible by the method proposed in this paper. The locations of these two dams are shown in Figure 18.2. The settings of these and several other small dams with strong seepage problems are given in Table 18.1.

LeCongueria Dam in Chichihualco, Guerrero State, was built on an alluvial fan in transition to an alluvial plain. Runoff into the dam may be ponded but persists only temporarily before it all infiltrates into the highly permeable alluvium. Compounding the problem here is the fact that the reservoir is now filled with sediment. Espanita Dasm in San Jose Iturbide, Guanajuato State, was constructed on an alluvial plain. Like the LaCongueria Dam, it rarely has any water in its reservoir, since it is built on highly permeable sediments. In both cases the dams have resulted in artificial recharge of the groundwater, but in neither case has the local landowner drilled wells to utilize the additional water. Indeed, the landowners have complained to the writer that before the dams were constructed they had water to irrigate their crops, but after the dams were built they did not have enough water. The writer has inspected both dams regularly since 1976: by 1985, the reservoirs were nearly full with sediment, but seepage persisted at both sites.

There are several factors which must be considered in choosing sites for groundwater recharge. They include the following:
- adequate reservoir storage capacity, determined by geomorphology and stratigraphy;
- suitable textural characteristics in the aquifer, controlling porosity and permeability;
- adequate precipitation;
- flood runoff volume and frequency;
- water quality;
- sediment load of the runoff water.

There are two general techniques for achieving artificial recharge: surface spreading and recharge wells. The water commonly used for artificial recharging includes natural streamflow, high flows diverted from a nearby water course, or

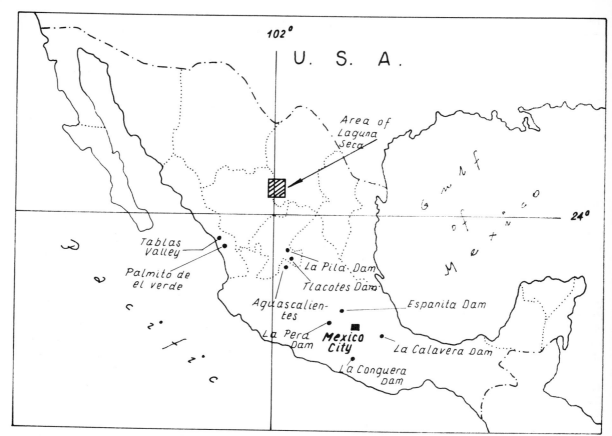

Figure 18.2. Map of Mexico, showing the location of preliminary dam seepage and recharge studies, and the Laguna Seca region

industrial cooling waters. In some places, storm drainage water has been collected and allowed to infiltrate through basins or wells. In many parts of the world during the past several years, sewage effluent, either treated or untreated, has served as a major source of recharge water (Asano, 1985).

Injection or recharge wells might consist of shallow, relatively large pits or shafts, or small diameter, screened wells. Such wells permit direct access from the surface water source to the groundwater reservoir (Pettyjohn, 1974, 1985: see Figure 18.3A). Urban (1986) has recharged a bedrock aquifer at a rate 2.5 times greater than the daily evaporation rate, introducing one million gallons of water during the first week.

Water spreading is defined as the release of water over the ground surface for the purpose of increasing the quantity of water infiltrating into the ground and percolating to the watertable (Todd, 1959: p. 252) (Figure 18.3B). It can be accomplished by flooding or spraying a tract of land, by impounding water in shallow excavations, or by damming and widening a natural stream channel. Essentially, water spreading supplements natural recharge by increasing the quantity of water available for recharge or extending the time that water is in contact with the soil. The effectiveness of water spreading as a means of artificial recharge is largely dependent upon the permeability of the materials through

Table 18.1. Principal characteristics of inspected sites of small dams suitable for artificial recharge of aquifers

Name of small dam and place	Drainage pattern	Longitudinal profile and gradient of streams	Transverse valley profile/ Geomorphology of dam site	Sediments in reservoirs
LaCongueria, Leonardo Bravo, Chichihualco, State of Guerrero	Rectangular, on jointed rock	Rugged and steep, with rapids and falls, and dissolution caves at the limestones; 100 m km^{-1}	V-shape; alluvial fan	Cobbles and gravel; good permeability
Espanita, San Jose Iturbide, State of Guanajuato	Dendritic, on more or less flat-lying strata	High gradient; lateral sides ungraded for stream erosion: 50 m km^{-1}	U-shape; small alluvial plain	Cobbles, gravels, sand; volcanic and rhyolitic; very good permeability
LaPera, Acambar, State of Mexico	Radial, on a volcanic cone	Abrupt, with rapids and sharp curves; 80 m km^{-1}	V-shape; alluvial cone	Cobbles, basaltic fragments, volcanic ashes and scorias; very good permeability
Boquilla de las Perlas, Calera, State of Zacatecas	Dendritic, on Jimulco scarp	Long slope with high gradient; 50 m km^{-1}	U-shape; alluvial fan	Sand and gravel, some silt; good permeability
LaPila; Calera, State of Zacatecas	Dendritic, in flat-lying strata	High gradient; lateral sides diverge downstream; 25 m km^{-1}	U-shape; small alluvial plain	Sands and gravels; excellent permeability
Tlacotes; Ojo Caliente, State of Zacatecas	Trellis, in folded sedimentary rocks in an elongated landform	High gradient; lateral sides ungraded for stream erosion; 60 m km^{-1}	V-shape; narrow creek	Sand and gravel; excellent permeability
LaCalabera; Izucar de Metamoros, State of Puebla	Parallel, on a moderate to steep slope	High gradient; 70 m km^{-1}	V-shape; small alluvial fan	Sand and gravel, some silt; good to excellent permeability

which the water must pass in its downward movement to the watertable.

From the quantitative standpoint, however, the size of the spreading area, length of time that water is in contact with the soil, and the amount of suspended material in the water and its chemical quality are also important factors. The least permeable layer through which the water must pass will generally control the rate of downward movement, and the dimensions of the spreading basins and period of inundation will determine the volume of water that reaches the watertable (Pettyjohn, 1985).

Laguna Seca

Laguna Seca is located approximately 45 km southeast of Torreon, in the state of Coahuila, at 25°14′N and 103°12′W (Figure 18.2). It consists of alluvial fans, playas, and a desert plain which are built out to the northeast from the Sierra de Jimulco (Figure 18.4). The dynamic groundwater level (the average level with pumping) here has declined from 27 m below the surface in 1950 to 80 m in 1976, and to more than 100 m in 1987 (Figure 18.5). The static level (without pumping) dropped from about 45 m below the surface in

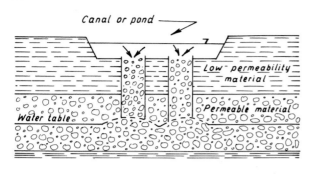

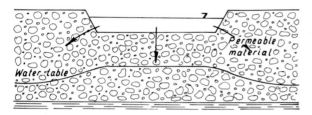

Figure 18.3. Artificial groundwater recharge methods (after Pettyjohn, 1985): (A) Injection wells; (B) Surface spreading into shallow trenches or basins

1963 to 100 m in 1987. The average yield decreased from nearly 40 to less than 30 l s^{-1} over approximately the same period. Because the Laguna Seca zone has the typical water problems of rural arid and semiarid regions of Mexico, it was chosen as the location of the first comprehensive experimental massive recharge project.

The annual precipitation in the Laguna Seca region is between 250 mm and 350 mm, which comes in only 15 to 25 rainfall events. The temperature ranges from a minimum near $-4°C$ during some days in the winter season in January and February to more than 40°C maxima in June, July, and August. The writer has recognized, on the basis of size, abundance, and type of vegetation, that some microclimates with lower temperatures and higher precipitation exist within the field area.

The lands of the Laguna Seca region are worked by a few landowners and several cooperative groups. The total area under cultivation is estimated to be between 3500 and 3600 ha. The cultivated area is irrigated with water drawn from 100 to 120 deep wells in the alluvium in the valley. The most important cash crop is cotton. Production is six to seven tonnes per hectare (communication from agronomy staff at the Banco Nacional de Credito Rural del Norte). In addition to cotton, the main crops include alfalfa, wheat, beans, tomatoes, potatoes, sorghum, grapes, and lemons. Livestock such as cattle (beef and dairy), pigs, sheep, goats, and donkeys are raised in the region. Poultry and egg production are also important to the local economy.

Mayer (1965) discussed the general geology of the region. Topographic and geological maps have been published by the Direccion de Estudios del Territorio Nacional (DETENAL), Secretaria de Programacion y Presupuestos (1974). The principal rocks which can be recognized in the Laguna Seca region are Palaeozoic and Mesozoic limestones, sandstones, shales, and mudstones, which are marine sediments deposited in a geosyncline (Kellum, 1932; Mayer, 1965). These sediments make up most of the Sierra de Jimulco (Figure 18.6A).

For the purpose of this work, the most important deposits were formed during the late Pliocene, Pleistocene, and Holocene Epochs. They are the alluvial fans, playas, and plains which are built out to the northeast from the Sierra de Jimulco. Along portions of the mountain front, recent uplift has caused streams crossing the fans to become entrenched and this has resulted in a downfan shifting of the area of deposition (*cf.* Bull, 1977). Materials which make up the fans display the size gradation which is typical of alluvial fans—very coarse at the head and fine at the toe. Proximal deposits include large, semirounded boulders embedded in poorly-sorted gravels (Figures 18.6B, C). Silts increase in abundance toward the toes of the fans, where occasional clayey mud also can be found. The arroyo channels are sometimes anastomosed or braided, and run radially away from the fanhead trenches. The writer has observed the accumulation of gravel and some boulders in the channels during heavy rains.

The recent arroyos permit one to study the stratigraphy of the fans. They are the result of buildup by both fluvial and debris-flow deposi-

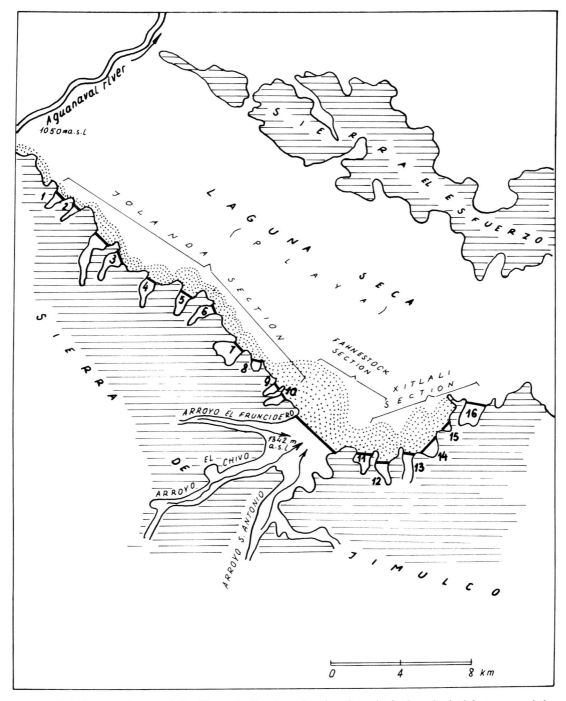

Figure 18.4. Laguna Seca and the Sierra de Jimulco, showing the principal geological features and the major structures of the proposed project

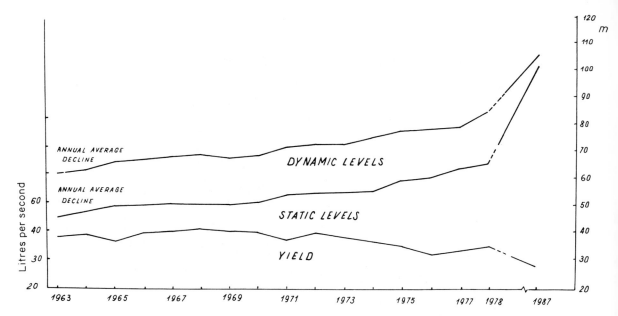

Figure 18.5. Static and dynamic surface levels and average yield from a series of pilot wells in Region Lagunera, including Laguna Seca (data from Secretaria de Agricultura y Recursos Hidraulicos, Distrito No. 17, Region Lagunera)

tion. The different thicknesses of the beds indicates that varied sedimentation events contributed to fan construction. One can also see the backfilled channels of former arroyos, so there have been episodes of entrenchment and filling. These are most common near the head of the fan. The alluvial deposits are fairly well sorted and in some instances cemented by caliche. The debris-flows are poorly sorted and are also cemented in places.

Artificial Recharge at Laguna Seca

The total water available for recharge has been estimated using data available from the Mapa Geohidrologico de Mexico (Instituto de Geologia de Mexico, 1960) and precipitation maps from DETENAL (1974a). Using the basic water balance equation and some reasonable approximations, it is possible to derive some value for the amount of water available for recharge: i.e.,

$$P = I + ET + R$$

where P = precipitation, = 300 mm annually;
I = infiltration, assumed to be $0.18P$;
ET = evapotranspiration, assumed to be $0.50P$;
R = runoff, assumed to be $0.32P$;

the data and fractions being derived from information given on the geohydrological map. The

Figure 18.6. Characteristics of alluvial fans at Laguna Seca. (A) 'Fahnestock' section in the El Alto des Palomillo. Here, Arroyos El Chivo, El Frundidero, and San Antonio converge to form a large alluvial fan with estimated 18 to 20 million cubic metres recharge per annum. Behind, folded and faulted limestones of the Sierra de Jimulco, with peaks at 3000 m; (B) Yolanda section: view toward the fan apex, showing the bouldery gravels which would form the recharge zone; (C) cobble gravels at the Fahnestock site in Laguna Seca, with a small clay dam ('Bordo') intended for flood control. This illustrates the style of construction intended for the recharge dams. Photos by Dr. Luis Maeda Villalobos, Instituto Cientifico de la Region Lagunera, A.C

Table 18.2. Principal characteristics of individual drainage basins of proposed dams and the estimated annual recharge*

Drainage basin	Drainage area (km^2)	Gradient (% slope)	Sediment on floor of proposed reservoir	Dam length (m)	Approx. recharge volume (m^3 yr^{-1})
Yolanda section					
1	2.6	20	Conglomerate, sand and gravel, coarse debris; good permeability	300	263 000
2	1.3	30	Conglomerate, sand and gravel, coarse debris; good permeability	250	125 000
3	2.3	30	Conglomerate, sand and gravel; good permeability	300	233 000
4	6.4	Very steep no datum	Conglomerate, sand and gravel; good permeability	500	642 000
5	6.0	Very steep no datum	Conglomerate, sand and gravel, silt; good permeablity	700	598 000
6	5.9	Very steep no datum	Conglomerate, gravel, sand and silt, good permeability	650	585 000
7	2.8	Very steep no datum	Conglomerate, sand and gravel; good permeability	300	275 000
8	1.1	Very steep no datum	Conglomerate, sand; good permeability	500	105 000
Xitlali section					
9	2.0	20	Conglomerate, sand, coarse debris; good permeability	300	200 000
10	1.3	30	Conglomerate, sand, coarse debris; good permeability	250	125 000
11	2.0	30	Conglomerate, sand; good permeability	300	200 000
12	6.0	Very steep no datum	Conglomerate, sand and gravel, good permeability	500	600 000
13	5.0	Very steep no datum	Conglomerate, gravel, sand, silt; good permeability	700	500 000
14	5.0	Very steep no datum	Conglomerate, gravel, sand, silt; good permeability	400	500 000
15	2.0	Very steep no datum	Conglomerate, sand and gravel; good permeability	300	200 000
16	1.0	Very steep no datum	Conglomerate, sand; good permeability	250	100 000

*Location of each basin is shown in Figure 18.4. See text for basis of recharge figures. These data are presented as preliminary figures: prior to construction, it will be necessary to make detailed surveys of the individual sites.

Table 18.3. Characteristics of the proposed dams*

Dam	Crest width (m)	Maximum height (m)	Slope of dam surface upstream	Slope of dam surface downstream	Dam cross-sectional area (m^2)	Foundation area (m^2 m^{-1})	Total length (m)	Expected recharge (m^3 yr^{-1})	Estimated cost ($US: 1986)
Fahnestock	2.0	7.0	2:1	15:1	65	21	2500	16×10^6	1.2×10^6
Yolanda	2.0	3.0	3:1	2:1	26	19	3500	2.8×10^6	0.8×10^6
Xitlali	2.5	3.0	3:1	2:1	27	8	3000	2.4×10^6	0.5×10^6

*See Figure 18.7 for terminology.

total area of El Fruncidero, El Chivo, and San Antonio watersheds is 106 km^2. Therefore:

$P = 106$ km$^2 \times 300$ mm yr$^{-1} = 32 \times 10^6$ m^3 yr^{-1}

Similarly,

$ET = 16 \times 10^6$ m^3 yr^{-1}
$I = 6 \times 10^6$ m^3 yr^{-1}
$R = 10 \times 10^6$ m^3 yr^{-1}

The total volume estimated to be available for recharge, including the present infiltration portion is 16×10^6 m^3 yr^{-1}. This amount will, of course, vary from year to year.

The writer considers this recharge volume to be a minimum average, hence a conservative estimate of the potentially available water, because vegetation observed growing in the mountains is known to require more than the estimated 300 mm annual precipitation suggested on the DETENAL isohyet map. Therefore, it is inferred that the total precipitation in the basin may be greater than the 32×10^6 m^3 estimated. Similar calculations have been made for all of the drainage basins in the Laguna Seca zone and recharge estimates are given in Table 18.2.

Because of the geomorphology and permeability characteristics of the alluvial fans, it is proposed to capture all of this water by constructing recharge dams at locations shown in Figure 18.4. In addition, it is suggested that a variety of recharge ditches, trenches, ponds, and recharging wells be excavated. In order to prevent sediment loading of the recharge dams and trenches, it is proposed to build inexpensive gabion sediment traps upstream of the dams. Figure 18.7 is a sketch showing typical structures for the gabions and dams (see also Figure 18.6C). Different structures must be designed to meet the specific needs of each site. The proposed project consists of three major sections: 'Fahnestock' (Figure 18.6A), Yolanda (Figure 18.6B) and Xitlali. The Fahnestock section is to consist of a single structure about 2500 m long. The other two sections will each consist of several shorter segments. Some features of the major sections are given in Table 18.3

The major construction costs were based on prices obtained from the Direccion de Construccion de la Secretaria de Agricultura y Recursos Hidraulicos, converted to dollars. The figures given in Table 18.4 are only approximate because the engineers have had no experience in con-

Table 18.4. Estimated construction costs

Activity	Unit cost* ($US: 1986)
Excavation and earth movements for small dams: earth roads	1.22/m^3
Construction of earth dams: movement of talus and spoil, construction of terraces	1.15/m^3
Compaction of dam: compacted terrace formation	0.54/m^3
Purchase and transportation of riprap materials: concept A-7-A materials for terraplain construction	0.58/m^3 km
Installation of riprap protective rock cover on face of dam: concept A-5-7 compaction and terraplain formation	1.11/m^3

*Based on unit prices obtained from civil engineers at the Department of Construction of the Water Resources and Agriculture Ministry. The construction phase is given first followed by the engineers' best approximation of a process closely resembling that phase.

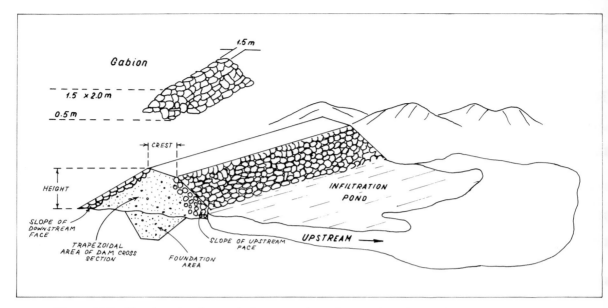

Figure 18.7. Sketches showing typical construction of an earthen dam and gabion for the project

structing recharge works and, therefore, have provided estimates based upon what they thought was most nearly like the required work. Based on these estimates, the construction costs for each of the three sections of the proposed recharge works have been calculated separately and are contained in Table 18.3. The total cost of building all three sections is estimated to be 2.5 to 2.7 million United States (1986) dollars and they will recharge approximately 21×10^6 m^3 of water annually.

Summary

Geohydrological studies indicate that all the areas in Mexico using groundwater have shown significant declines in their piezometric surfaces. This emphasizes the great importance of applying the techniques of massive artificial recharge of groundwater in Mexico as soon as possible. The Region Lagunera, within which Laguna Seca lies, is no exception.

The total irrigated land in the Region Lagunera is approximately 1.5×10^5 ha, which produces crops with an average annual value of more than \$US400 million per year. The land irrigated with water from 2400 deep wells is about 0.6×10^5 ha (the remaining 0.9×10^5 ha being fed from two dams). All of the deep wells show a strong decline in their piezometric surface. All of them urgently need to be recharged.

At the present time in the Laguna Seca region approximately 20×10^6 m^3 of potentially useful water goes unused for a variety of reasons each year. Much of this unused water could be captured for domestic or agricultural use if the project were constructed.

The main technique of artificial recharge in the Laguna Seca region will be surface water spreading behind small dams which the writer proposes be constructed for the purpose. These dams would retain the runoff that currently flows through the creeks and arroyos and is lost largely to evaporation from the surface of the playa. Several wells would be installed downslope from the dams to permit monitoring of changes of water level to determine the effectiveness of the recharge experiment.

The infiltration works proposed here have a cost lower than 15 cents per cubic metre of water (U.S. currency, 1986) in the first year. If one considers that the project will remain useful for about 10 years, each cubic metre will cost only about 1.5 cents.

References

Asano, T. 1985. *Artificial Recharge of Groundwater*. Stoneham, MA., Butterworth. 767 pp.

Böckh, E. 1977. *Note on potential measures on artificial groundwater discharge under arid–semiarid conditions*. Hannover, Federal Republic of Germany, Bundesanstalt für Gewisschaften und Rohstoffe. 4 pp.

Bull, W. B. 1977. The alluvial fan environment. *Progress in Physical Geography*, **1**, 221–260.

DETENAL (Direccion de Estudios del Territorio Nacional) 1974a. *Mapa de Climas de Durango–Coahuila con Isohyetas*. Balderas 74, Mexico D. F.

DETENAL 1974b. *Hojas Hidalgo y Jimulco, Topograficas y Geologicas*. Balderas 74, Mexico D. F.

Heath, C. R. 1981. *Introduction to Groundwater Hydrology*. Dublin, Ohio, Water Well Journal Publications. 285 pp.

Instituto Geologica de Mexico 1960. *Mapa Geohidrologico de Mexico*. Mexico D. F.

Kellum, L. B. 1932. Reconnaissance studies in the Sierra de Jimulco, Mexico. *Geological Society of America Bulletin*, **43**, 541–564.

Mayer, P. R. F. 1965. *Resumen de la Geologia de la Hoja Viesca*. Publicado por el Instituto de Geologia de la UNAM. Un Mapa. Ciudad Universitaria, Mexico 20 D. F. 27 pp.

Pettyjohn, A. W. 1974. Artificial recharge: a potential solution for many water problems. *Water Well Journal*, **28**(9), 1–17.

Pettyjohn, A. W. 1985. *Introduction to Artificial Groundwater Recharge*. Stillwater, Oklahoma State University Press. 44 pp.

Todd, D. K. 1959. *Ground Water Hydrology*. New York, Wiley. 336 pp.

Urban, L. 1986. Artificial recharge in Ogallala Formation. *Cross Section Bulletin (Lubbock, Texas)*, **32**(9), 3 pp.

CHAPTER 19

Geomorphic Appraisals for Development on Two Steep, Active Alluvial Fans, Mt Cook, New Zealand

I. E. Whitehouse
Department of Scientific and Industrial Research, Christchurch

and

M. J. McSaveney
University of Canterbury, Christchurch

Abstract

Geomorphic hazards put developments on the active alluvial fans of Black Birch and Twin Streams, near Mt Cook, central Southern Alps, at significant risk. Aggradation at the fanhead and channel avulsion from the subsequent overtopping of the artificially raised banks are the major hazards to development at Black Birch Stream. The concern at Twin Stream is how often, and in what quantities, floodwater might cross the developed part of the fan. Three major floods with aggradation have occurred at both sites in the last 80 years. Only the last of these is well recorded. Magnitudes and frequencies of unmeasured events can be assessed crudely from historical photographs and anecdotes, interviews with witnesses, and personal observations. Following the appraisal for Black Birch fan, the primary artificial levee was enlarged at the fanhead to increase the level of protection to safely contain at least the 50-year flood and aggradation. In addition, a secondary artificial levee was constructed down the mid-line of the fan to reduce the consequences of primary levee failure. At Twin Stream fan, the risk to the development area is most simply reduced to an acceptable level by blocking the supply of floodwater to distributary channels on the northern side of the fan. By upgrading existing artificial levees downstream of a highway bridge at the apex of the fan it should be practicable to safely contain the 100-year flood discharge. The solutions to manage, mitigate, or avoid the hazards and risks on these fans are conservative to allow for the uncertainties in the information, and for the need to test the adequacy of the solutions by trial-and-error.

Introduction

Alluvial fans are common in the Southern Alps of New Zealand. Often they provide comparatively flat, well-drained sites with splendid views, and are favoured for residential and tourist development, but invariably they lack hydrological data, and have a brief observational record. For development where the fans are considered active, there is increased recognition of the need for appraisals of geomorphic hazards and risks as part of the planning process or for designing protective works. This paper outlines such appraisals for two fans near Mt Cook in the central Southern Alps: Black Birch Stream fan and Twin Stream fan (Figure 19.1). In both studies, the lack of hydrological measurements and objective observational record is partially overcome through use of

Alluvial Fans: A Field Approach edited by A. H. Rachocki and M. Church
Copyright © 1990 John Wiley & Sons Ltd.

geomorphic evidence of past activity, historical photographs and anecdotes, and interviews of witnesses to past events.

Impetus for the appraisal for Black Birch fan came in 1985 when an application for permission to build a Youth Hostel led to a review of the adequacy of flood protection on the fan (Williman et al., 1986). From the lessons of a large storm in 1979, the review sought to objectively appraise safety on the fan, the adequacy of existing protection, the consequences of a supra-design event, and how the placement and design of buildings or prearranged evacuation programmes could reduce the risk. The appraisal reported in this paper was part of the review. For Twin Stream fan the impetus was the application to increase the number of tourist cabins on the fan. Advice on the adequacy of existing flood protection for the area was sought by the developers. Our appraisal for them forms the basis of the account given here.

Alluvial Fans and their Hazards

Alluvial fans are depositional landforms that develop incrementally where sediment-laden mountain torrents (as flood-flows or debris-flows) exit from steep, confined channels to flow onto more gently sloping surfaces where their channels are no longer constrained to a fixed position. The decrease in stream grade, and increase in channel width which usually accompanies it, decrease the stream's power to transport sediment and the excess of sediment is deposited. The coarsest deposits and the fastest rate of aggradation generally occur near the fanhead, where the rate of change in stream grade with distance down the fan tends to be greatest. Deposition in the channel causes the stream to episodically abandon one course to seek new courses. This often sudden abandonment of a channel, and creation of a new one which may lead water to an entirely different area, is termed an avulsion.

The significant geomorphic hazards on alluvial fans are the unpredictable avulsions and the nature of the events in which avulsions occur. The traditional floodplain flood hazard, with decreasing risk of brief episodic inundation with increasing height above normal river level does not occur on steep alluvial fans. Simple geometry dictates that any point on the fan below the apex is at a lesser height than the river no great distance upslope, and thus is at risk from escaping floodwater. On an active fan, too, the hazard is less one of temporary inundation by water, than one of longer-term incorporation into a new water course. Geometry dictates that, in general, the probability of flooding of any point on the fan decreases systematically downfan (as discussed by Dawdy, 1979). The uncontrolled course of floodwater across an alluvial fan, however, usually is braided, and not a single broad sheet of water, so the probability of a point away from the watercourse being flooded is not simply that of the stream breaching its banks in a flood of given magnitude: the latter must be reduced by the proportion of the floodplain likely to be occupied by water during the flood. Although fan geometry causes hazard severity and risk to vary systematically down the surface of an alluvial fan (Dawdy, 1979, 1981), their distributions across the fan are seldom purely random. During the processes of channel avulsion, the stream abandons often well-defined and sometimes deeply scoured channels which remain to direct the course of future flooding in the event of water entering them. Likewise, some forms of development on the fan surface, such as roads, embankments for bridge approaches, and excavations to provide level ground for buildings, may guide new courses by directing water or localizing deposition. Thus the detailed assessments of hazards and their risk should take into account both the general geometry of alluvial fans and the particular topography of the fan under investigation.

In addition to the hazards of temporary and permanent inundation by water, there also is the hazard of inundation by gravel, termed deposition or aggradation. Two types of depositional hazards may be present. The more serious one, with greater potential to be a hazard to life, is that of debris-flow deposition. Characteristics of the drainage basin supplying water and sediment to the fan determine whether a fan surface will be at risk from debris-flows (for information on recognizing debris-flow hazards, see Kellerhals and

Church, this volume). Deposition from the traction load of streams also occurs. It is a lesser hazard to life, but it may be no less damaging to property than debris-flow deposition. For both types of deposition, the severity of the hazards and the risks of their occurrence at any point generally decrease downfan as a function of fan geometry. The developments on the two fans discussed in this paper are principally at risk from avulsion and floodwater inundation. The significant debris-flow hazard in one of them stems only from the height of the artificial levee and the saturated loose gravel that will be impounded by it. As a consequence, this paper does not discuss the assessment of debris-flow hazards on alluvial fans.

Appraisal Techniques

The appraisals for the two fans reviewed the geomorphology of the fans and their historical floods and channel changes. Because of their proximity, the two fans share a common history of major flooding. Our previous work in the area (Whitehouse, 1982) had given us a historical record of these, gained from reviews of the longer rainfall and river-flow records from climatically related regions of New Zealand, newspaper accounts from library archives, and historical anecdotes from books. Witnesses to the more recent events were sought and interviewed. We also had access to departmental files concerning construction and repair of earthworks and river-control measures in the two areas. Our aerial photographic index and archive were used to obtain all aerial photographs of the areas (spanning 1954–1986). These were viewed stereoscopically. Although we had detailed general knowledge of the two sites from previous work in the area, each site was visited and reviewed.

A prediction of the magnitude of the 100-year flood was obtained for each fan using regional flood estimates of Beable and McKerchar (1980) as modified by Williman (written communication, 1982) for the Southern Alps. These estimates are based on regional relationships between discharge and basin area, and between flood flows and mean annual discharge.

Where possible data were presented quantitatively but, as on many active fans, data are limited and largely qualitative. All information was presented in a form that could be used in planning and engineering design of developments on the fans.

Black Birch Fan

PHYSICAL SETTING

The southern half of Mt Cook Village is built on this geologically young fan. On its 0.55 km^2 area there are about 45 residences, a school, a Youth Hostel, a mechanical workshop, water storage tanks, and the sewerage ponds for the village. The fan is composed of coarse bouldery alluvium. Gradient on the symmetrical fan decreases from 3.8° at the fanhead, to 2.3° at its southern toe where it buries terraces of Hooker River (Figure 19.1). Until river-control work was initiated in the late 1960s, the whole fan was subject to irregular flooding, aggradation, and course changes (Figure 19.2). Black Birch Stream has been in its present position on the fan since a natural course change in the early 1960s. Within a few years of the change, it had to be encouraged to stay there. To this end, large boulders were raked from the stream bed to induce scour. Later, more substantial artificial levees were required.

Above the fan, Black Birch Stream is a perennial mountain torrent draining a steep, 4.8 km^2 basin on the eastern flanks of the Sealy Range, The basin is underlain by highly fractured sandstone and siltstone, and has a rugged topography. Altitude ranges from 755 m at the fan apex, to 2230 m at Mt Annette. The broadly U-shaped valley was sculpted by glacier ice, but little permanent snow remains in the basin today. Slopes average 38° (Lewandowski, 1975). The steep upper slopes generally are rocky bluffs, while the lower slopes are mantled with a generally thick regolith consisting of screes and deposits from glaciers, landslides, and snow avalanches.

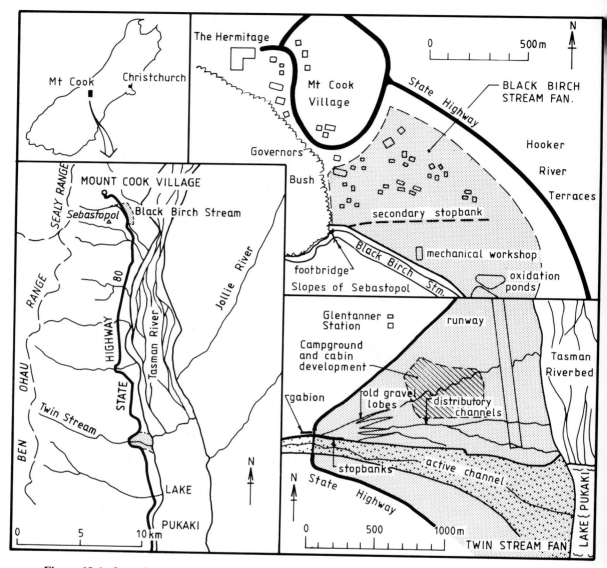

Figure 19.1. Location of Black Birch Stream and Twin Stream fans, South Island, New Zealand

Erosion within the basin dominantly is by frequent debris-flows and debris avalanches, primarily triggered by rain but also by snowmelt. These mass movements are common on the valley sides and in tributary channels, but we saw no evidence that they travel far along the main channel. This channel is steep (6° or greater), bounded by large boulders (about 2.0 m diameter), and undercuts most of the fans built by its tributaries. The basin is snow covered for most of winter, but vegetation trimlines indicate that snow avalanches do not commonly reach closer than about 1 km from the fanhead. Three snow avalanches are reported to have reached the fanhead between 1920 and about 1950 (D. Darroch, written communication, 1956). Erosion of the bed and banks of the main channel is the major direct source of sediment reaching the fanhead. The

GEOMORPHIC APPRAISALS FOR FAN DEVELOPMENT

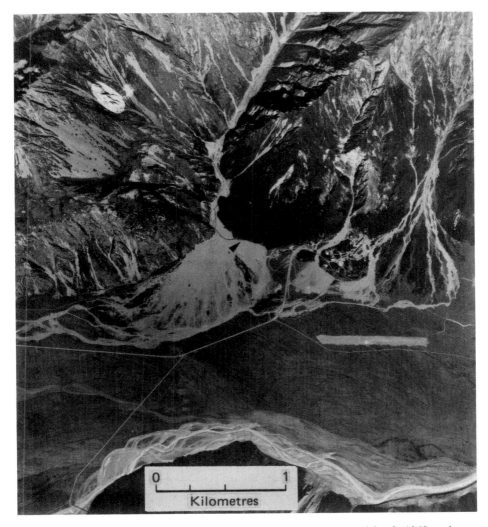

Figure 19.2. Vertical aerial view of fan at Black Birch Stream, March 1960, prior to expansion of Mt Cook village onto the northern flanks of the fan. Arrow marks stream channel passing to northern edge of fan, alongside Governors Bush (aerial photograph 2716/1, Department of Survey and Land Information)

channel provides temporary storage for most material reaching it from the valley sides. This storage can be as brief as during part of a single storm. At least several million cubic metres of sediment are stored in the wider reaches of the stream above a confining, narrow rock gorge at the fanhead.

No hydrological measurements have been made within the basin of Black Birch Stream. Daily rainfall records from Mt Cook Village, adjacent to the fan, provide a basis for estimates for the basin. Mean annual precipitation here is 3985 mm (New Zealand Meteorological Service, 1983). From likely rainfall gradients in the region, the mean for the basin is probably about 4500 mm. Evaporation losses may be as little as 500 mm. Thus, mean annual discharge of the stream is likely to be about 0.6 m^3 s^{-1}. The

100-year flood may be about 100 m³ s⁻¹ (Williman, written communication, 1984).

HISTORY OF AGGRADATION AND DEVELOPMENT

Aggradation at the fanhead and channel avulsion from the subsequent overtopping of the artificial levees are the major hazards to development. In March 1913, a major storm is reported to have brought two feet (610 mm) of rain in 20 hours to the site of the first Hermitage (du Faur, 1915). Photographs of Sebastopol, on the south side of Black Birch Stream, taken around 1911–1914 record the formation of a large gully, probably in this storm (Whitehouse, 1982). Considerable aggradation is likely to have occurred on Black Birch fan then, although it went unrecorded. Such a storm would have a return period of about 250 years (Whitehouse, 1982) but because the accuracy of the 'measurement' is in doubt, this estimate of return period is only speculation. Nevertheless, the storm was severe and had a long return period.

A photograph from the late 1920s shows Black Birch Stream occupying a broad swath in the middle of the fan, with recent deposition and scour over most of the fan surface. Aerial photographs show it in a similar location in 1954, and this is believed to reflect the likely pattern of behaviour throughout the first century of European occupation of the region.

On 26 December, 1957, 491 mm of rain was recorded at Mt Cook Village. The return period of this 24-hour rainfall is about 30 years (Whitehouse, 1982). Aggradation during this storm shifted Black Birch Stream from the centre of the fan to the extreme northern edge, adjacent to Governors Bush (Figure 19.1). Aerial photographs taken in March, 1960, show more than ten large erosion scars within the lower drainage basin of Black Birch Stream which appear fresher than the gully attributed to 1913. About four may relate to the 1957 storm, having appeared since October, 1954, when the previous aerial photographs were taken. Between 1954 and 1960 the character of the mountain torrent above the gorge changed from a deeply-scoured channel strewn with large boulders to a broad channel of low relief with few large boulders. This is a result of aggradation within the channel. On the fan itself, however, there is little apparent change other than the shift in channel course: undoubtedly there was deposition as well.

Aerial photographs taken in March, 1965, show the stream flowing along the southern margin of the fan below Sebastopol, in the course it maintains today. It was still in this position when photographed on 10 December, 1968, having maintained it apparently unassisted, incising some 3 to 4 m below the upper surface of the fanhead.

The first river protection works appear to have been constructed in about 1969 when sewerage oxidation ponds were built on the fan. Large boulders were 'raked' (with bulldozers) from the centre of the channel to its edge to form an artificial levee on its northern side. This extended from the fanhead to below the oxidation ponds (Figure 19.1). The first few houses were moved onto the fan in 1971. Development has increased steadily since then. By May, 1976, there was considerable development on the fan, and substantial artificial levees had been built (Figure 19.3). The safety of this development has been questioned on various occasions. Previous hazard assessments and recommended solutions are reviewed in Williman et al. (1986).

Flooding scoured the artificial levee at the Sebastopol footbridge and by the mechanical workshop (Figures 19.1 and 19.3) in May, 1978. It was repaired and a second one constructed inside of it (Figure 19.4). This was completed in early 1979 and was designed to provide a subsidiary floodway should the inner protection be overtopped.

On 2 December, 1979, 537.2 mm of rain fell at Mt Cook Village. The return period of this 24-hour rainfall has been estimated to be 55 years (Whitehouse, 1982). During the associated flood, there was considerable aggradation on the fan. At the height of the storm, the channel bed level rose about 7 m and water washed over the Sebastopol footbridge (Figure 19.4). From a 1975 contour plan and eyewitness reports, about 93 000 m³ of gravel accumulated within the confines of the

GEOMORPHIC APPRAISALS FOR FAN DEVELOPMENT

Figure 19.3. Oblique aerial view of fan at Black Birch Stream in May, 1976, showing the nature of the fan surface on which development has proceeded. Channel is confined by an artificial levee on the southern (left on photograph) extremity of the fan against the foot of Sebastopol slopes. The levee-enclosed building, left–centre, is the mechanical workshop. (Photograph V. D. 24, Mt Cook National Park, Department of Conservation)

channel. Fretting of the artificial levee presented increasing danger to villagers on the fan and they were taken to safer ground. Many debris-flows were active in the basin. On 3 December, after the peak of the storm had passed, a large gully of about 70 000 m^3 volume formed on the lower northern slope of Sebastopol (Figure 19.5), building a cone of debris in the gorge at the fanhead. This threatened to lift the stream out of its channel. A bulldozer was used to direct the stream away from the levee and to make it scour a deeper channel. This averted the threatened flood.

When the damaged levee was repaired following the flood, the channel fill was used to increase its height (Figure 19.4). Storage capacity for aggradation was increased to 114 000 m^3 (Williman et al., 1986). In the storms which have occurred since December, 1979, no net aggradation has occurred. There have been no major slope failures within the basin, and the upper channel has scoured and become armoured with large boulders.

RETURN PERIODS OF MAJOR AGGRADATION EPISODES

The major historical aggradations of 1913, 1957, and 1979 were ranked subjectively in order

Figure 19.4. Oblique aerial view of Black Birch Stream immediately after the December, 1979, storm (Photograph S. A. R. 115, Mt Cook National Park, Department of Conservation), showing the fanhead and gorge infilled by debris-flow fan (a). Source of debris-flow is shown in Figure 19.5. About 7 m of aggradation occurred at the footbridge (b). Bulldozer continues to assist degradation of the channel and build up inner levee (c). In the following months, the enlarged inner levee was amalgamated with the outer levee (d)

Figure 19.5. Oblique aerial view of Black Birch Stream fan in March, 1987. The recently constructed secondary artificial levee (a) is visible in the middle of the fan. The large gully (b) was eroded in the storm of December, 1979: adjacent gully (c) probably formed in a storm in 1913

of increasing volume: 1957 < 1979 < 1913. The data for this ranking are meagre: there are our own and other observations for the 1979 event; aerial photographs bracketing the 1957 event; and almost nothing for 1913. We assess the 1913 event as the largest aggradation, despite the lack of evidence, because this assumption is the safest alternative. The occurrence of three major events in 75 years indicates a recurrence interval of about 25 years for massive aggradation. We now introduce the subjective ranking: the 1957 event is likely to be closer to the 25-year recurrence interval on volume than the other two. The 1913 event is likely to be closer to the 75-year event, and the 1979 one closer to the 50-year event. This reasoning is fraught with pitfalls of uncertainty, but we believe it is an appropriate and conservative (in regard to safety in engineering design) use of very meagre data. The 1979 storm rainfall at the Mt Cook raingauge is much more reliably estimated as a 55-year event. We view this as support for our estimate, even though recurrence intervals of rainfalls only rarely coincide with recurrence intervals of associated river flooding, and there are no substantive data yet on recurrence intervals of bedload yields for aggradation to know how these might relate to flood or rainfall frequency. The estimated return periods for the 1913 and 1957 events however, are used only to place the more reliably estimated 1979 event in perspective. The perspective is that the 1979 flooding and aggradation was unusual, but not so exceptional that a similar occurrence should not be accommodated in the design of protective measures.

OUTCOME OF APPRAISAL

The post-1979 flood protection works should safely cope with an event at least as big as the 1979 one, but the hazard involved in their failure is significant. In a major flood, aggradation could raise the bed level at the Sebastopol footbridge, causing floodwater to overtop the artificial levee. Following initial overflow, there could be rapid scour and channelization, and the overflow could quickly capture the full flood flow of the stream. As the levee scoured to its full depth, the loose, saturated, fine gravel filling the channel behind it could be mobilized easily, possibly slumping into the rapidly growing channel with the consequent mobilization of a debris flow at the fanhead. Because of the relatively low gradient on the fan, it is unlikely that a debris-flow would travel more than a few hundred metres, but could guide the course of the new channel and assist in overtopping low obstacles. Thus the course of a breakout from near Sebastopol bridge is unpredictable. It will not necessarily follow existing channels at the fanhead, or be readily controlled by sandbagging.

To accommodate the uncertainties of defining probabilities of aggradation events from the available data, and to reduce the consequences of levee failure, the level of protection was increased in late 1986 by strengthening and raising the levee at its upper end, adjacent to Governors Bush, and by constructing a secondary artificial levee down the crest of the fan (Figure 19.5). This created an alternate floodway in the event of failure of the primary protection and in effect allocated a larger portion of the fan for Black Birch Stream. In addition, to further safeguard life, a flood-warning system was installed on the Sebastopol footbridge and evacuation procedures set in place.

The designation of an alternate floodway limits future development to within the presently developed area (Williman *et al.*, 1986). The intensity of development within this area will be limited principally by the capacity of recommended evacuation procedures.

Twin Stream

SETTING

This larger fan of approximately 2.2 km^2 area is located about 22 km south of Black Birch Stream (Figure 19.1). A highway crosses the fan near its head. Tourist cabins and a campground are located at mid-fan on the northern side. A small commercial airport is located below this. The gradient of the fan decreases from 3.2° at the fanhead to 1.8° at its base. Twin Stream occupies a 150 to 350 m wide, braided channel slightly to the south of the centre of the fan (Figure 19.6). It

GEOMORPHIC APPRAISALS FOR FAN DEVELOPMENT

Figure 19.6. Oblique aerial view of Twin Stream fan in March, 1987 (photograph by J. Barringer). The proposed area for cabin development is adjacent to the campground and cabins on the northern (right in this photograph) side of the fan

has occupied this position at least since 1931 when the first plans of the area were prepared, and probably for much longer. At least in part, this stability is due to river-control work at the apex of the fan. The bedlevel of the stream has degraded 2 m in the last 27 years at the site of the highway bridge (T. D. Bird, written communication, 1985). This primarily is due to the raking of larger boulders from the bed to form crude artificial levees, which has exposed finer bed material in the channel to erosion. At the fanhead, the present fan is entrenched at least 25 m into an older (early Holocene) and much steeper fan. Remnants of this fan occur along the lateral margins of the present fan which buries them at their lower ends. The stream is presently scouring a low (about 1 m high) terrace on the south side of the active fan.

From the fanhead to the lower end of the protection works, the active stream channel is well armoured with larger boulders. Many have median diameters of about 0.6 m, with some up to 1.0 m. Scouring of the stream channel over this reach appears to be largely a consequence of these larger boulders being raked from an otherwise stable channel. Because of the river works, the locus of deposition on the fan is several hundred metres downstream of the bridge. The finer gravel load of the stream is sluiced through the narrow bridge section, and deposition begins where the channel is unconfined. Below the limit of levees, the stream regularly extends its active channel when in spate. On the south bank it spreads gravel onto cultivated fields where it is not confined by a low terrace edge. On the north side, it spreads gravel into dense scrub, to enter old channels which run parallel to the main active channel. A number of these distributary channels cross the northern half of the fan. Many of these have not carried water since the active channels of the stream moved away from this sector, possibly as late as the early part of this century. Several of the major channels, however, have carried floodwaters at various times since—at least into the 1960s.

Above the fanhead, Twin Stream drains a steep, approximately 26 km^2 basin on the eastern flanks of the Ben Ohau Range. The basin has rugged topography ranging in altitude from 620 m at the fan apex to 2675 m at the head of the basin. Slopes are steep. The geology, glacial history, erosion processes, and sediment sources are very similar to those in Black Birch Stream. Twin Stream does not have a rock-bound gorge at its fanhead.

Rainfall has been measured within the basin by Archer and Cutler (1983). They estimated a mean annual precipitation of 1700 mm for a site at an altitude of 850 m and from shorter-term records suggested that precipitation may be 3500 mm at 1800 m. The flow of Twin Stream has not been measured. The flood of 100-year recurrence interval at the bridge is estimated to be about 100 m^3 s^{-1}, with velocities of up to 3.5 m s^{-1} on the 3.4° slope of the fan in this reach (E. B. Williman, personal communication, 1986). Average water depth on the fan below the bridge is unlikely to exceed 1 m during floods.

FAN HISTORY

The concern on Twin Stream fan is how often, and in what quantities, floodwater might cross the northern part of the fan. Very high flows would have occurred in Twin Stream in 1913, 1957, and 1979, as at Black Birch Stream. We could find no written record of the 1913 event at Twin Stream. The later events were witnessed by the present land-holder and developer, and their effects show on aerial photographs. It is likely that the 1979 flow had a recurrence interval of about 50 years, as the gauged Jollie River across Lake Pukaki (Figure 19.1) had about a '50-year' flood (G. L. Davenport, written communication, 1980) and rainfall at Mt Cook had a 55-year recurrence interval.

The first flood protection work on Twin Stream was a gabion groyne built in the 1930s on the northern bank at the apex of the fan. Its purpose was to prevent flooding and aggradation on the road and fields in front of the Glentanner Homestead. This comparatively minor work was remarkably successful: the now 50-year old groyne has not been breached or overwhelmed in a number of floods since its construction. Bedlevel inside the groyne closely matches the ground level beyond it, attesting to the long-term stability of the channel bedlevel at the fanhead. Since 1955, when the highway bridge was constructed, this groyne has been redundant for its original purpose, and has served to protect the northern approach to the bridge. Some river protection would have been built following bridge construction but its extent is not precisely known. Since the initial construction of these works, there has been continual *ad hoc* work in the river channel about the bridge, involving raking large boulders from the active river bed. These boulders have been pushed to the side adding to the artificial levee. This levee has been damaged by scour in high flows, but has grown through repeated additions following damage.

The first aerial photographs of the fan, taken in October, 1954, show fingers of fresh gravel splaying north from the main channel at the fanhead just below where the highway bridge now is located. These extend 800–1000 m down the fan to four slightly entrenched, narrow, sinuous channels that continue to the base of the fan (Figure 19.7).

In the high flows of December, 1957, scour and aggradation occurred within the gravel fingers on the northern side of the fan apex and on the lower fan, particularly north of the main channel. Floodwater flowed in at least some of the distributary channels on the northern side of the fan to cover a field near Glentanner Homestead. Using handtools during the flood, the land-holder and an assistant were able to turn the stream back to its former course.

Artificial levees were extended after 1957 to cut off the northern distributary channels and they do

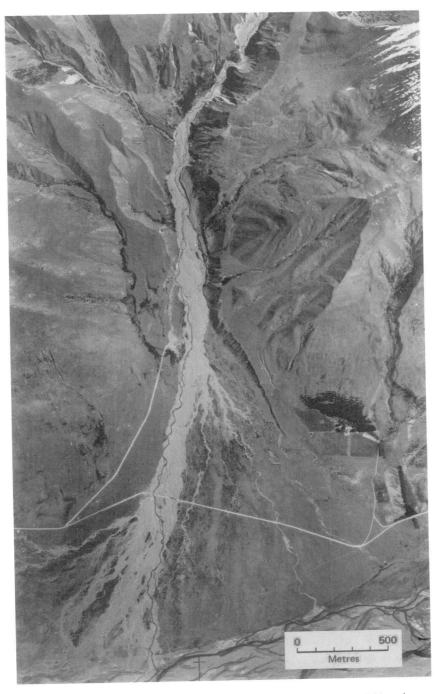

Figure 19.7. Vertical aerial view of Twin Stream fan, October, 1954, prior to development. Preparation for the southern abutment of the present highway bridge has begun at the fanhead. Subsequent to this photograph, river-control work associated with the bridge has substantially blocked floodwater and stream avulsion from entering the northern side of the fan (right of photograph) (aerial photograph 2311/16, Department of Survey and Land Information)

not appear to have carried significant water in the 1979 flood. Then, aggradation was primarily within and adjacent to the main channel, particularly on the lower fan. Following the 1979 event, levees were built by raking boulders from the channel for approximately 400 m above and 200 m below the highway bridge.

A campground, tourist cabins, shop, and tearooms were built on the northern part of the fan in 1978 (Figure 19.1). They were not flooded in 1979. An airport runway built in 1957 was extended for commercial use in 1980 and 1986 across the lower northern part of the fan (Figure 19.6).

PRESENT FLOOD RISKS

The cabins and campground are afforded considerable flood protection by the embankment for the northern approach to the highway bridge. A significant flooding hazard originates only at the downstream side of the bridge where inactive channels lead directly, through dense scrub, to the campground and cabin area. Fan geometry elsewhere directs floodwaters away from this area. River-control work in the vicinity of the bridge has greatly decreased the risk of flooding into the channels that lead to the development area. The level of protection exceeds that required to accommodate the 50-year flood. A flow of about this magnitude was passed safely in 1979 with minimal damage to the flood-protection structures, and without flooding the development. In addition, the levees were added to in channel maintenance after this event. They lack a coherent form, however, and presently consist of several loosely constructed banks of boulders and gravel. These may interact to aid in their mutual destruction by scour in the event that floodwater passes between them.

Scour of the levee immediately below the bridge would lead water directly into channels which pass through the cabin area. The exact nature of the flood hazard to the cabin area is uncertain. Flow through the scoured levee is not likely to be the full flood flow, and the breach must occur within a short, 20 to 50 m distance below the bridge for floodwater to enter channels that pass through the cabin area. Allowing for these factors reduces the magnitude of the flood overflow through these channels to probably less than half that passing the bridge. The channels do not have even this capacity, and the floodwater is likely to flow widely and shallowly, and be slowed significantly by the dense scrub. Some local scour, as occurred in 1957, is likely, with resultant deposition immediately downstream of scour sites. The hazard is unlikely to be a danger to life, but certainly will be damaging to property.

Downstream of the existing artificial levee, there is no current risk of bank overflow reaching the cabin area. There is, however, a long-term risk that continued deposition will lead the stream ultimately into this region. The deposition required is large and likely to take decades. The threat is more immediate to the southern end of the airport runway below the cabins, and protective work will be needed for this long before there is any threat to the cabins from this direction.

PROPOSED SOLUTION

The risk to the present cabin area and proposed extension can be minimized most simply at the point of origin of the hazard, at the downstream side of the bridge. The existing, roughly pushed-up levee, about 200 m long and 1.5 m high, downstream of the bridge on the northern side, is of about the right dimensions and appropriately positioned to prevent flood breakout into the channels of concern. Although it has survived a flow of about a 50-year recurrence interval, its construction is *ad hoc* without design, and it was built to protect the highway rather than the campground. In our report to the developers, we recommended the design of an artificial levee to contain the estimated 100-year flood within the river channel. This would entail upgrading the existing bank to a consistent width and height, and facing it with large boulders. We believe that this would allow for the uncertainty of the data used to assess the flood risk, and provide a conservative level of protection. After considering the likely consequences of a flood exceeding this design level, we saw no need to provide for flood-warning or for secondary floodways, although it is prudent that development is not proceeding in existing former channels, which are being left for natural drainage of stormwater on the fan.

Discussion

In determining site suitability for development and in designing any necessary protection work for a development on an alluvial fan, it is invaluable to have realistic appraisals of the hazards to which the particular fan is exposed, and their likely magnitudes and frequencies of occurrence. Avulsion is the principal hazard to development on a fan, but to design protection from this, it is more important to determine the magnitudes and frequencies of stream discharges and volumes of aggradation in the stream channel, than it is to determine the frequency of past avulsion. Although magnitudes and frequencies may not be able to be established with scientific rigour, it is essential that all available information is presented in forms that can be used in planning and engineering design. This may mean that some magnitudes and return periods of events must be estimated from qualitative information gleaned from historical photographs and anecdotes, newspapers, and interviews of witnesses.

The efficacy of such designs can be tested only by trial. Because of the element of uncertainty as to the outcome of such a trial, it is particularly important to assess the likely consequences of failure when avulsion leads the stream into the 'protected' area. If the consequences of failure are unacceptable, development should not proceed, or some provision should be made to reduce the consequences to an acceptable level.

The nature of the geomorphic hazards on Black Birch Stream and Twin Stream fans are very different despite their close proximity and superficial similarities, and consequently so too are the solutions to minimize them. Houses or tourist developments were already present on both fans before the appraisals were carried out, and the options of zoning their entire surfaces as unsuitable for development were considered not viable. Such a zoning solution, or other solutions also not considered on Black Birch or Twin Stream fans, may be more appropriate on other fans. These may involve other hazard-mitigation structures (using protective works), hazard management (using warning systems and landuse zoning), hazard avoidance (by zoning), or a combination of these solutions.

At Black Birch Stream, large changes in bed-level on the fan were the principal hazard to consider in the design of protection. This, and the uncertainties in estimating their magnitude and return periods, have required a massive artificial levee at the fanhead, an alternate flow path for floodwater and debris when the levee is overtopped, and a flood-warning system to allow villagers time to seek safer ground. This combination of protective works and procedures has been arrived at after nearly two decades of trials. Although there have been successive major improvements in design, at no stage has there been a disastrous failure. The latest scheme is purposefully developed to avoid this possibility.

At Twin Stream fan, such dramatic changes in bedlevel are not a possibility, and flooding alone is the principal hazard. With the embankment of the approach to a highway bridge already offering substantial flood protection, it is practical to look at containing the 100-year return-period flow of the river to the existing channel with relatively modest levees. The consequence of levee breaching is not such a serious hazard as to be a substantial danger to life. Thus, no secondary protection appears to be necessary. As at Black Birch Stream fan, however, the situation should be reassessed as the scheme is tested.

The objects in hazard management on alluvial fans are to maintain public safety and to safeguard property by separating zones intended for human use from those at significant risk of use by the river. Depending on the rate and nature of the alluvial processes, these may be achieved by notional separation (zoning), or by constructing physical, separating barriers. On the two fans discussed here, physical barriers are used because the risks to life and property were judged (by others, acting in part on our recommendations) not to preclude continued human use. These barriers are intended to shift and confine the areas in which avulsion may occur, and significantly reduce the risk of avulsion into regions wanted for 'urban' use. Here, the theoretical difficulties in accurately assessing risk on alluvial fans (see discussions between Dawdy, 1979, 1981; and McGinn, 1980) present no difficulty in practice, because when barriers are used to contain the floodwater and aggradation, and inhibit avulsion, they conventionally are designed to some specified and arbitrarily low probability of exceedence. Thus, the practical problem is not to calculate the risk to each dwelling, but to nominate some arbitrarily long recurrence interval whose risk is judged to be on the threshold of acceptability, and then to design structures to *contain* the events of this probability. This expedient approach determines the risk to the protected region as a whole, and ensures a lower degree of risk to any dwelling within the protected zone.

For success in approaches which rely on trial and error there is need to periodically reappraise the adequacy of hazard mitigation. In such reappraisal, all available information on the nature of the hazards should be reviewed, and requantified if newer information or techniques require it. The approach adopted should be conservative with respect to safety, with solutions taking both the information, and its uncertainty, into consideration. The solutions implemented should also recognize that it may be necessary to add to, or change the mode of protection as new information comes to light. Thus, continuing geomorphic appraisal is an integral part of hazard management on steep, active alluvial fans.

Acknowledgments

We thank the Water and Soil Directorate, Ministry of Works and Development, and Department of Conservation, Christchurch, for use of information collected as part of the review of flood protection at Mt Cook. We thank Mr Ian Ivey, Glentanner Station, for use of information collected as part of the investigation of flood protection on Twin Stream fan. Review and comment from Mr Brin Williman, Dr M. Paul Mosley, and Dr Les Basher are appreciated.

References

Archer, A. C. and Cutler, E. J. B. 1983. Pedogenesis and vegetation trends in the alpine and upper subalpine zones of the northeast Ben Ohau Range, New Zealand. 1 Site description, soil classification, and pedogenesis. *New Zealand Journal of Science*, **26**, 127–150.

Beable, M. E. and McKerchar, A. I. 1980. New Zealand flood study. *Water and Soil Technical Publication*, **20**. Ministry of Works and Development, Wellington, New Zealand.

Dawdy, D. R. 1979. Flood frequency estimates on alluvial fans. *American Society of Civil Engineers Proceedings, Journal of the Hydraulics Division*, **105**, 1407–1413.

Dawdy, D. R. 1981. Flood frequency estimates on alluvial fans: closure of discussion. *American Society of Civil Engineers Proceedings, Journal of the Hydraulics Division*, **107**, 379.

Du Faur, F. 1915. *The Conquest of Mt Cook*. London: George Allen and Unwin. 250 pp.

Lewandowski, R. J. 1975. *The Geology of the Black Birch Fan and Catchment, Mt Cook National Park, and its Engineering Structures*. Unpublished M.Sc. thesis (Geology), University of Canterbury, Christchurch, New Zealand. 145 pp.

McGinn, R. A. 1980. Discussion of 'Flood frequency estimates on alluvial fans' by D. R. Dawdy. *American Society of Civil Engineers Proceedings, Journal of the Hydraulics Division*, **106**, 1718.

New Zealand Meteorological Service 1983. Rainfall normals for New Zealand 1951 to 1980. *New Zealand Meteorological Service Miscellaneous Publication*, **185**, 36 pp.

Whitehouse, I. E. 1982. Erosion on Sebastopol, Mt Cook, New Zealand, in the last 85 years. *New Zealand Geographer*, **38**(2), 77–80.

Williman, E. B., McMillan, B., and Alexander, D. 1986. *Mount Cook Village Review of Flood Protection Black Birch Fan*. Ministry of Works and Development, Christchurch, New Zealand. Report WS 1129. 47 pp. (unpublished).

Index

Abandoned channels, 75, 158, 161, 170–173, 217–219
Accretion, vertical and lateral, 187
Accumulation–removal cycle, 82
Aconcagua, 57
Aeolian deposits, 28, 221
Aeolian erosion, 277
Aeolian sands, 166
Aeolian silts, 41
Africa, 20
Agassiz Lake, 179, 190
Aggradation, 164, 168, 179, 218, 247, 249, 272, 291, 293, 347, 351, 374
  –entrenchment cycle, 18
  episodic, 250
  return period of, 375
Alaska, 198
Alberta Creek, 341
Alleröd, 242
Allocyclic control, 272, 298, 299
Allometric analysis, 16
Allometric models, 8
Alluvial apron, 5
Alluvial fan(s)
  aggradation, 296
  artificial, 20
  definition, 4, 27
  dissected, 92, 259
  distribution control factors, 94
  environments, 5
  exhumed, 272
  expansion, 105
  faulted, 91, 100
  geometry of, 326
  hazards, 370
  longitudinal profile, 137–139
  morphology, 135, 259
  as piedmont landforms, 126
  problem, 3, 8, 20
  research, 3
  studies, paradigm in, 7
  surface deformations, 92
  tectonically initiated, 146
  terraced, 92
  transverse profiles, 137

Alluvial fan-like accumulation, 239, 250
Alluvial megafan, 151
Alluvial ridges, 183
Alluvial sediments, 186
Almeria region, 250
Alpine glaciation, 98, 104
Alps, 28, 55, 231, 336, 351
Alternate bars, 159, 162
Alternative paradigm, 6
American Great Basin, 69
American South West, 16, 19, 69
American West, 3, 6, 12
Anastomosing rivers, 172
Andes, 28
Antecedent drainage, 153
Appalachian Mountains, 109–128
  piedmonts, 113
Aquifers
  artificial recharge of, 356
  texture of, 357
Arizona, 12, 19
Arroyo channels, 360
Artefacts, 289
Artificial embankments, 154–155
Arun river, 151
Assiniboine river, 179, 181, 189
  sediment transport rate, 182
Autocyclic control, 272, 296–298
Avulsion, 175, 187, 272, 295–296, 298, 336, 341, 345, 347, 351, 370, 383
Axial streams, 137–138, 146

'Badlands', 251
Baffin Island, 207
Bagrot Valley, 35, 41
Baltic Sea, 311
Ballistic ripples, 165
Bankful discharge, 293
Barchan, 165
Base-level change, 12
Basin–fan area ratio, 191
Basin–Range Province, 12, 19
Batkor Valley, 35, 41

INDEX

Beaches
   constructional, 198, 217, 220–221
   destructional, 198, 217, 220–221
Bedload streams, 195
Bedrock, weathering of, 75
Bengal North, 321
Betic Cordillera, 249
Bhimnagar barrage, 159
Bifurcation, 164
Bioturbated sediments, 223
Biwa Lake, 93
Black Birch Stream, 369
Black-box, 19
Blue Ridge Mountains, 110, 113, 115, 121
Bonneville Lake, 196, 205
Bottomset beds, 205
Boulder conglomerates, 49, 133
   levees, 119, 177
   lobes, 118
Bouldery diamictons, 32
Brahmaputra Valley, 135
Braid islands, 169
Braided channels, 16, 161
Braided stream, 166, 378
Braided stream gravels, 125
Braiding
   degree of, 272
   index, 166
   intensity of, 166–168
Bristol Channel, 198
British Columbia, 207, 335
Buddist petroglyphs, 34
Bölling, 242, 308

Calcite cement, 293
Calcrete crusts, 257
Canterbury Plains, 28
Caribbean Sea, 198, 202, 214
Carpathian Mountains, 231
Catastrophic events, 15, 17, 30, 50
Catastrophism, principle of, 52
Cayman Trough, 214
Chandigarh Dun, 131
Channel
   cross-section of, 158
   incision, 49
   meandering, 157
   plugging of, 337
   primary and secondary, 156
Chemical weathering, 28
China, 20
Chri Mountains, 102
Chubu, 93, 100
Clast size, 283, 289
Closed system, 9
Cloudburst floods, 81
Coastal dunes, 217

Compaction, degree of, 35
Compound bars, 161
Conveyer-belt transport system, 85
Copper river, 198, 199, 204
Core drillings, 234–235
Costa Rica, 16
Cottonwood Canyon, 81, 83
Crevasse, 189
Critical stream power, 248

Dams, 357
   in canyons, 81
Danube, 229–239
Darjeeling district, 321
Davis cycle of erosion, 8, 18
Davisian concept, 6–7
Davis's evolution paradigm, 6
Dead ice, 314
Dead Sea Rift, 196
Death Valley, 13, 16, 18, 19
Debris
   avalanches, 110, 119
   chutes, 351
   cones, 336, 341
   deposition, 84–85
   flow, 4, 7, 11, 28, 37, 255, 325, 338, 340, 348–351, 375
     behaviour, 84
     deposits, characteristic of, 41, 345
     dominated fans, 16, 110
     fan formation and, 73
     fans, 16, 41, 96–97, 110, 115, 255
       geometry of, 115
       monitoring system, 96
       sedimentology, 118
     frequency of, 94, 115
     hazard, 349, 351–353
     indicators, 341
     lateral ridges, 74
     leveed, 16
     linear, 50
     lobes, 74, 96–97
     physical characteristics of, 73
     processes, 72–87
     recurrence intervals, 85, 117
     size distribution of material, 74–75
     surges, 339
     thunderstorm rain and, 72–73
   jam, 344
   levees, 118
   removal volume, 82
   supply and precipitation, 104, 112
   torrents, 338
Deflation, 50, 243, 287, 290, 294–296
Deforestation, 328
Dellwood, 115

INDEX

Delta
  front, 195, 201
  lobes, 164
  slope of, 201
Denudation, 28
Deposition, rate of, 9
Depositional chronologies, 18
Depositional hazard, 370
Depositional processes, 15
Desert varnish, 287
Diagonal bars migration, 172
Diamond-shaped bars, 173
Diamictons, 34
Diapiric disturbances, 254
Diapiric uplift, 249, 253
Distal trenching, 263
Distributary channels, 282, 295
Diversion dykes, 351
Du Boys equation, 291
Dynamic equilibrium, 9, 169
Dynaric Mountains, 231

Earthquake, 77
Earthworms, 163
East African Rift, 196, 201
East Anatolian Transform Fault, 196
Eastern Oman Mountains, 272
Empirical approach, 18, 19
Entrenched streams, 133
Epeirogenic uplift, 249
Ephemeral lakes, 198
Equilibrium
  concepts, 6, 8
  line, 57
Erosion, rate of, 9
Erosional energy, 10
Erratic boulders, 58
Eustatic changes of sea-level, 99, 104, 299
Evapotranspiration, 293
Evolutionary hypothesis, 8

Faceted boulders, 60
Facies, types of, 283, 287–288
Fan
  aggradation, 11, 12, 15, 247
  apex, 4, 12, 140, 187, 216, 277, 327, 339, 379–380
  –basin relationship, 8, 16
  building, hiatus in, 124
  channels, anastomosing of, 140
  cross-profile, 4, 140
  delta, 195, 213
    depositional models, 202–207
    Gilbert type, 202, 205–206
    lacustrine, 198
    subaqueous, 200
    wave dominated, 198
  deposits in geologic past, 13
  matrix free, 118
  fabric of, 145
  thickness of, 142, 183, 232–233
  development model, 35, 248
  dissection, 17, 123
  dynamics paradigm, 7, 8, 10, 18, 20
    crisis in, 14
  entrenchment, 7, 8, 137, 293, 379
  evolution, 7
  experimental, 12, 20
  formation
    by Gilbert, 71
    mechanism, 294
    tectonics and, 253, 299
  gravel aquifers, 126
  humid tropical, 219
  ideal, 135
  literature, 15
  littoral, 198
  segmentation, 12
  segments, 138
  subaerial, 197, 217, 221
  submarine, 202, 217
  subsidence, 175, 230, 356–357
  surface gradient, 137, 183, 191, 217, 259, 309, 327, 371
  terrace(s), 28, 92, 257
  toe
    edge, 138
    truncation, 28, 41, 49
Fan as relict feature, 11
  coalescent, 5
  humid, 110
  humid-temperate, 16, 197
  relict, 17
  segmented, 115
  steep, 79
  wet-tropical, 16
Fanhead trenching, 4, 11, 12, 17, 41, 49, 137, 249, 255, 257, 263
Faults, 133
  active, 100
  Plio-Pleistocene, 250
  reverse, 100
    slip, rate of, 100
Filltop terrace, 92, 98
Fish Lake Valley, 71
Flash flood, 275, 324
Flat lobe, 96
Flood
  deposits, 96
  risks, 382
Floodplain, 152, 157, 159, 163, 217, 324
  deposits, 218–219
  fan, 179–192
  loess, 236, 242
  surface, topography of, 119

Floodplain—*cont.*
   valley, 58
Flow, duration of, 293
Flow-duration curve, 155
Fluvial deposits, 63
Fluvial dissection, 58
Fluvial erosion, 41
Fluvial fans, 121–126
   dissection, 123
   drainage, shift of, 123
   longitudinal profile, 123
   sediments, 123–126
Fluvial incision, 28
Fluvial megaclasts, 142
Fluvial sediments, 28
Fluvial theory, 3
   and fans, 18
Fluvially dominated fans, 16, 110, 220, 255
Foreset beds, 205
Fossil meanders, 295
Freeze–thaw cycles, 105
Fresno County, 12, 19
Frost shattering, 28
Froude number, 291, 292

Gabion sediment trap, 365, 370
Gang Benchen, 57, 61
Ganga River, 152, 158, 176
Geomorphic thresholds, 16
Gilgit River Valley, 28, 35–46
Ginseng Hollow, 119
Glacial diamictons, 34
Glacial erosion, 28
Glacial polishing, 58
Glacial sequence, characteristic of, 58
Glacial spillway, 179
Glaciodeltaic sediments, 182
Glaciofluvial debris, 56
Glaciofluvial deposits, 28, 37
Glaciofluvial sands, 49
Glaciolacustrine clays, 180
Graben, 154
Grain-size curves, 182
Granular desintegration, 28
Gravitational transfer, 76
Groundwater, pumping of, 356
Groundwater-fed streams, 170–173

Harvey Creek, 341
Hazar Lake, 196, 206
Hazard
   avoidance, 345
   identification, 340
   problem, 336
Hidaka Range, 92, 98
Himalayan Front Fault, 153
Himalayan piedmont, 131, 133, 135

Himalayan streams, 138
Himalayas, 28, 131
   Darjeeling, 323
   Eastern, 321
   Lesser, 153
   Lower, 133
   Nepal, 151, 153
   Outer, 144
Hindu Kush Mountains, 28
Historical perspective, 15
Hokkaido, 92, 98
Holocene, 99, 103, 245–246
Horst, 154
Hunza River, 28–37
Hurricane
   Agnes, 118
   Camille, 113, 117, 118
   Flora, 216
   Juan, 118–119, 123
   landfall, 113
Hyperconcentrated flood-flow, 126
Hyperconcentrated flood sediments, 110, 255
Hyperconcentrated sediments flow, 118

Ice jam, 187, 189
Ice marginal ramps, 55–66
   slope of, 59
Ice marginal spillway, 305
Ina Valley, 93, 100
Indogangetic Plain, 151
Indus
   River, 28
   Valley, 35–37
Intellectual crisis, 6
Interchannel areas, 173
Interference ripples, 165
Intergranular cementation, 258
Intermittent flow, 140
Intermontane basins, 27, 249
Intersection point, 4, 11, 15, 257, 259, 263
Intertidal zone, 199
Inverse grading, 96
Iron oxide, 175
Irrigation, 356

Japan, 196
Japanese Alps, 93

Kames terrace, 311
Kanto Plain, 92
Kaolinite, 330
Karakoram Mountains, 27–52
   climate, 30
   glaciations, 35, 37, 46
   uplift rate, 28
   vegetation, 30
Kiso Range, 100

Kitakami Valley, 92
Kosi megafan, 151–153
Kosi river, 151
　hydrology of, 154–155
　bank retreat rate, 159
Kuen Lun Mountains, 57
Kuh-Rud Mountains, 57
Kuh-i-Jupar Mountains, 57–60
Kurye Basin, 97

Lacustrine sediments, 28, 40, 60, 183, 198
Lag deposits, 119
Laguna Seca, 359
Landslides, 216
　frequency of, 348
Large magnitude events, 17
La Salle palaeochannel, 189, 191
Last glaciation, 97
Late Glacial, 306
Lateral moraines, 58
Leba river, 305
Levee(s), 158, 163, 183, 189, 374, 380
Lichen cover, 32
Linguoid bars, 159, 161, 164
Littoral processes, 216
Loess, thickness of, 240
Lone Tree Creek, 81, 83
Longshore currents, 198, 202
Lower Derwent Valley, 16

Macrotidal setting, 198–199
Main Boundary Fault, 133
Malawi Lake, 196, 201
Manitoba Lake, 180, 183, 189
Manning equation, 291
Meander wavelength, 291
Mexico, 355
Microtidal setting, 198, 216
Misfit streams, 183
Mount Cook, 369
Multiple distributary channels, 191

Nanga Parabat, 28, 30, 57
Nelson County, 115, 119
Neotectonics, 131
Neogene tectonic, 62
Nepal, 151–152
Nevada, 19
New England, 109
New Zealand, 196
Non-cohesive sediment gravity flow, 15
'Normal science', 6
North Bihar, 151–152
North Carolina, 109, 121
North European Lowland, 317

Older Dryas, 242

Ontake Volvano, 100
Open system, 9
Outwash plains, 56, 59
Overbank flow, 187, 189
　deposits, 197
Owens Valley, 52, 71, 83
Oxbow lakes, 152, 176

Palaeochannels, 183, 272, 277, 290
　generations of, 277, 295
　sinuosity of, 279
Palaeocurrent analysis, 309
Palaeohydrologic analysis, 291
Palaeodischarges, 291
Palaeohydraulic models, 290
Palaeovalley, 35, 37, 40
Pamir Mountains, 28
Pannonian Lake, 230–232
Parabolic dunes, 243
Paradigm, 6
Paraglacial sedimentary fans, 52
Paraglacial sediments, 46
Parallel gullies, 78
Pasu area, 46–50
Peak-flow, 164
Pediment, 5, 12, 121, 133, 135
Perennial streams, 80, 170
Periglacial environment, 92, 104, 133
Periglacial soil creep, 76
Petrocalcic horizon, 258
Piedmont, 5, 110, 275, 321, 332
　glaciation, 56, 58
Piracy process, 115
Plate tectonic, 196
Playa, 9
Pleistocene, 11, 55, 57, 97, 98, 119, 124, 134, 147, 232, 249, 299
Plio-Pleistocene, 250, 272
Point bars, 158, 159, 162
Poland, 306
Polygenetic landforms, 27
Porosity, 357
Portage la Prairie, 179, 180
Pradolina, 306–314
Precipitation, mean annual, 360, 373, 380
Process-form relationship, 4
Prodelta, 200
Proglacial sediments, 46
Pumice, 100

Qilian Mountains, 61
Qinhai–Xizang Plateau, 57, 61
Quantitative approach, 6
Quaternary, 28, 119, 229–240, 249, 251, 257
Queen Dicks Canyon, 73

Radial channel pattern, 74

Radial profile, 4, 12
Radiocarbon datings, 15, 32, 117, 119, 242
Rainfall
  intensity, 86, 87
  recurrence interval, 216, 374, 380
  threshold of, 113
Recessional moraine, 41
Recharge, volume of, 365
Red River, 189
Red Sea, 196
Rejuvenation, 13
Resedimentation, 41
River
  capture, 294, 298
  terraces, 306
Roan Mountains, 119
Roaring River, 15, 17
Roches moutonnées, 58
Rock varnish, 18, 32, 49
Rocky Mountains, 15, 122, 338

Sand
  hummocks, 243
  mounds, 172
Scientific revolution, 6, 14
Scroll
  bars, 183, 187
  cones, 50
Sediment
  availability, 298
  fans, 28, 30, 37, 41
  load rate, 28–30
Sedimentary sequences, 14
Sedimentary structures, 283–287
Sedimentological analysis, 32
Seismic activity, 201
Shear fracture, 159
Shear stress, critical, 291
Sheetflooding, 15, 255, 336
Sheet-flow, 16, 137, 164
Sheetwash, 117
Shelf-fan delta, 202–204
Shellmouth Reservoir, 191
Shenandoah River Valley, 110, 115
Shighar Valley, 37
Shikoku, 93
Shiretoko Mountains, 92
Shisha Pangma, 57, 61
Shoreline processes, 195, 198
Side bars, 158, 161
Sierra de Jimulco, 359–360
Sierra Nevada, 71
Sieve deposits, 197
Sirsa River, 140
Siwalik Hills, 131, 133, 322
Siwalik sediments, 133–134, 153
Skardu Valley, 37

Slope-fan delta, 202–203
Smoky Mountains, 121
Snow avalanches, 372
Snowmelt flood, 86
Sociology of science, 6
Soils, 245, 330
Splays, 163
Steady-state equilibrium, 9
Stochastic process, 348
Storm frequency, 11
Stream
  gradient, 146
  incision, 28
  piracy, 11
Subaqueous processes, 195
Subaqueous sliding, 207
Subchannels, 156, 164
Submarine canyons, 202, 223
Subtropical monsoon, 133
Superimposed channels, 277
Surface
  boulders, 41
  calcretes, 257
  retention, 256
Surf-zone, 199
Swamps, 158, 175, 219
Systems theory, 8

Talus streaming, 77, 79
Tamur River, 151
Tanganyika Lake, 196
Tea gardens, 328–329
Tectonic uplift, 12, 272, 306
Tectonic valley, 100
Tectonically active areas, 5, 133, 200, 249
Tephrochronology, 92
'Terrain locked' storms, 113
Thermoluminescence dating, 34
Threshold slope, 12
Tibetan Plateau, 61, 153
Tidal currents, 196
Tidal flats, 158, 199
Tidal lagoon, 204
Tisza River, 222–237
Thalweg, 146
Till deposits, 28
  resedimented, 37
Timber line, 104
Tokachi River, 98–99
Tohoku, 92
Topological parameters, 166
Topset beds, 205
Traction load, 205
Tractive force, 11
Transfluence cols, 58
Trans-Himalayas, 38
Transport competence, 80

INDEX

Transverse bars, 161
Transverse dunes, 165
Tropical climate, 215
Trunk canyon, 79
Turbidite deposits, 202
Twin Stream, 191

Uniformitarianism, 17
Unit bars, 161
Uplift of mountains, 294
   rate of, 12
Upper flow regime, 164
Urstromtal, 306
U-shaped valleys, 58, 371
Uspallata basin, 57, 60–61

Valley fill sediments, 28, 35
Virginia, 16
Volcanic ash, 92, 99

Wadi, 275
Wahiba Sand Sea, 282
Water-laid sediments, 4
Wave
   action, 165
   convergence of, 222
   diffraction, 222
   heights, 217
   energy, 199, 221
   erosion, 221
   refraction, 202, 222
   ripples, 223
Wedge-shaped deposits, 13
Westgard Pass, 69
Wet mass movement, 77
White-box, 9
White Mountains, 9, 11, 16, 19, 69
Width/depth ratio, 158
Wind
   activity, 165
   furrows, 243
Wintertime flood, 86
Wisconsin Glaciation, 119
Würm, 236–237, 239, 258, 311

Yallahs River, 214, 216
Yallahs Submarine Basin, 215
'Yazoo' drainage pattern, 158

Zagros Mountains, 57, 196